Genetics and Molecular Biology of Rhythms in *Drosophila* and Other Insects

Advances in Genetics, Volume 48

Serial Editors

Jeffery C. Hall
Waltham, Massachusetts

Jay C. Dunlap
Hanover, New Hampshire

Theodore Friedmann
La Jolla, California

Genetics and Molecular Biology of Rhythms in *Drosophila* and Other Insects

by
Jeffrey C. Hall
Department of Biology
Brandeis University
Waltham, Massachusetts

ACADEMIC PRESS

An imprint of Elsevier Science

Amsterdam   Boston   London   New York   Oxford   Paris   San Diego
San Francisco   Singapore   Sydney   Tokyo

Academic Press
An imprint of Elsevier Science.
525 B Street, Suite 1900, San Diego, California 92101-4495, USA
http://www.academicpress.com

Academic Press
84 Theobald's Road, London WC1X 8RR, UK
http://www.academicpress.com

International Standard Book Number: 0-12-017648-3

PRINTED IN THE UNITED STATES OF AMERICA
02   03   04   05   06   07   MM   9   8   7   6   5   4   3   2   1

"The crackling of the last embers in the grass: stubborn insects
. . . the god that is time passes through the branches"

—Octavio Paz

Contents

Genetics and Molecular Biology of Rhythms in *Drosophila* and Other Insects

Jeffrey C. Hall
Department of Biology
Brandeis University
Waltham, Massachusetts 02454

ABSTRACT

Application of genetic variants (Sections II–IV, VI, and IX) and molecular manipulations of rhythm-related genes (Sections V–X) have been used extensively to investigate features of insect chronobiology that might not have been experimentally accessible otherwise. Most such tests of mutants and molecular-genetic experiments have been performed in *Drosophila melanogaster*. Results from applying visual-system variants have revealed that environmental *inputs* to the circadian clock in adult flies are mediated by external photoreceptive structures (Section II) and also by direct light reception that occurs in certain brain neurons (Section IX). The relevant light-absorbing molecules are rhodopsins and "blue-receptive" cryptochrome (Sections II and IX). Variations in temperature are another clock input (Section IV), as has been analyzed in part by use of molecular techniques and transgenes involving factors functioning near the heart of the circadian clock (Section VIII). At that location within the fly's chronobiological system, approximately a half-dozen–perhaps up to as many as 10—clock genes encode functions that act and interact to form the *circadian pacemaker* (Sections III and V). This entity functions in part by transcriptional control of certain clock genes' expressions, which result in the production of key proteins that feed back negatively to regulate their own mRNA production. This occurs in part by interactions of such proteins with others that function as transcriptional activators (Section V). The implied feedback loop operates such that there are daily variations in the abundances of products put out by about one-half of the core clock genes. Thus, the normal expression of these genes defines circadian rhythms of their own, paralleling the effects of mutations at the corresponding genetic loci (Section III), which are to disrupt or apparently eliminate clock functioning. The fluctuations in the abundance of gene products are controlled transcriptionally and posttranscriptionally. These clock mechanisms are being analyzed in ways that are increasingly complex and occasionally obscure; not all panels of this picture are comprehensive or clear, including problems revolving round the biological meaning of a given features of all this molecular cycling (Section V). Among the complexities and puzzles that have recently arisen, phenomena that stand out are posttranslational modifications of certain proteins that are circadianly regulated and regulating; these biochemical events form an ancillary component of the clock mechanism, as revealed in part by genetic identification of factors (Section III) that turned out to encode protein kinases whose substrates include other pacemaking polypeptides (Section V). *Outputs* from insect circadian clocks have been long defined on formalistic and in some cases concrete criteria, related to revealed rhythms such as periodic eclosion and daily fluctuations of locomotion (Sections II and III). Based on the reasoning that if clock genes can regulate circadian cyclings of their own products, they can do the same for genes that function along output pathways; thus clock-regulated genes have been identified in part by virtue of their products' oscillations (Section X). Those studied most intensively have their expression influenced by circadian-pacemaker mutations. The clock-regulated genes discovered on molecular criteria have in some instances been analyzed further in their mutant forms and found to affect certain features of overt whole-organismal rhythmicity (Sections IV and X). Insect chronogenetics touches in part on naturally occurring gene variations that affect biological rhythmicity or (in some cases) have otherwise informed investigators about certain features of the organism's rhythm system (Section VII). Such animals include at least a dozen insect species other than *D. melanogaster* in which rhythm variants have been encountered (although usually not looked for systematically). The chronobiological "system" in the fruit fly might better be graced with a plural appellation because there is a myriad of temporally related phenomena that have come under the sway of one kind of putative rhythm variant or the other (Section IV). These phenotypes, which range well beyond the bedrock eclosion and locomotor circadian rhythms, unfortunately lead to the creation of a laundry list of underanalyzed or occult phenomena that may or may not be inherently real, whether or not they might be meaningfully defective under the influence of a given chronogenetic variant. However, such mutants seem to lend themselves to the interrogation of a wide variety of time-based attributes—those that fall within the experimental confines of conventionally appreciated circadian rhythms (Sections II, III, VI, and X); and others that consist of 24-hr or nondaily cycles defined by many kinds of biological, physiological, or biochemical parameters (Section IV).

I. INTRODUCTION

The purpose of this monograph is to describe and evaluate essentially all the genetic and molecular studies of biological rhythms in insects that have ever been performed. It may be valuable to have all this information in one place, even though some of the findings to be mentioned or discussed are obscure. Certain of them may be regarded as weird. Nevertheless, it seems as if chronogenetic studies of insects have spread out, often usefully so, into more biological phenomena than has occurred in analogous genetic investigations of other organismic forms. Therefore it seems warranted—if only this one time—to mine every nook and cranny of insect rhythm genetics, large and small. As to the major phenomena, they will be treated in particular detail, for it is possible that some of the specifics of these findings have been glossed over in older or contemporary reviews. Which reviews? It seems impossible to single out any of them (*Gott sei Dank*): I have lost track because such summaries are so numerous that is barely possible to list, let alone keep up with, all the rhythm reviews.

A fair fraction of the recent summaries of chronogenetics and molecular chronobiology take a comparative approach with respect to different systems: for example, how the *Drosophila* and the mammalian ones are at once akin and divergent in terms of interspecifically homologous genes whose functions underlie circadian pacemaking. However, the present work, devoted almost exclusively to the rhythm variants and rhythm-related genes of insects, takes a somewhat different tack, which is tacitly to chronicle this *history* of this business for one group of related invertebrates. What might be the value of this quasi-historical approach? Well, it is arguable that one of, if not the, most important lines of inquiry involving biological rhythms is that which started more than 30 years ago with the discovery of circadian variants in *Drosophila*. Even though these seminal "clock mutants" lay fallow for several years after they were sytematically searched for and isolated, the eventual molecular ascertainment of the key gene so identified proved to be of enormous heuristic value for delving into circadian processes in this insect species *and* in many other kinds of organisms, well beyond invertebrates or even animals. Actually, mutating an equally significant clock gene in the fungus *Neurospora* followed closely on the heels of pioneering the chronogenetic approach in *Drosophila*. As the fly and fungal factors took off during the mid to late 1980s, it began to sink in to an (up until then) indifferent scientific "community" that rhythm geneticists-cum-molecular biologists were onto something. By the early oughts (this century), investigators studying cyclical phenomena from gene-based perspectives in every conceivable type of organism have turned the one-time cottage industry implied in these passages into heavy industrial research. It encompasses biological and molecular rhythms running in forms that range from microbes to mammals, along the way sweeping through invertebrates well beyond *Drosophila*, higher plants, so-called lower verebrates, and culminating if you will at human chronogenetics. A long time ago, the writer of a general-genetics textbook (whose name escapes me) said that someone should erect a large statue of a fruit fly and place it near some prominent scientific institution or the other. Well, at a somewhat smaller level of homage-paying, a similar event might well occur in the vicinity of ground zero for investigative chronobiology, wherever that might be.

How to organize the particulars of this current, claiming-to-be-comprehensive treatment of insect chronogenetics? Such studies of insect rhythms can nicely be summarized in the following order of topics: applying mutations in analyses of *inputs* to the animals' clockworks, searching for mutants that might be defective in *central pacemaking* and analyzing functions encoded by the corresponding genes, and looking for other genetic factors whose functions are involved in *outputs* from the clock. However, telling this story will require description of genetically and molecularly defined clock factors before elements of the input subtopic can be completed. This out-of-order feature of the piece will nevertheless permit the overall story to be told by way of another potentially useful organizing principle: genetic approaches to various components of this insect's

rhythm system described roughly in historical sequence, in terms of the identification and application of chronobiological mutants, followed by isolation and
manipulation of the genes defined by such mutations.

Insect chronogenetics wallows in genetics itself just as much as it deals
with the chronobiology of a limited number of insect species. In truth, most such
studies involve the fruit fly, *Drosophila melanogaster*. In order to cope with the au-
thentic genetic gibberish that routinely revolves round biological considerations
of that dipteran system, it might be useful to peruse the Glossary appended to this
work, which explains the genetic and molecular-biological terms, comprehension
of which may be necessary for the more biologically oriented person who has an
interest in rhythms. Geneticists and molecular biologists may find other entries
within the Glossary valuable for decoding the chronobiological jargon with which
this review is necessarily replete.

II. MUTANTS USED TO IDENTIFY CELLS AND TISSUES THAT MEDIATE INPUTS TO CIRCADIAN PACEMAKERS IN *DROSOPHILA*

A. Eye-removing mutations

The first genetic study of rhythms in *Drosophila* was aimed at asking whether
the photoreception that synchronizes cultures of developing *D. melanogaster* is
"extraocular." It was found that mutant *sine oculis* (so^1) animals are entrainable by
light so that their eventual eclosion is periodic (Engelmann and Honegger, 1966).
That such an insect's (or other invertebrate's) "circadian photoreception" would
not require its eyes was not surprising (reviewed by Hall, 2000). Nor was this particular application of mutant-as-tool to ask this kind of question very meaningful:
A *Drosophila* culture can have the various animals in it synchronized for rhythmic
eclosion at very early developmental stages—not the embryonic one, but as early
as the first-instar larval stage (Brett, 1955). This is well before nonformation of
the imaginal eye in an so^1 mutant culture is at issue, and indeed photic stimuli
that induce certain acute behavioral responses are unaffected by *so* mutations
(Hotta and Keng, 1984; Sawin-McCormack *et al.*, 1995). That said, it may be
valuable to register that another *sine oculis* allele, so^{mda}, causes more severe early-
developmental effects, in that larval photoreceptors fail to differentiate (Serikaku
and O'Tousa, 1994); the so^{mda} mutant has been applied in a *neuro*-chronobiological
context (Malpel *et al.*, 2002), but not yet in one that involves the functional
meaning of circadian photoreception.

B. Neural mutants, both anatomical and biochemical

For that kind of functional test, it was more useful to ask whether eyeless so^1
adults would be light-entrainable for locomotor-activity rhythms, which they are.

A Daily locomotor cycles of wild-type *D. melanogaster*

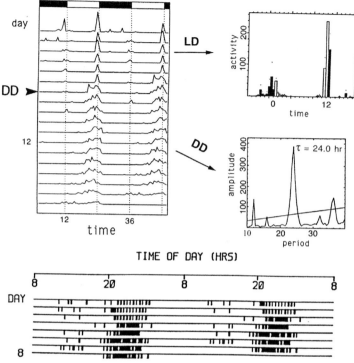

Figure 1. Locomotor activity of *Drosophila melanogaster* adults. (A) On the left, an actogram showing intraday and day-to-day fluctuations of general locomotion for an individual wild-type fly. It was placed in a food-containing glass tube (4 mm inside diameter ×6 or 7 cm), flanked by an infrared-light emitter/detector pair; when the fly walks down the tube and breaks the light beam (operating at a wavelength outside the spectrum to which this species's visual or rhythm system responds), a locomotor event is recorded by a computer [see Hamblen *et al.* (1986) for a description of such an automated activity monitoring setup, including a drawing of the glass-tube/light-beam arrangement]. Days 1–2 of such activity are plotted on the top line, with the height of a bump or squiggle representing the relative magnitude of the locomotor movements recorded; such behavior on days 2–3 is plotted on the next line down; etc. This double-plotted record thus depicts successive days of locomotor fluctuations horizontally and vertically as well. For the latter, it is evident that peaks of locomotion occurring at times 12 and 24 (for example) are stacked right on top of one another within the top half of the actogram. For this, the fly was in 12-hr light:12-hr dark (LD) cycles (medium-level white light, marked by the white horizontal bars at the very top), and it "entrained" to this environmental condition, that is, behaved in synchrony with the LD cycles. The essentially perfect vertical stacking of the locomotor peaks means that they occurred at the same time every morning and evening. Note that the morning peak involves less activity than the evening one, which is typical of this species's behavior. After about 1 week of LD monitoring the lights were turned off; the beginning of this

continues

B   Locomotor arrhythmicity of the *disco* mutant

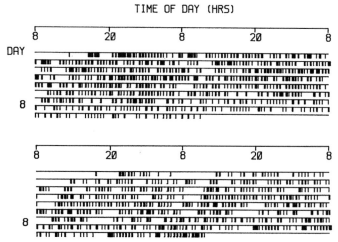

Figure 1. (*continued*) constant-dark part of the test is denoted by DD. The robustness of the locomotor peak near the beginning of the "subjective day" (chronobiological jargon) diminishes or disappears in DD; the evening peak (or plateau), however, persists in this condition and broadens to encompass a higher proportion of a daily cycle compared with LD-entrained behavior (again, both such DD-behavioral attributes are typical). The time of the subjective-evening peaks drifts slightly to progressively earlier onsets in DD toward the bottom of the actogram because this behavior is "free-running," that is, endogenously controlled, with no light-induced daily reset. On the top right such LD resetting is revealed by a different kind of plot, in which the average activity per time bin (30 min in this case) per day is indicated by a given histogram bar (above each one is a dot, denoting the standard error associated with that per-day mean); locomotion during the dark phase, black histogram bars; during the light, white bars. Behavioral anticipations of lights-on and of lights-off are evident in this average-activity plot. In the middle right a periodogram is plotted, which is the result of analyzing DD behavior in this manner—actually by applying chi-square periodogram functions (Sokolove and Bushell, 1978), such that a given ordinate value ("amplitude," in dimensionless units) which crosses the "significance line" ($\alpha = 0.05$) indicates significantly periodic behavior for the corresponding time value on the abscissa. The weak persistence of the morning peak in DD is such that this fly's behavior is defined by meaningful (peak) period values (τ's) of 12 and 36 hr; however, the most robust periodogram peak is clearly that corresponding to 24 hr, reflecting the major evening plateau of locomotion in DD (a high-amplitude harmonic peak at 48 hr would be evident as well had the periodogram analysis been extended out to that abscissa value). This 24-hr value is taken to be the best estimate of this fly's free-running locomotor rhythm, a typical result for adults of this *Drosophila* species. However, there is no canonical such value for *D. melanogaster* in the literature; the average reported τ's for a given group of wild-type controls drift around in a range between ca. 23.5 and 24.5 hr; other organisms, in contrast, seem to given more standard periodicities of free-running behavioral rhythmicity [e.g., the value for wild-type *Mus musculus* is typically about $23\frac{3}{4}$ hr, as exemplified by Vitaterna *et al.* (1994) and Antoch *et al.* (1997)]. At the

Flies expressing so^1 along with a *small-optic-lobes* (*sol*) mutation were also found to behave in synchrony with light:dark (LD) cycles (Helfrich and Engelmann, 1983; Helfrich, 1986). Application of optic-lobe-deranged mutants was not very definitive in terms of a circadian-photoreceptive structure, a pacemaker one, or both being in these visual-system ganglia or not, because *sol* mutants and the like retain a fair fraction of these tissues. In any case, extraocular photoreception is sufficient to entrain this imaginal behavior. By the way, would-be ocular reception that might feed into the rhythm system from the ocelli seems unnecessary because all so^1 individuals lack these "simple" eyes. (Most of them are devoid of the compound eyes as well, and those that form a vestige of these structures are discarded before their throughly eyeless siblings are monitored for locomotor behavior.)

A potential bonus from behaviorally testing eyeless and optic-lobe-defective mutants was that some such flies exhibited free-running rhythms with complex patterns (Mack and Engelmann, 1981; Helfrich and Engelmann, 1983; Helfrich, 1986). In this constant-dark condition (DD), locomotor maxima—which are extensions of the evening peaks normally observed in prior (LD) cycles (Figure 1A)—sometimes split into two free-running components. It follows that one such component in an so^1 record of this type has a different circadian period τ from the other. These anomalies are not observed with other externally blind mutants (discussed later), which implies that *sine oculis* flies are brain-damaged as well as eyeless. They are, as is discussed in Section X.B.1.

Complex rhythms were also observed in higher-than-normal proportions in *sol* flies and those expressing another optic-lobe mutation, *lobula-plateless* (Helfrich, 1986). Therefore, on the one hand, it may be useful to delve into these mutant brains, as they could help identify tissue structures involved in the generation of rhythmic behavior (Section VI and the follow-up minidiscussion within Section X.B.1). On the other hand, the unexpected abnormalities of free-running rhythms in these supposed visual-system-specific mutants leads to a warning about applying such mutations as tools. They may cause *pleiotropic defects* that could undermine interpretability of the desired main effect.

Another form of anatomical eyelessness can be caused by the combined effects of *eyes-absent* (*eya*) and *ocelliless* (*oc*) mutations. This genotype allows

bottom, a raster-plot actogram for a wild-type fly's behavior in DD is double-plotted, with the density of vertical markings reflecting numbers of light-beam crossings during a given 30-min time bin; such marks blend together into horizontal bars' worth of activity during times of heavy locomotion during the subjective evening. (B) Actograms showing aperiodic behavior of the *disco* mutant in DD. Most such individuals (like the two whose locomotion is plotted here) behave in an arrhythmic manner in this condition (see text): Activity events occur arbitrarily with respect to time, so a bin chosen from any portion of a *disco* mutant's daily cycle could just as easily contain locomotor markings or not, whereas the wild type's behavior is such that a bin at hour 8 (or even 14) has a low chance of being marked, but one in the 20- to 24-hr range is quite likely to be filled (see A).

for cleaner free-running behavioral rhythmicity compared with that affected by so^1 (Frisch et al., 1994). Flies homozygous for either eya or so^1 were tested further in constant darkness (DD), but here there was imposed a 12-hr: 12-hr high-temperature:low-temperature cycle; in that varying condition the mutants were initially monitored in constant light (LL), then shifted to DD (Tomioka et al., 1998). The purpose was to examine photic modulation of behavioral rhythmicity as it is entrainable by temperature cycles (cf. Section IV.C.1). The locomotor changes observed after the LL → DD transition (roughly speaking, relatively more activity during the cryophase compared with the thermo one) in flies with eyes were not observed in the eyeless mutants (Tomioka et al., 1998). It was concluded that the compound eyes are the principal input route for light as it enters the system to "cooperate" (as these authors would have it) with the effects of temperature on Drosophila's locomotor rhythmicity.

To enhance the interpretability of visual-variant effects on rhythm entrainability, mutations beyond the externally eye-removing so, eya, and oc ones were applied. Thus, physiologically blind no-receptor-potential-A (norpA) and optic-ganglia-blind disconnected (disco) mutants were found to entrain, such that all individuals behaved in synchrony with 24-hr LD cycles and with one another (Dushay et al., 1989; Hardin et al., 1992a; Wheeler et al., 1993). disco mutants are also arrhythmic in free-running conditions (Figure 1B), which will be taken up in more detail within the neural-substrates subtopic (Section VI.A). Another complexity associated with these visual-system mutants is that norpA, which encodes a phospholipase C (PLC) involved in phototransduction within the compound eye and ocelli (Montell, 1999), causes slightly faster-than-normal free-running rhythms when mutated (Dushay et al., 1989). Perhaps this is a brain effect, by analogy to what is arguably the case for so^1. Indeed, the significance of the norpA gene is not limited to the visual system: This gene is expressed in the central nervous system (CNS) as well as in the external eyes (Zhu et al., 1993), and mutants involving this locus are defective in chemosensory responses as well as light-mediated ones (Riego-Escovar et al., 1995). An alternative to the notion that norpA's short-period phenotype results from the mutation's pleiotropy is that attenuated light input, in advance of testing the flies in DD, affects eventual function of the pacemaker. Consistent with this supposition is that norpA can severely diminish larval vision with regard to various features of the animal's light-induced behavior (Hotta and Keng, 1984; Busto et al., 1999; Hassan et al., 2000), if not all such photomodulated locomotion (Sawin-McCormack et al., 1995). Yet the PLC encoded by this gene is found in Bolwig's nerve, which projects centripetally from the one known larval photoreceptive organ and contacts putative circadian-related neurons within the larval brain (Kaneko et al., 1997; Malpel et al., 2002; see Section VI.A.1). Based, therefore, on the reasoning that norpA may diminish light input to the clock system during development—if not throughout larval life, then, say, during metamorphosis—the following environmental manipulation suggests itself: Constant-light rearing, compared to the consequences of DD or LD conditions, might be expected to cause lengthened behavioral periods (the opposite of *norpA*'s effect on adult rhythmicity). This suggestion connects with further ambiguities, which stemmed from LL-rearing tests performed on wild-type flies. These puzzles arise in the context of mutations that affect circadian pace-making itself (Power *et al.*, 1995b; Tomioka *et al.*, 1997). Such light-manipulation experiments will be discussed (Sections IV.C.1 and IX.B.) after *Drosophila*'s clock mutants have been introduced (Section III.A).

C. Rhodopsin's role as a circadian-input factor and inferences about contributions from other light-absorbing substances

None of the results of applying visual-system mutants in tests of the adult fly's entrainability to LD cycles indicate that the circadian photoreception subserving this process is entirely extraocular—instead, that external photoreceptors are not necessary. A relevant object lesson came from visual-system manipulations of *D. pseudoobscura* in terms of light-induced phase shifting of eclosion peaks. For this, the results of vitamin A deprivation suggested not only that the photoreceptive structures and functions can be extraocular, but also that that input route to the eclosion clock is the only pertinent one: The severely retinal-depleted animals exhibited no decrement in sensitivity to the effects of light on phase shifting (Zimmerman and Goldsmith, 1971). Many years later, this issue was taken up with regard to adult locomotor rhythms. It was found that the mutants *so*[1] or *norpA*, doubly variant adults expressing the *oc* and *eya* mutations, or retinal-depleted (wild-type) flies each exhibit much reduced sensitivity in terms of entrainment in LD cycles, including the ability to re-entrain to shifted such cycles (Helfrich-Förster, 1997a; Stanewsky *et al.*, 1998; Ohata *et al.*, 1998).

Therefore, it seems as if circadian photoreception is mediated partly by external photoreceptors and partly by putative internal ones. This could provide versatility to the system. Retinal photoreceptors contain opsins that absorb maximally in the blue-green or ultraviolet (UV), depending on the cell type (Montell, 1999). Perhaps the extraocular circadian photoreceptors possess substances absorbing most robustly at wavelengths intermediate between those just implied. This suggestion does not come out of the blue, but is based on the following: The nominal peak for light-induced eclosion phase-shifting is in the <500-nm range (Frank and Zimmerman, 1969; Klemm and Ninnemann, 1976), as if the (solely) extraocular photoreceptive functions underlying that process involves a flavoprotein functioning as a blue-light receptor. In adults, the spectral-sensitivity peak for locomotor-rhythm entrainment was shifted to shorter wavelengths (towards blueness, around the mid-400-nm range) when *norpA*, *so*, or retinal-depleted flies were so tested (Helfrich-Förster, 1997a; Ohata *et al.*, 1998).

However, the results of applying different colors in behavioral experiments are complex: First, note that behavioral phase shifts induced by *short* light pulses exhibit maximal sensitivity in the range of 400–500 nm and a sharp drop-off beyond 500 nm (Suri *et al.*, 1998); a subsequent, more refined experiment revealed peak sensitivity at 475 nm (Suri, 2000). Second, certain behavioral tests involving adults monitored in LD cycles showed that re-entrainment could be mediated by low-intensity light in the 650- to 700-nm range, and such red light (throughout the 12-hr L phase) was effective even for eyeless or *norpA* mutants (Zordan *et al.*, 2001). These results are superficially inconsistent with those of Ohata *et al.* (1998), which were based on application of the *so¹* mutation and retinal depletion. However, it is difficult to compare the criteria in these two studies in terms of their assessments as to whether a given fly entrained. The extremely weak behavioral phase shifting induced by 650-nm light pulses (Suri *et al.*, 1998) compared with the effective role of red light in one study of entrainment regimens (Zordan *et al.*, 2001) may reflect differences between the effects of photic manipulations in light-pulse versus LD-cycling experiments. This issue will reappear when the input subject is revisited (Section IX.B.3) after molecular chronogenetic factors and phenomena are in hand.

    Whatever light-absorbing molecules may be functioning in extraocular photoreceptors, where might such structures lurk? They could include pacemaker cells themselves, as will necessarily be dealt with later (Section IX.B.3), based in part on the expression of clock genes in certain brain neurons (Section VI). Previous to those studies, attention had been drawn to an entity known as the Hofbauer–Buchner (H–B) eyelet, named after its discoverers, who found it to be labeled by application of a photoreceptor-specific antibody; this reagent led to staining of all known photoreceptors in the fly head and of the novel structure (Hofbauer and Buchner, 1989). The H–B eyelet is situated underneath the retina, between it and the distalmost optic lobe (Hofbauer and Buchner, 1989; Robinow and White, 1991; see Figures 11 and 17 in the color insert, as discussed in Sections VI and IX). Subsequently, cells within the eyelet were found to contain at least one opsin subtype (Yasuyama and Meinertzhagen, 1999; Malpel *et al.*, 2002), but it is unknown whether the H–B structure also possesses blue-absorbing material. However, it is intriguing that, during the midpupal stage, these cells send projections into a provocative region of the adult CNS (Hofbauer and Buchner, 1989; Yasuyama and Meinertzhagen, 1999; Malpel *et al.*, 2002). As the molecular and the neural-substrate sections (Sections V and VI) will reveal, the external eyes and the H–B eyelet do not form the entirety of the circadian-photoreceptive system. One result that created this scenario can be described now, and it involved application of yet another visual-system mutation. This is a *glass* allele (gl^{60j}), which eliminates all known external photoreceptors on the fly's head and the H–B eyelet as well (Helfrich-Förster *et al.*, 2001). Yet, these mutant adults still entrained in light: dark conditions re-entrained to shifted LD cycles (Helfrich-Förster *et al.*, 2001). The reason for this result is that the gl^{60j} mutant flies are only doubly defective. The most interesting application of this mutation involved combining gl^{60j} with a mutation in a *D. melanogaster* gene that encodes a blue-light-absorbing protein called cryptochrome, a full discussion of which (in Section IX.B) must wait until treatments of this organism's chronogenetics and molecular chronobiology have been plowed through.

III. MUTANTS APPARENTLY DEFECTIVE IN CENTRAL-PACEMAKING FUNCTIONS UNDERLYING *DROSOPHILA'S* CIRCADIAN RHYTHMICITY

A. Presumed clock mutants

1. *X*-chromosomal mutations at two loci

The philosophy that underpinned systematic searches for rhythm mutants in *Drosophila*, as an entrée into the mysteries of circadian clocks, led to one of the triumphs of "forward-genetic" analysis of complex processes. R. J. Konopka sensed that mutating functions acting at or near the heart of central-pacemaking functions might be the only way to identify such factors. In contrast, others thought that working inward from clock-input functions or working back from clock outputs might arrive, or even collide, at the central pacemaker (e.g., Johnson and Hastings, 1986). Few to no meaningful steps were taken in either such experimental direction, each of which necessarily involved starting with known entities: for example, a light input that might cause, as its first change, an alteration in the quantity or quantity of a particular piece of cellular biochemistry; or, from the opposite end of the overall process, some final feature of clock output being a known biochemical parameter. If, however, one were to induce mutations in a gene that encodes a core element of the clock, one makes no *a priori* assumptions about what kind of molecular factor will be mutated.

Thus, in the summer of 1968, R. J. Konopka began testing the descendants of chemically mutagenized *D. melanogaster* for strains in which eclosion-rhythm abnormalities manifested themselves in LD cycles. (Normal free-running eclosion and abnormalities of it caused by a more recently identified *per* mutation are shown in Figure 2.) Konopka's original search for rhythm mutants resulted in an aperiodic strain during that fateful summer. Cultures of this genetic line, found among the first 200 lines monitored by the investigator's eye (as recounted by Greenspan, 1990; Weiner, 1999), produced no apparent daily peaks of adult emergence in constant darkness. After approximately 1,800 additional strains were screened (Konopka and Benzer, 1971), two further variants were isolated: one with eclosion peaks spread only about 19 hr apart ($per^S = period^{Short}$), the other with ca. 29-hr free-running periods ($per^L = period^{Long}$).

As a result of this successful mutant hunt, three chronogenetic object lessons came rapidly to the fore: (i) It was neither impossible nor too easy to induce and isolate rhythm variants, in these cases, ones exhibiting rather strikingly

A Normal free-running adult-emergence rhythm of D. melanogaster and aperiodic eclosion of disco

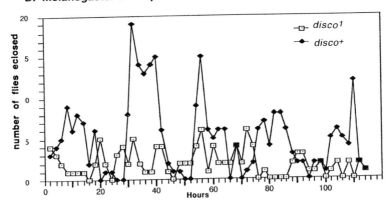

B Eclosion defects caused by a period mutation

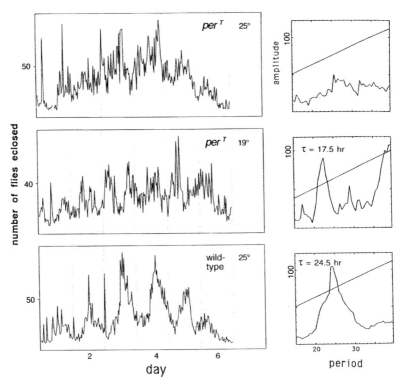

abnormal circadian rhythmicity. If something like one-fifth or one-third of the candidate strains exhibited rhythm defects, this might imply that almost "any" gene could cause such abnormalities when mutated, and that these genetic changes would usually be causing rhythm defects by indirect, epiphenomenological routes. (ii) The three original mutations all mapped to the same genetic locus on the X chromosome, which (as alluded to above) was called *period* (*per*). [See Figure 18 in the color insert (referred to out of order here because this diagram, discussed at the end of this work, is mainly meant to summarize most of its subjects).] These *per* mutations need not have been allelic, for each mutation was isolated independently—as opposed to, for example, per^S having stemmed from putting mutagenized X's in heterozygous condition with the arrhythmia-inducing per^0 mutation to search for new mutant alleles by noncomplementation with the original one. This was the first, but not the last instance of "repeat hits" occurring at a given rhythm-related locus; the implications of the eventual array of loci and their independently isolated mutant alleles will be considered (Section IV.B) once all such genes are verbally on the table (literally in Table 1; also see Figure 3, color insert). (iii) If one induces a mutation causing rhythm defects, it need not have identified a clock gene. For example, per^0 could be brain-damaged and arrhythmic for reasons

---

Figure 2. Emergence of flies from *D. melanogaster* cultures. (A) Normal eclosion in constant darkness and that affected by a *disco* mutation. Two sets of *Drosophila* cultures—one free of rhythm-affecting mutations, the other true-breeding for the *disco¹* mutation—were reared in parallel in LD, but transferred to DD as the animals proceeded into metamorphosis. Numbers of flies emerging from the pupal stage into adulthood were counted every 2 hr (by hand) for about 4.5 days. *disco⁺* eclosion [which did not involve wild-type cultures, but instead those in which females bore an "attached-X" chromosome (don't ask)] is defined by emergence peaks broadly distributed over the subjective-morning hours (cf. Figure 1) on successive days. In contrast, *disco*'s eclosion involves jagged peaklets and short-duration troughs that occurred in a temporally arbitrary manner. This figure was adapted from Dushay *et al.* (1989), who formally analyzed these eclosion profiles and found significant 24-hr periodicity for the control, none for *disco*. (B) Free-running eclosion affected by a fast-clock *period* mutation. These three plots stemmed from automated eclosion monitoring in DD. By a method reported first in Konopka *et al.* (1994), emerging flies drop from a disc onto which pupal cases were glued into a funnel whose stem is surrounded by a ring of infrared emitter/detector pairs (cf. Figure 1); beam breakages lead to digitizable signals that are accumulated for several days' worth of 30-min time bins. As shown in the top panel, cultures true-breeding for the per^T mutation (cf. Table 1) reared at 25°C exhibited sloppily peaky eclosion that was not significantly periodic—indicated by the chi-square periodogram on the right (cf. Figure 1). When this mutant was reared at 19°C (middle panel), eclosion defined by 17.5-hr periodicity resulted. This period value is longer than that for a per^T locomotor activity rhythm, as shown in Figure 4, reflecting the slower clock pace for this mutant at lower temperatures (Konopka *et al.*, 1994). Genetically normal (wild-type per^+) cultures (bottom panel) led to 24.5-hr-periodic eclosion (cf. control profile in A). These plots were adapted from Hamblen *et al.* (1998).

Table 1. Rhythm Mutants in *Drosophila* Stemming from Searches for Genetic Variations Causing Abnormalities of Periodic Eclosion, Adult Locomotion, or Clock-Gene Expression[a]

	Mutated alleles	Fully mutant phenotypes	Phenogenetic details/remarks
Major mutants			
Clock			
	Clk^{Jrk}	Arrh.: E, B	Over +: ca. 50% arrhythmic; homozygous: anomalous responsiveness in LD (no morning B peak)
	Clk^{1-7}	Arrh.: B	Fully recessive; E not tested
cryptochrome			
	cry^b	*per, tim* expression arrh.	Free-running (f-r) B rhythmic, but reduced sensitivity to re-entraining light stimuli; by itself, blind to phase-shifting (B) effects of short light pulses; when combined with *glass* mutation: ca. no re-entrainability; molecular and odor-sensitivity rhythms antenna absent
cycle			
	cyc^{01}	Arrh.: E, B	Over +: variously found to give 1-hr-longer-than-normal period or wild-type-like durations of behavioral cycles; homozygous: anomalous responsiveness in LD (see Clk^{Jrk})
	cyc^{02}	Arrh.: B	See cyc^{01}
double-time			
	dbt^S	18: B; 20: E	Over +: 22 (B)
	dbt^L	27: B (over *Df* or dbt^P)	Over +: 25 (B)
	dbt^P	Lethal	Over +: normal B; a.k.a. *dco*, defined by several allelic lethals in addition to dbt^P
	dbt^{PexlX}	27 or arrh.: B	Imprecise excision of transposon in dbt^P; ca. 60% of individuals tested arrh.
	dbt^g	Arrh.: B (over *Df* or dbt^P)	Over +: 29; not homoygosable because linked lethal was coinduced
	dbt^h	29: B	Over +: 25
	dbt^{ar}	Arrh.: B	Over +: 29
lark	sic	Early phase: E	E effect occurs in *lark*/+ cultures; mutation is recessive lethal; B: normal; see Table 6 for more
period			
	per^{01}	Arrh.: E, B	Over +: ca. 24.5 (B)
	per^{04}	Arrh. or >30: B	Most hemi/homozygotes arrh.; over +: ca. 24.5 (B)
	Df-per/Df-per	Arrh.: B	Hyperactive in f-r B tests
	per^S	19: E, B	Over +: 21.5
	per^{Clk}	22: E, B	Over +: 23
	per^T	16–17: E, B	Eclosion arrh. at 25°C, ca. 17 at 19°C; over +: 20 (B)

Table 1. *continued*

	Mutated alleles	Fully mutant phenotypes	Phenogenetic details/remarks
	per^{LI}	29: E, B	Over +: 26 (B)
	per^{Lvar}	Arrh. or ca. 30: B	Most hemi/homozygotes arrh.; over +: 25 (B)
	per^{SLIH}	27: E, B	Over +: 26 (B); serendipitously found in laboratory strain
shaggy			
	$sgg^{EP(X)1576}$	21: B	Dominant effect of UAS-containing *EP* transposon inserted at *sgg* locus and driven by *tim-gal4* transgene (thus, *sgg* overexpression)
	sgg^{MII}	25: B	Lethal allele rescued to adult survival (for B testing) by activation of *heat-shock–sgg* transgene
	sgg^{D127}	26: B	See sgg^{MII}
Timekeeper			
	Tik, a.k.a. CK2α^{Tik}	Ca. 26.5: B	Isolated behaviorally as a dominant lengthener of per^S; homozygosity for the (third) chromosome on which Tik induced: lethal (likely due to Tik itself); an intragenic revertant (R) strain was identified in which TikRl+ flies exhibited ca. 24.3-hr periods (compared with the 23.6-hr controls)
timeless			
	tim^{01}	Arrh.: E, B	Fully recessive
	tim^{03}	Arrh.: B	Isolated by causing *per* noncycling
	tim^{04}	Arrh.: B	Isolated by causing *per* noncycling
	tim^{S1}	21: B	Over +: 22 (B)
	tim^{S2}	22: B	Over +: 23 (B)
	tim^{L1}	28: E, B	Over +: 26 (B)
	tim^{L2}	26: E, B	Over +: 25 (B)
	tim^{L3}	28: B (over *Df*)	Over +: 26 (B); not homozygosable because linked lethal was coinduced
	tim^{L4}	28: B	Over +: 26 (B)
	tim^{UL}	33: B	Over +: 26 (B)
	tim^{rit}	26: B	Over +: 25 (B); naturally occurring amino-acid substitution
	tim^{SL}	Normal: B	Isolated by causing per^L to be ca. 26
Minor mutants			
AC-72	sic	Arrh.: B	Ca. one-third of individuals arrh.
Andante			
	And	26: E, B	A.k.a. *Andante*; over +: ca. 25
	And-like dy^{n1}	25: B	Isolated as newly induced *dusky* (*dy*) wing mutant (other such *dy*'s are rhythm-normal)
	And-like dy^{n3}	26: B	See dy^{n1}
	And-like dy^{n4}	25: B	See dy^{n1}

continues

Table 1. *continued*

	Mutated alleles	Fully mutant phenotypes	Phenogenetic details/remarks
clock-7			
	cl-7	Arrh.: E, B	Induced in *D. pseudoobscura* and isolated as E arrh. in LD or DD; also arrh. B
	cl-7^{10}	Arrh.: E, B	Recessive effects of this mutation and of *cl-7* on B permitted complementation test, which showed the former to be allelic to the latter (nee *cl-10*) in terms of arrh. B; however, *cl-7/cl-7^{10}* LD-rhythmic for E (late phase) and weakly E rhythmic in DD (long period)
clock-8			
	cl-8	Arrh.: E, B	See *cl-7*, and note that the second of the *cl-8*
	cl-8^9	Arrh.: E, B	mutations (née *cl-9*, isolated as E arrh. in DD, though E rhythmic in LD) proved to be allelic to *cl-8* in terms of arrh. B; however, *cl-8/cl-8^9* LD-rhythmic for E (late phase) and weakly E rhythmic in DD (long period)
E64	sic	Arrh.: B	90% of individuals arrh.
EJ12	sic	Arrh.: B	Ca. two-thirds of individuals arrh.
gate	sic	Arrh.: E	Goes arrh. after 2–3 f-r E cycles; may be allelic to *psi-2*
hpa	sic	Quasi-arrh.: E	Identified in *D. jambulina*; E temperature-entrainable (in constant photic conditions) and f-r thereafter in LL
hra	sic	Arrh.: E	Identified in *D. jambulina*; E temperature-entrainable, but only in DD; no f-r or thermal condition arrhythmicity following any photic
linne	sic	Arrh.: E	Identified in D. subobscura; E temperature-entrainable and f-r thereafter in DD or LL
ml20	sic	Early LD phase: E	B: normal f-r period
psi-2	sic	Early LD phase, f-r 25: E	B: anomalous LD phases and lengthened f-r period
psi-3	sic	Early LD phase, f-r 25: E	B: normal
restless	sic	Arrh.: B	Ca. half of individuals arrh.
Toki	sic	25: B	Over +: ca. 24.5

[a]B, behavioral rhythm phenotype; E, eclosion phenotype. Such phenotypes involve: (a) free-running (f-r) circadian periods (τ's) that are appreciably different from the normal value: ca. 24 hr (h) for *D. melanogaster* (Figure 1), in which all of these mutants were identified except for *cl-7* and *cl-8* (*D. pseudoobscura*) *hpa* plus *hra* (*D. jambulina*), and *linne* (*D. subobscura*); (b) arrhythmicity (arrh.); or (c) in a few cases altered phases in light:dark (LD) cycling conditions. τ's are in most instances rounded to the nearest hour, except for mutant types that are routinely found to be intermediate between two hourly values. Under "Mutated alleles," sic means that the term in the lefthand column is the designator for the only mutation induced. In some of these cases (except for *dy*, *gat*, the *psi*'s, and *Toki*), the mutated loci have not been mapped. Most mutations listed have had their chromosomal positions

similar to the effects of ablating a mammal's suprachiasmatic nucleus (SCN) in the hypothalamus (reviewed by Weaver, 1998)—not that such a neuroanatomical variant in *Drosophila* would be uninteresting. Another possibility is that an arrhythmia-inducing mutation alters a factor involved in ongoing function of the system (not its morphological features), but that its duty is to operate within an "output pathway." In terms of per^0's isolation phenotype, the pathway would end at the proximate regulation of periodic eclosion.

As the further implications of point (iii) are considered, this section is not spinning off into an overly discursive treatment of the subject: These implications establish various core tenets of the genetic approach to the clock, ones that will hold the subject in good stead with respect to further mutations and additional clock genes. Thus, regarding per^0 and brain damage, consider that the same gene can also mutate to cause a fast clock and a slow one in the per^S and per^L mutants. This suggests that the original arrhythmic mutant suffers from a clock-functional problem, possibly a pacemaker collapse. Moreover, the putative function in question was quickly shown to be potentially a broad one, for all three *per* mutations cause parallel defects in the rhythmicity of adult locomotion (Konopka and Benzer, 1971). Yet it should also be possible to find an eclosion and behavioral mutant in which anatomical structures that underlie both rhythms are disrupted (see Figures 1 and 2, and further discussion of the *disco* mutant, e.g., Section VI.A.2).

R. J. Konopka searched for newly induced rhythm mutants for about 20 years, choosing to track X chromosomes transmitted by mutagenized males. Three further arrhythmic *per* mutants were induced and isolated (Konopka, 1988; Hamblen *et al.*, 1986; Hamblen-Coyle *et al.*, 1989). Two turned out to be

---

determined (Figure 18) in part by applications deficiencies, a.k.a. *Df*'s. This abbreviation is standard nomenclature (in *D. melanogaster*) for designating a deficiency (intrachromosomal deletion) of the pertinent locus—here, *dbt* or *per* (for the latter, certain partly overlapping *Df*'s, each of which removes the locus, allow for viable *Df-per/Df-per* flies that have been B-tested). The per^{02} and per^{03} mutants are not listed because their nonsense-mutation-creating nucleotide substitutions and known phenotypic effects are identical to those of per^{01}. "tim^{02}" is not listed because it is a recessive lethal second-chromosomal *Df* that is deleted of more than the *tim* locus. For several mutants, the B phenotype only is indicated, which does not imply that E is necessarily normal; instead: not tested. One of the minor mutants (*psi-3*) has its phenotype designated by E only; in this case, only that rhythm was found to be aberrant. This mutant and others in this category are denigrated as "minor" because, in the first place, they cause rhythm abnormalities that are mild, erratic, or both; and in the second place, most such mutants have not been extensively analyzed, even in terms of their isolation phenotypes (usually involving abnormal B), let alone by subsequent study (e.g., neurobiological, molecular). Most of the mutants listed in either category have appeared in primary reports (see main text); exceptions: Clk^{1-7} (R. Allada and M. Rosbash, unpublished); tim^{04} (R. Stanewsky and J. C. Hall, unpublished); AC72, *cl-7, cl-8, EJ12, restless* (see Hall, 1998a); *m120* (see Sehgal *et al.*, 1991). Also, the case of *And* (whose incipient molecular genetics is on the way and could boost this factor into the top section of the table) has only just begun to be clarified in terms of its relationship to wing-affecting mutations in the "*m-dy* complex" (DiBartolomeis *et al.*, 2002).

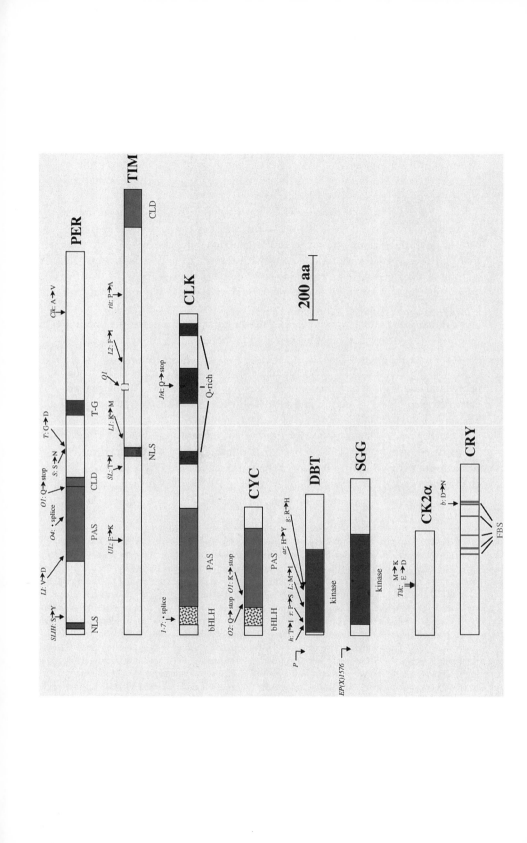

Figure 3. Proteins encoded by major rhythm-related genes in D. *melanogaster*. The products of genes originating from mutagenesis and screening for circadian-rhythm abnormalities are diagrammed. The eight genes chosen are those that can cause substantial such phenotypic changes when mutated and encode products deemed to play major roles in circadian pacemaking. The lengths of these proteins in numbers of amino acids (aa) are indicated by the scale bar; the longest such polypeptide consists ca. 1420 of residues (long form of TIMELESS; Rosato *et al.*, 1997b), the shortest ca. 335 (CK2α). Some of these protein sizes should be regarded as approximate or average values, owing to polymorphic forms identified in different D. *melanogaster* strains (e.g., PER, nominally 1,224 aa according to http://flybase.bio.indiana.edu/) or different protein isoforms that result from separate translation-initiating codons [within the open-reading from (ORF) of a given gene], possibly in conjunction with alternative splicing inferred from cDNA characterizations. For example, CLK nominally consists of 1,023 aa as shown here, but truncated forms of this protein may be produced (Allada *et al.*, 1998; Darlington *et al.*, 1998), and a putative (full-length) isoform was inferred to be produced by a transcript type containing an mini-exon that encodes an extra 4 aa (Bae *et al.*, 1998). The SGG kinase exists in multiple forms, ranging, for the three examples reported by Siegfried *et al.* (1992), from proteins containing ca. 500 to 750 aa (also see below). In this diagram, the sites of the induced mutations, usually nucleotide substitutions, are indicated above the rectangles by aa number (starting from the N-terminal methionine, #1) and either the resulting aa change; or a stop codon; or the alteration of a splice-acceptor site in *per*^{04} or of a splice-donor one in *Clk*$^{1-7}$ (Δ-splice: at the 3′ end of *per*'s third intron and at the beginning of *Clk*'s second intron); or an intragenic deletion [bracketed] in *tim*^{01} (starting at the codon specifying long-form aa 746 and missing ca. 21 residues' worth of codons; but in an out-of-frame manner such that 35 additional aa are encoded, followed by a premature stop codon that would be encountered by the translational machinery). The often cryptic single-letter aa abbreviations, resorted to for space-saving, are S = serine, Y = tyrosine, V = valine, D = aspartate, N = asparagine, G = glycine, A = alanine, E = glutamate, K = lysine, T = threonine, I = isoleucine, M = methionine, F = phenylalanine, Q = glutamine, P = proline, R = arginine, H = histidine. The transposon (P-element) insertion at the *double-time* locus that resulted in the lethal *dbt*^{P} mutation (which also causes molecular abnormalities associated with clock factors in larval-brain neurons) is shown to the left of DBT's N-terminus because this element is inserted in the intron between *dbt*'s second and third exons (which are noncoding, the entire open reading frame being contained in the fourth exon). Similarly, the *EP(X)1576* mutation is shown to the left of the apparently relevant open reading frame within the *shaggy* gene because this *EP* transposon is inserted ca. 1 kb upstream of the translation-start site for a 496-aa SGG-kinase isoform (known as SGGY or ZW3-C; Siegfried *et al.*, 1992) that is at higher than normal levels in the *sgg*$^{EP(X)1576}$ mutant whose embedded (yeast) UAS sequences are driven by brain-expressed, transgene-encoded (yeast) GAL4.

With regard to the CK2α polypeptide near the bottom, previous identification of sequences encoding this enzymatic catalytic subunit caused the mutationally defined gene name *Timekeeper* (*Tik*) to be renamed by an italicized version of the abbreviation for the casein kinase 2α subunit (Table 1); the CK2α^{Tik} mutant was found to contain two aa-substituting base-pair changes, as shown. Note that some of the mutations listed in Table 1 have not been molecularly characterized (e.g., several of the *tim* variants, the two *sgg* mutations other than the *EP* allele). Most of the intrapolypeptide domains referred to in the text are indicated by rectangles or vertical lines within the main rectangle; abbreviations: NLS = nuclear-localization sequence; PAS = PER/ARNT/SIM residues that are similar in sequence among relatively N-terminal segments of these three proteins and others, including CLK and CYC as shown (not all of PER's PAS is delineated in this diagram because molecularly identical to the original such mutation (Hamblen-Coyle *et al.*, 1989), which is now called *per*^{01}. The fourth (*per*^{04}) is molecularly unique (Figure 3; see color insert) and allows for barely appreciable and extremely long period (>30-hr) free-running rhythms in a small fraction of the flies monitored (Hamblen-Coyle *et al.*, 1989). As well as isolating *per*^{02} through *per*^{04}, Konopka induced additional period-altering mutations at this locus (Table 1 and Figure 3): (i) *per*^{L2}, which turned out to be caused by the same molecular change harbored by *per*^{L1}, née *per*^{L} (Gailey *et al.*, 1991); (ii) *per*^{Lv} (Konopka, 1988), "*v*" for variegated, in that most individuals expressing this mutation are arrhythmic, but the remainder exhibit ca. 30- to 36-hr behavioral cycles; (iii) *Clock*, a mutation causing 22- to 23-hr periods in constant conditions (Dushay *et al.*, 1990); this mutant is not to be confused with an autosomal one that came to be known as *Clock*^{Jrk} (next section); the *X*-chromosomal mutation originally seemed to map near, but not at, the *per* locus, but the original *Clk* variant was subsequently shown to be a *per* allele after all (Table 1 and Figure 3) and is now called *per*^{Clk} (Dushay *et al.*, 1992; Hamblen *et al.*, 1998); and (iv) *per*^{T}, a supershort mutant (Figures 2B and 4A) with approximately 16-hr periods (Konopka *et al.*, 1994). Thus it seemed as if *per* is a key *X*-chromosomal locus, which, when mutated, results in substantially altered circadian rhythms (Table 1). (v) One further *per* mutation, *per*^{SLIH} (Figure 4C), occurred spontaneously in a laboratory strain and was noticed because of its ca. 27- to 28-hr periods (Hamblen *et al.*, 1998).

The second major gene on *D. melanogaster*'s *X* chromosome that influences circadian rhythmicity was found in a screen that exploited molecularly manipulated transgenes. Transposon derivatives called EPs have been scattered around the fly's genome. Many of the landing sites could identify genetic loci of interest as follows: EPs have the potential for mediating misexpression (including overexpression) of nearby genes, for they contain *cis*-acting regulatory

---

Figure 3. (*continued*) ca. 40 aa in the C-terminal region of this motif overlap most of the ca. 60-aa CLD); CLD = cytoplasmic localization domain; T–G = threonine–glycine repeat; bHLH = basic α-helical aa followed by a loop and another helix; Q-rich = high proportions of glutamine (including stretches of poly-Q) in relatively C-terminal regions of CLK, presumed to be involved in activation (per se) of transcription as influenced by this protein; kinase = core components of two enzymes inferred (from analyses of kinases analogous those diagrammed here) to mediate their activity as such (including an ATP-binding site located within relatively N-terminal regions of the DBT and the SGG kinase domains; and a catalytic region farther downstream, which is proximately involved in attachment of phosphate groups to the substrates of these enzymes); FBS = flavin-binding sites inferred (by similarity to related photolyase proteins) to be contained within the CRY polypeptide (the six vertical lines designate, from left to right, a single aa that touches the cofactor; a quintet of adjacent such aa; a nearby pair; another such; an aspartate residue mutated to an asparagine in *cry*^{b}; and a trio of clustered aa, the C-terminal one of which is six residues downstream of the mutationally defined D residue).

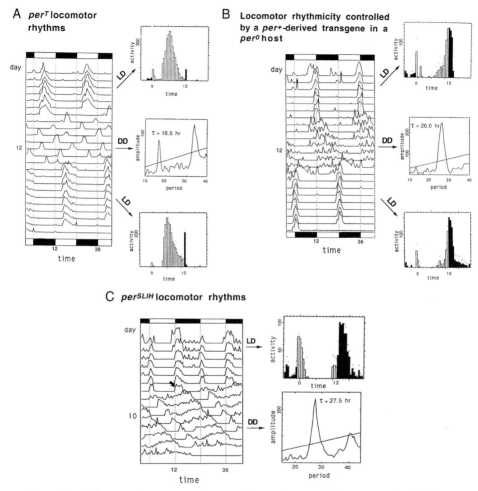

Figure 4. *period* gene variants causing shorter- or longer-than-normal locomotor cycles. (A) Actogram, average-activity plots, and a periodogram resulting from monitoring the behavior of a short-period *per^T* mutant *Drosophila* (see Table 1 and Figure 2) in LD, then DD (see Figure 1), then in LD again (the latter segment of this test involved a "dark:light" cycle that was reversed with respect to the initial one). This mutant individual entrained in LD such that the morning peaklets (indeed) occurred at that time on successive days; the "evening" peaks (with reference to normal locomotor cycles) were also produced in a vertically stacked manner, but at times that preceded lights-off by several hours. During the middle portion of this test, the evening peaks free-ran such that they occurred 7.5 hr earlier on successive days, hence the 16.5-hr periodogram peak (and its harmonic at 33 hr). These plots were adapted from Konopka et al. (1994). (B) Locomotor test of a fly carrying the arrhythmia-inducing *per^{0l}* mutation (see Table 1 and Figures 3 and 6) along with a 7.2-kb piece of DNA cloned from *per^+*. The 5' end of this transgene is within the "first intron" of the *per* gene (see Figure 9); thus, it lacks 5'-flanking regulatory sequences, but encodes a

sequences taken from *Saccharomyces cerevisiae* (Rørth, 1996). These stretches of DNA are called upstream activating sequences (UASs), in particular ones that are targets of this yeast's GAL4 transcription factor. These soluble and *cis*-acting gene-regulatory factors are nicely active in *Drosophila* in a wide variety of biological and experimental circumstances (Brand and Perrimon, 1993). Thus, an EP/misexpression screen includes sources of GAL4 (Rørth *et al.*, 1998), and to ground a search in chronobiology, a transgene was applied in which regulatory sequences from the autosomal *timeless* (*tim*) gene (see next section) were fused to *gal4* ones (see Section X.B.3 for details of this molecular construct). A series of 2,300 doubly transgenic (EP; *tim-gal4*) types was screened for locomotor rhythm abnormalities (Martinek *et al.*, 2001). One EP line was followed up (among the seven rhythm-abnormal ones identified), owing to the 21-hr free-running periods exhibited by the EP;*tim-gal4* flies (Table 1)—in this case females carrying one copy of each transgene (Martinek *et al.*, 2001). The EP transposon was found to be inserted near the *X*-chromosomal *shaggy* (*sgg*) locus. It happens to be very near *per* (Figure 18; see color insert) and was previously identified by lethal *sgg* (a.k.a. *zw3*) mutations (e.g., Young and Judd, 1978), which cause developmental pattern abnormalities (e.g., Perrimon and Smouse, 1989; Siegfried *et al.*, 1992; Blair, 1994).

2. Autosomal mutations at the putatively most important quintet of loci

The second gene identified in searches for rhythm variants that turned out to be mutated in an actual clock factor is *timeless*. By analogy to the screen that resulted in *sgg*, the original tim mutant came from a mutant hunt involving transposon mobilization (Sehgal *et al.*, 1991). In this case, *Drosophila* P-elements were made to stray from their original genomic location on the *X* chromosome

---

(*continued*) full-length PER protein. The locomotor monitoring of this transgenic individual (carried out as for the mutant fly in A) assessed the extent to which the "7.2" transgene is able to "rescue" the effects of *per^{01}*—one of which is to allow for only environmentally "forced" locomotor changes in LD (see Figure 5B). In contrast, the *per^{01}*-with-7.2 fly entrained in LD (top and bottom portions of the actogram, with accompanying average-activity plots); however, the evening peaks occurred at later-than-normal times (being shoved into the early nighttime with reference to the wild-type behavior shown in Figure 1A). A corollary is that the free-running locomotor rhythmicity of this 7.2 transgenic type is defined by longer-than-normal cycle durations, as shown in the middle portion of the actogram and its accompanying periodogram. These plots were adapted from Frisch *et al.* (1994). (C) Activity rhythms exhibited by a long-period *per^{SLIH}* mutant (see Table 1 and Figure 3). For this test, the fly was monitored in LD, then DD (the onset of the latter condition is marked by a heavy, curved arrow). Its entrained behavior involves later-than-normal morning and evening peaks of locomotion and longer-than-normal free-running behavior. This actogram, average-activity plot, and periodogram were adapted from Hamblen *et al.* (1998).

such that they might land in or near autosomal genes. Unlike the EP situation, the elementary P-element transposons would be likely to cause—if any genetic defect—subnormal expression of an adjacent gene that could result in a rhythm defect. The autosomal genes in question would have molecular tags that facilitate their initial cloning, by analogy to $sgg^{EP(X)1576}$ (Figure 3), a transposon-tagged mutant that led to reisolation of *shaggy*, in this case as a rhythm-related gene (Table 1), revealing its functions to be broader than those devoted to body-plan formation during development.

The most salient autosomal mutant that stemmed from the P-mobilization screen of Sehgal *et al.* (1991) was an arrhythmia-inducing variant on the second chromosome, which causes both eclosion (the isolation phenotype) and adult behavior to be aperiodic. Originally this variant, then called *aj42*, also exhibited anomalous night-active behavior in LD cycles, a locomotor abnormality (cf. Figures 1A and 5B; see color insert) that curiously disappeared when the mutant was reported more formally and rebaptized as *timeless* (Sehgal *et al.*, 1994). The additional rhythm-related behavioral anomaly, beyond *aj42*'s basic free-running arrhythmicity, seems to have been caused by a genetic-background effect. The *tim* mutation per se also turned out to have nothing to do with the P-element transposon that was newly inserted on this chromosome. Nevertheless, this *timeless* mutant (now called tim^{01}) seemed worth pursuing because of a key molecular abnormality caused by it: Expression of the *per* gene was found to be substantially altered (Sehgal *et al.*, 1994), as will be described in Section V.

Meanwhile, it is important to register that additional *tim* mutations were induced by chemical mutagenesis: two arrhythmic variants (Stempfl *et al.*, 2002; R. Stanewsky and J. C. Hall, unpublished; see Section IX.B.3); two short-period mutants (21–22 hr); and four long-period (26–28 hr) ones (Rothenfluh *et al.*, 2000a). An additional slow-clock *tim* mutant (tim^{rit}; Matsumoto *et al.*, 1999) had been previously isolated by extracting chromosomes from natural populations; this variant was originally named *Ritsu* (Murata *et al.*, 1995). All of the period-altering *tim* mutations (Table 1) lead to behavioral cycle durations that are 1–2 hr closer to normal when heterozygous with the normal allele (Matsumoto *et al.*, 1999; Rothenfluh *et al.*, 2000a).

One final mutation at the *timeless* locus that must be mentioned, the first one identified by virtue of free-running period anomalies, is $tim^{Suppressor-of-Long}$ (tim^{SL}). It was isolated by deliberate mutagenesis of the per^L mutant; a dominant, autosomal suppressor mutant was found that substantially ameliorates the *per* mutation's effects (Rutila *et al.*, 1996). Thus, the per^L, $tim^{SL}/+$ double mutant exhibited free-running periods 3 hr closer to normal than in the case of per^L alone. A further shortening of 1 hr (down to 25.5 hr) was effected when tim^{SL} was rendered homozygous, but the tim^{SL}/tim^{SL} genotype influenced the rhythm defect of per^S only slightly, *lengthening* the latter's period-altering effect by 1 hr (Rutila *et al.*, 1996). This strategy of mutating a mutant is commonly applied in microbial and invertebrate genetics, aimed at identifying factors that interact with those defined by the starting mutations. It might even be that the interacting, mutated factor would cause no abnormalities by itself; this is the case for tim^{SL} in a per^+ genetic background (Rutila et al., 1996). More generally, the matter of interactions between clock factors in *Drosophila*—including those involving the *per* and *tim* gene products—is a key feature of this organism's molecular chronobiology (as elaborated in Section V).

Further screenings for autosomal rhythm mutants led to induced changes at three chronobiologically important loci: (i) *Clock,* and its *Jrk* allele, which causes locomotion or eclosion to be arrhythmic when the mutation is homozygous. At least half of a group of $Clk^{Jrk}/+$ flies tested for locomotor behavior were arrhythmic. This is as good place as any to note that most apparent clock mutations are semidominant one way or the other [reviewed by Hall, 1995, 1998a); dominant or semidominant mutations are typically named by words and abbreviations beginning with upper-case letters as in the case of the autosomal *Clock* (*Clk*) mutant, but this custom seems just as frequently honored in the breach (Table 1)]. Indeed, all period-altering mutations at the major loci cause shorter– or longer-than-normal cycle durations when heterozygous with a given normal allele (Table 1); these periods are intermediate between ca. 24 hr (Figure 1A) and the value associated with a fully mutant effect (e.g., per^S hemizygous in a male, per^S homozygous in a female; tim^{rit} homozygous on the second chromosome of either sex; and so on). Having said this, I note that a second, independently isolated *Clk* allele (Table 1) is fully recessive arrhythmic type in that all $Clk^{1-7}/+$ individuals are rhythmic and exhibit normal locomotor periods (R. Allada and M. Rosbash, unpublished). (ii) *cycle* mutations, of which two were induced in separate screens (Rutila *et al.*, 1998b; Park *et al.*, 2000a), are also recessive arrhythmics in terms of both behavior and eclosion. However, (iii) the isolation of *double-time* mutants was based squarely on a screen for dominant autosomal mutants (Price *et al.*, 1998), assuming that semidominance of rhythm mutants is common, if not the rule. It is also less laborious to test the effects of mutagenized autosomes when they are heterozygous with an unmutagenized chromosome *2* or *3*, rather than rendering the treated chromosomes homozygous in subsequent generations (as was done to isolate the *Clk* and the *cyc* mutants).

Screening for dominant mutants is more forceful than matters of time-saving. What if a newly induced rhythm mutation were lethal when homozygous? That strain would be discarded in screens for recessive rhythm mutants. Indeed, *dbt* is a vital gene (Price *et al.*, 1998). This was revealed in two ways: (iiia) Not long after the original rhythm mutation at the locus was isolated, by virtue of causing ca. 22-hr periods in *dbt*/+ flies, a recessive-lethal transposon that happened to be inserted at the locus was characterized (Table 1 and Figure 3); this strain had arisen prior to any knowledge of the gene's relationship to rhythms (Price *et al.*, 1998); (iiib) other investigators were independently studying lethal *dbt* alleles—which they called *discs-overgrown* (*dco*)—with regard to the extensive developmental abnormalities caused by *dco* mutations, none of which seems connected with the animal's circadian system (Zilian *et al.*, 1999). Less severe damage to the gene (Figure 3), caused (for example) by the original period-altering mutation dbt^S (Table 1 and Figure 3), allows for viable adults. These mutations, which lead to abnormally fast or slow clock paces, depending on the allele, result in more substantial period alterations in homozygotes. For instance, the original *dbt* mutation produces ca. 18-hr locomotor periods in flies homozygous for dbt^S, although the corresponding effect on eclosion rhythms was less dramatic (Price *et al.*, 1998; Table 1). The next mutation induced at the locus (the L allele shown in Figure 3) was identified as a ca. 25-hr dominant (the behavioral phenotype of $dbt^L/+$ flies); dbt^L/dbt^L homozygotes exhibit 27-hr periods for rest–activity cycles and eclosion (Table 1). The three most recently isolated mutations at this locus are also (in part) long-period variants; two of them were found in screens (Suri *et al.*, 2000) completely independent of those that led to the seminal such variants, and those were found independently of one another. (This apparently random remark connects with the point that "anyone" seems likely to reidentify the same set of genetic loci in a screen for rhythm mutants.)

One of the two *dbt* mutants induced after recovery of the first set was detected as homozygous-viable mutation with ca. 29-hr free-running locomotor periods (subsequently this dbt^h mutation turned out to be semidominant); the companion mutation was recognized in a strain for which the previously muta-genized third chromosome was homozygous lethal (Table 1), but the heterozy-gous types were saved, tested, and found to give ca. 29-hr periods (Suri *et al.*, 2000). This strain was shown to carry a gratuitous lethal mutation induced else-where on the third chromosome (which reinforces the value of testing heterozy-gotes before discarding such a putative mutant) because, when this dbt^g allele was placed in heterozygous condition with a *dbt* deletion or the aforementioned transposon mutation (the P allele in Figure 3), the results were viable flies that behaved arrhythmically (Suri *et al.*, 2000). The same locomotor phenotype was demonstrated for yet another *double-time* allele, dbt^{ar} (Rothenfluh *et al.*, 2000b), a homozygous-viable mutant that was initially identified as a dominant 29-hr mu-tant in conjunction with induction of the most recently isolated set of *tim* mutants (Rothenfluh *et al.*, 2000a).

Similar to the manner by which tim^{SL} was isolated, an autosomal dom-inant "suppressor" of per^S was induced by Lin *et al.* (2002a). However, the new mutation, which caused the per^S behavioral period to be 1.5 hr closer to normal, led to longer-than-normal cycle durations in a per^+ genetic background (same when combined with per^L). In other words, the novel *Timekeeper* (*Tik*) mutation would have been isolable in a straight screen that did not incorporate the possi-bility of gene interactions. The *Tik*-bearing third chromosome carries a recessive lethal mutation that has so far been inseparable (in recombination tests) from the factor causing lengthened behavioral periods (Lin *et al.*, 2002a). Provisionally, *Tik*, like *sgg* and *dbt*, is a vital gene.

Why were the *Clk, cyc, dbt,* and *Tik* mutants followed up such that the four corresponding genes were revealed almost certainly to be bona fide clock factors? Other variants resulting from mutagenesis and screening for rhythm mutants were not chased down vigorously (as is described in Section III.B). The four genetic cases just noted, however, were shown—by molecular analyses that closely accompanied the mutant isolations—to include effects of these genetic changes on *period* gene expression (see Section V). Thus, following the lead created by the first chapter of the *timeless* story, this additional quartet of autosomal genes seemed worth delving into at levels deeper than elementary phenogenetics.

3. Chronogenetic principles stemming from isolation of the major mutants and from elementary analyses of the altered loci

Certain patterns can be distilled from the results of screening for rhythm variants, their initial phenotypic characterizations, and mapping of the genetic loci responsible for the chronobiological abnormalities. First, consider that chemical mutageneses were far more successful, in that one new rhythm variant was found among approximately 1,000 strains descended from a given set of the originally mutagenized flies. The original *tim* mutant, from the transposon-mobilization approach, required grinding through more than 10,000 of the relevant strains. A screen based on transposon-induced misexpression seems more efficient, however (Martinek *et al.,* 2001).

Nevertheless, a second advantage of chemical mutagens is that they typically cause changes *within* a given gene (transposons do not). The chemically induced intragenic hits will include loss-of-function mutants (such as those caused by premature stop codons) and ones with intrinsically altered proteins (a.k.a. missense mutants). If a given rhythm-related gene encodes a protein and the etiology of a missense mutation in it is an amino-acid substitution, that molecular abnormality can be rationalized to cause semidominant effects. Stated another way, this common chronogenetic property suggests that an inherently abnormal, but still functional, protein might interact with the normal gene product produced from the other chromosome that carries the wild-type (a.k.a +) allele. This interaction could derange the function of the overall product if it is naturally some kind of multimer. An analogous scenario would involve the amino-acid-substituted protein interacting anomalously with the product of another clock-related gene, once again altering the function of a putative protein complex.

Thus, these genetic results suggest that such complexes are a feature of the clockworks, and the force of this notion will be come to the fore in subsequent discussions. Given what we know so far about the phenogenetic results, including those stemming from applying deletions of certain mutated loci, an amino-acid substitution in a clock protein can have effects on the overall process that are worse than nothing: deletion-over-+ is less defective than missense-mutation-over-+. It follows that an intragenic null mutation, heterozygous for the normal allele, would cause no or only mild rhythm defects. This is just what is observed, in that, for example, *per*-null variants heterozygous with per^+ lead to solid rhythmicity, although slightly longer than normal periods are observed (e.g., Smith and Konopka, 1982; Hamblen-Coyle *et al.*, 1989; Cooper *et al.*, 1994). Conversely, an extra copy of the per^+ allele—achieved by introducing a chromosome aberration called a duplication into flies that are otherwise genetically normal—results in shorter-than-normal periods (Smith and Konopka, 1982). In fact, computational analysis of these "gene-dosage" alterations, and those inferred to occur in the period-altered *per* mutants, led to useful functions for which apparent levels of the gene's products are nicely related (logarithmically) to circadian-cycle durations (Coté and Brody, 1986).

That actual *per*-product concentrations can be correlated with period lengths will be noted in due course. Meanwhile, elementary gene-dosage effects were also revealed for the case of *Clk* by cytogenetics: a deletion of the locus heterozygous with Clk^+ causes significant period lengthening (Allada *et al.*, 1998). This may be an epiphenomenon because flies heterozygous for the recessive arrhythmia-inducing *Clk* mutation and Clk^+ exhibit normal periods (Table 1). Therefore, the *Clk* deletion, which is also missing many other genes, may have caused period lengthening because something other than the *Clk* gene is also absent (i.e., present in only one gene dose in the deletion/+ heterozygote). With regard to the other probable clock genes, *tim*-null or *tim*-deletion heterozygotes and analogous genotypes involving the *dbt* gene are such that one copy of the normal allele is sufficient for full rhythmicity and normal periodicity (Rothenfluh *et al.*, 2000a; Price *et al.*, 1998). Moreover, an increase in tim^+ dosage (effected by a duplication) was associated with a normal behavioral τ (Rothenfluh *et al.*, 2000a), unlike what happened when analogous manipulations were effected for the wild-type *per* allele. The gene-dosage effects of *cyc* are erratic among separate studies: The original mutant allele, cyc^{01}, resulted in fully rhythmic flies when it is heterozygous with cyc^+, but free-running locomotor periods were longer than normal, the same phenotype as that exhibited by *cyc*-minus deletion heterozygotes (Rutila *et al.*, 1998b). However, $cyc^{02}/+$ flies exhibited normal circadian periods (Park *et al.*, 2000a).

A retest of the effects of these genotypes, all monitored in parallel for locomotion in DD (O. Leclair, M. Mealey, and J. C. Hall, unpublished), led to a wild-type control (average) period of 24.4 $\pm$ SEM 0.2 hr ($n = 18$); $cyc^{01}/+$, 24.2 $\pm$ 0.1 hr ($n = 30$); $cyc^{02}/+$, 24.9 $\pm$ 0.1 hr ($n = 19$); *cyc*-deletion/+, 24.2 $\pm$ 0.1 hr ($n = 34$). On balance, then, these *cyc*-null variants (see Figure 3) are distressingly ambiguous in terms of their semidominant effects on the clock's pace.

A third genetic principle stemmed from the mutant hunts, and it will resurface in the clock-output subtopic (Section X): Semidominant rhythm variants permit the mutant hunter to tap into vital genes, which might go undis-covered in the execution of genetically limited mutant-induction strategies. Such genes are versatile in that they regulate essential biological processes that seem unrelated to rhythm control, but this does not diminish their heuristic value for chronobiological study. One still wishes to home in on the rhythm-related function that is one of the roles played by such a pleiotropically acting gene.

The fourth genetic point that seems to have surfaced as a result these entry-level chronogenetic studies is made in the same context as identifying vital plus putatively inessential genes in searches for rhythm mutants. Nowadays, screening for dominant or recessive variants has the potential to avoid missing the identification of any and all relevant factors. Therefore, one gingerly suggests that the majority of them may have already been identified. This supposition stems mainly from the fact that so many cases of repeat hits at a small number of genetic loci keep recurring. A given chemical-mutagenesis effort puts all of the organism's genes at risk, but when almost any of them is hit, no rhythm abnormality results. If a gene that is mutated such that a clear, perhaps striking, rhythm defect occurs, this locus can be thought of as special. The function it encodes may hover close to the clockworks, and there might be a small number of such factors because otherwise a given new mutation would often define a heretofore unknown gene. Instead, the emerging result is that a given new allele was induced at a locus which turned out to be in a gene already identified by previous mutageneses.

There are at least two problems with this sanguine view. One of them involves the fact that mildly defective rhythm variants accompany the isola-tion of others that cause complete arrhythmicity or substantial period alterations. Candidate-mutant strains in which only about one-half to two-thirds of the in-dividuals are arrhythmic (Table 1) tend not to be followed up (Hall, 1998a). Perhaps a bona fide clock variant lurks within such a strain, and it could de-fine a novel, chronobiologically important gene. This factor would be forgotten unless it got mutated later on to result in a more dramatic rhythm abnormality. Other material that the clock-gene hunters may have been mutating but fail-ing to identify would be at vital rhythm-relevant loci, which, on the one hand, happened to be hit such that only null mutations occurred and developmental death ensued, and, on the other, were such that these losses of function led to no noticeable change in rhythmicity when heterozygous with the normal form of the gene.

Whether or not *Drosophila's* chronogenetic system is "saturated" at the level of formal genetic analysis, it was warranted to begin studying the stretches of DNA identified by extant mutants, that is, in advance of mutagenizing into oblivion, whereby no new mutation would define a heretofore unappreciated genetic locus. But relating the molecular part of the story should wait until further features of rhythm mutants and their various biological phenotypes are described.

B. Other mutants found in screens for rhythm variants

1. Minor *X*-chromosomal mutations

The screens of Konopka and colleagues for chemically induced mutants on the *X* chromosome led to only two rhythm variants beyond the *per* ones. The others were *Andante* and *E64*. The former exhibits a moderately slow clock (Orr, 1982; Konopka *et al.*, 1991). The latter mutant is similar to *disco* (cf. Figure 1B), in that the majority of *E64/Y* individuals are arrhythmic (Konopka *et al.*, 1994). Moreover, *E64*'s viability is subnormal, although not as depressed as in the case of *disco*, and *E64/disco* flies exhibit normal rhythms (Konopka *et al.*, 1994), that is, these two factors are not allelic. That a rhythm mutant, such as *disco* (Steller *et al.*, 1987; Dushay *et al.*, 1989), can exhibit approximately 10-fold lower-than-normal viability means that Konopka may have hit genetic loci of this type in his screens, but did not recover the poorly viable flies that resulted.

The case of *Andante* is intriguing, mainly because it long dropped by the wayside for unknown reasons. The original such mutation leads to ca. 25- to 26-hr periods for both eclosion and locomotor rhythms (Orr, 1982; Konopka *et al.*, 1991). It was also noticed that *And* flies exhibit *dusky*-like mutant wings (these are small and present a dark hue, owing to smaller-than-normal cell sizes). Indeed, *And* mapped to an *X*-chromosomal site where *dusky* (*dy*) is located; this factor is also known as the "*miniature-dusky* gene complex" because *m* mutations at this locus lead to even more extreme reductions in wing size. However, a classic wing-defective *dy* mutant was found to have 24-hr rhythms (Konopka *et al.*, 1991). Subsequent screening for anatomically *dy* wing mutants led to a handful of new variants at the locus, about half of which turned out to exhibit long-period behavioral rhythms (Table 1); the others were rhythm-normal, as were the three *m* mutants that got tested in parallel (Newby *et al.*, 1991). However, it is now believed that the rhythm-abnormal, *dusky* mutants isolated post-*Andante* (Orr, 1982) may be erroneous, at least because the original *And* variant is doubly mutated at two separate, but very nearby genetic loci: One mutation (within a vital gene, by the way: Jauch *et al.*, 2002) leads to a "moderately slow clock," the other (within an adjacent transcription unit) to abnormal wing development (DiBartolomeis *et al.*, 2002; B. Atken, E. Jauch, G. G. Genova, E. Y. Kim, I. Edery, T. Raabe, and F. R. Jackson, unpublished).

In any case, the pleiotropic complexities revolving around *Andante* as it was originally isolated (by Orr, 1982) may have undermined the motivation to continue studying this gene with much celerity, in fact it did not get formally reported until several years later (Konopka *et al.*, 1991). The only subsequent study involved "mosaic mapping" of body regions within genetically mixed individuals (described in another context in Figure 6B; see color insert). Which tissues need to express an *And*-like *dy* mutation for a long-period rhythm to be manifested or for an appendage to take on an abnormal appearance? Answers: somewhere within the head or within wing tissue itself, respectively (Newby and Jackson, 1995). However, the case of *Andante* otherwise petered out from the literature, and the incipient molecular genetics of this locus seemed to die on the vine (Jackson and Newby, 1993). Abandoning these mutants was not necessarily wise, as was revealed when multiple biological defects associated with further rhythm variants were realized. Some of the corresponding genes, notably *sgg* and *dbt*, are regarded major players in the control of the *Drosophila* clock (see Figure 7, color insert). Lest one get carried away with the puzzles revolving around *Andante* and the object lessons that such a substory might provide (see Section III.B.2), recent stirrings seem to indicate that these mutants and this gene are in the process of being resurrected. In fact, the *And*-encoded function is a catalytic one ensconced in the *shaggy* and *double-time* ballpark (Atken *et al.*, unpublished work cited in previous paragraph). We will wait till Section V.B.2 to learn about the field on which these factors play. Meanwhile, see Jauch *et al.* (2002) for a preview of *Andante*'s molecular biology. It involves a kinase activity, whereby *And*'s product would appear palpably to cooperate with the polypeptide that turns out to be encoded by *Timekeeper* (Lin *et al.*, 2002a)

2. Minor autosomal mutations

The *double-time* case—along with those of *tim*, *Clk*, and *cyc*—indicates that hunters for rhythmic variants were not stuck on screening only the X chromosome for such mutations. Actually, a tentative autosomal screen for rhythm mutants had occurred several years earlier. Genetic tracking of approximately 500 strains carrying chemically mutagenized autosomes—chromosomes 2 and 3, which sum to ca. 80% of *D. melanogaster*'s genome (Figure 18, color insert)—resulted in the *psi-2*, *gate* (*gat*), and *psi-3* mutants (Jackson, 1983). The *psi* cases are (or at least were) phase-angle eclosion variants, with earlier-than-normal peak times of adult emergence in LD cycles; the *gat* strain exhibited poorly gated eclosion. Tests of these variants' free-running eclosion revealed both *psi* mutants to exhibit ca. 25-hr periods. Monitorings of locomotion showed that *psi-3* adults were normal in LD and DD; whereas *psi-2*'s morning activity peak was earlier than normal, its evening peak was late, and this mutant free-ran with a longer-than-normal period (Jackson, 1983). *psi-2* and *gat* were induced on chromosome 2, and *psi-3* on chromosome 3. The potentially mutated loci were never pinned down with precision (Figure 18), although it was gingerly suggested that *psi-2* and *gat* may have been allelic to each other (Jackson, 1982, 1983). However, mapping of *psi-2* and *gat* (Figure 18) made it impossible for that mutation to be at the second-chromosomal *tim* locus (Sehgal *et al.*, 1994); and *psi-3*'s approximate map position (Figure 18) is far from those of either *Clk*[Irk] or the *cyc*[0]'s (Allada *et al.*, 1998; Rutila *et al.*,

1998b). Thus, the original rhythm mutations on *D. melanogaster*'s autosomes could have defined chronobiologically important genes about which almost nothing is known because they were abandoned, as almost happened for *Andante*. Deserting these autosomal mutants may have occurred because *psi-2* and *psi-3* (if not *gat*) exhibited rather mild rhythm abnormalities.

Another autosomal mutant is *Toki*. It was induced on the second chromosome and initially recognized by virtue of more than the usual proportion of adult emergents eclosing during the subjective night in DD (Matsumoto *et al.*, 1994). Subsequent behavioral testing of flies similarly homozygous for the mutagenized chromosome revealed ca. 25-hr free-running periods. *Toki/+* adults gave locomotor cycle durations intermediate between that value and normal periods (Table 1). Mutant homozygotes and heterozygotes were also shown to be hyperactive in day-to-day monitorings of adult behavior (Matsumoto *et al.*, 1994). *Toki*'s map position (Figure 18, color insert) demanded the conclusion that it is a unique chronogenetic factor (that is, distinct from *tim*, *psi-2*, or *gat*), but this mutant has not been followed up.

There is no way to dismiss the rhythm-related genes, defined by the set of mutants just summarized, as uninvolved in *Drosophila*'s clockworks. However, it is not surprising that mutants with relatively gentle or inconsistent rhythm defects failed to stimulate further interest. In fact, it is possible that loci like *Andante* or *Toki* cannot be mutated to cause more extreme period changes or all-out arrhythmicity. The *Andante*-like mutations at the *m-dy* locus are potentially instructive in this regard. Mutations subsequently induced at it were not recognized chronobiologically, but by their abnormalities of wing morphology. It follows that the new *dy* alleles could have led to "anything" in terms of rhythms; instead, subsequent tests showed all of them with abnormal rhythmicity to exhibit only 1-hr period lengthenings (Newby *et al.*, 1991), the same phenotype as the original *And* mutant (Orr, 1982; Konopka *et al.*, 1991). [This point, if any, leaves aside the severe caveat about the relationship between *And* and *m-dy*, as divulged by DiBartolomeis *et al.* (2002).]

The final group of mutants detected in dedicated searches for rhythm defects came from transposon mobilizations (as in the case of tim^{01}). One such variant, whose phenotypes were documented to a reasonable degree, was called *m120*; this strain exhibited a higher-than-normal degree of eclosion during the nighttime, but normal behavioral rhythmicity (Sehgal *et al.*, 1991), reminiscent of the *psi-2* phenotypes (Jackson, 1983). In fact, both that mutation and the putative *m120* one were mapped to the same chromosome (Jackson, 1983; Sehgal *et al.*, 1991). Other putatively transposon-tagged mutants were identified behaviorally. Their phenotypes were not documented in detail (see Hall, 1998a), but all four such types gave arrhythmicity for approximately one-half to three-fourths of the individuals tested from a given strain (Table 1). Such interindividual "variegation" is analogous to that of the per^{Lv} mutant (Konopka, 1988), although in the case of the transposon variants the rhythmic individuals exhibited essentially normal periods. This property also reminds one of the variably arrhythmic versus normal locomotor behavior of the chemically induced $E64$ mutant (Konopka et al., 1994). None of these five cases of "impenetrant" mutants stimulated further inquiries.

IV. ADDITIONAL MUTANTS WITH DEFECTS IN DAILY CYCLES AND OTHER TIME-BASED PHENOTYPES

A. Rhythm mutants originally identified by nontemporal criteria

1. Anatomical variants

Recall the *disco* mutants. These flies are largely devoid of the optic lobes (as introduced above) and in fact were isolated on that anatomical criterion (Steller et al., 1987). Most locomotor tests of flies hemizygous for an X-chromosomal *disconnected* mutation (Figure 18, color insert) showed that ca. 90–95% of individually tested $disco^1$ (Dushay et al., 1989; Dowse et al., 1989; Helfrich-Förster, 1998) or $disco^{1656}$ (Hardin et al., 1992a) males are arrhythmic (Figure 1B) in free-run (although Blanchardon et al., 2001, reported a lower proportion of behavioral rhythmicity for $disco^1$). Also, *disco* cultures are completly arrhythmic (Figure 2A) in terms of eclosion (Dushay et al., 1989). Thus, one infers that this mutation causes defects deeper than the eye–brain disconnection that gives the gene its name and results in general blindness. CNS defects in *disco* will be taken up in a later section (Section VI). Meanwhile, consider that *disco* causes a behavioral abnormality that parallels the mutant's eclosion arrhythmicity. These are like the phenotypes of the original *per* and *tim* mutants—eclosion variants that turned out to be locomotor-defective in the same ways. As a flip side of the same coin, the *Clk*, *cyc*, and *dbt* mutants were isolated as behaviorally arrhythmic and subsequently shown to exhibit the same kind of eclosion phenotype (Allada et al., 1998; Rutila et al., 1998b; Price et al., 1998). Mutants with these properties smack of variants that are generally defective in circadian rhythms, as if they might be devoid of or defective in a global clock function, not an output one devoted to a particular feature of the animal's cyclically varying phenotypes. As was oft-mentioned above, *per*, *tim*, *Clk*, and *cyc* can be considered to cause abnormalities of pacemaker function, in part because a mutation in any one such gene affects the expression of the other three (a substory that is elaborated later in conjunction with Figure 8; see color insert). In sharp contrast, *disco* mutants were found to exhibit normal *per* expression (Hardin et al., 1992a). Nonetheless, given *disco's* brain damage, its dually defective rhythmicity suggest that it lacks neural pacemaker structures underlying both eclosion and locomotor phenotypes. That *disco* mutants do so lack will be dealt with when the spatial expressions of rhythm-related gene products are taken up (Section VI).

Testing flies from other strains isolated as brain-damaged mutants (Heisenberg *et al.*, 1985) for those that exhibiting behavioral rhythm abnormalities led to the setting aside of *disco* alone: In the screening that uncovered this mutant as arrhythmic for adult locomotion and eclosion, other neuroanatomical mutants were found to be essentially normal (Table 2). Ultimately a bakers' dozen worth of these mutant types, assessed for periodic locomotion (Helfrich, 1986; Dushay *et al.*, 1989; Vosshall and Young, 1995), was found to exhibit marginal or no decrements in behavioral rhythmicity (Table 2).

Among the anatomical mutants that might have exhibited defects for this phenotype are those with aberrant morphology of a brain structure called the mushroom bodies (MBs; see Zars, 2000). One such variant, *mushroom-bodies miniature* (*mbm*), was noted to cause the flies' behavior frequently to degrade to arrhythmicity after they proceeded from LD-cycling conditions into constant darkness, but this effect disappeared after outcrossing *mbm* to a wild-type strain and reextraction of the mutant (Helfrich-Förster *et al.*, 2002). Consistent with the ultimately (but somewhat ambiguously) normal rhythmicity of *mbm*, three further MB-affecting mutations (Table 2) and hydroxyurea (HU)-induced ablation of these brain structures during development had minimal effects on free-running rest–activity cycles (Helfrich-Förster *et al.*, 2002). Locomotor monitoring of a fifth mutant—*mushroom-body defect* [tested cursorily by Vosshall and Young (1995) and more extensively by Helfrich-Förster *et al.* (2002)]—showed approximately 15–20% of *mud* individuals to be arrhythmic (Table 2). However, the minimal effects of other MB mutations or of HU-treatment led Helfrich-Förster *et al.* (2002) to speculate that *mud* causes quasi-arrhythmicity because of pleiotropic effects of this mutation on other brain structures. Please peruse the MB-mutant entries in Table 2 to appreciate certain behavioral complexities associated with some of these brain-damaged variants (overall, it cannot be said that the mutants are thoroughgoingly locomotor-normal). In any case, these genetic and chemical manipulations of the mushroom bodies are meaningful because this part of the fly's brain (Figure 11, see color insert) became a candidate for forming part of the neural substrate of rhythmic locomotor behavior (as is discussed in Section X.A.1).

Additional mutants that might have exhibited rhythm defects were surveyed in a similar spirit (Vosshall and Young, 1995). These variants had been found mainly on the basis of abnormal structure of the visual system, but almost none of them was abnormal for locomotor rhythmicity (Table 2). This reinforces the notion that circadian locomotor variants are relatively rare and special. In other words, if the mutants listed in Table 2 were involved in *Drosophila*'s circadian system, newly induced alleles at several of these loci would presumably have fallen out of the screens dedicated to isolating rhythm mutants.

Table 2. Rhythm Testing of Mutants Originally Identified on Other Criteria[a]

Mutant	Basic defect(s)	Rhythm defect(s) remarks
	Appreciably defective mutants and those repeatedly analyzed	
Creb2	$Creb2^{S162}$ mutation causes near-lethality; rare adult survivors exhibit reduced body size; gene encodes two isoforms of cAMP response element-binding protein, which polypeptides are undetectable in this mutant	Ca. 40% individuals f-r arrh.; rhythmics exhibited ca. 1-hr-shorter-than normal τ; luc-reported per expression as well as PER reduced and exhibit weak daily cyclings
dfmr l, a.k.a dfxr	Gene encoding a putative RNA-binding protein originally identified in its normal form; subsequently mutated to the null state by transposon excision, resulting in healthy, normal-appearing flies in most studies, poorly viable and sluggish ones in one investigation	Largely arrhythmic for locomotion in DD; eclosion rhythmitcity weak; certain qualitative anomalies accompany both kinds of circadian defects, although one study reported eclosion of mutant cultures to be like that of controls; overexpression of the gene leads to longer-than-normal free-running periods
disconnected	Eyes disconnected from brain in most individuals (thus optic lobes absent); adult viability very low; LNs (Figure 11) absent from most individuals; disco gene encodes transcription factor	Tends to be largely arrh. for B in DD [ca. 95% such in Dushay et al. (1989), although Blanchardon et al. (2001) reported only ca. 70–80% arrh.]; the most detailed scrutiny revealed ca. 80% arrh., ca. 19% complex B rhythms, ca. 2% clearly rhythmic (Helfrich-Förster, 1998); LD entrainment sloppy; E arrh.; several reports indicate (among the separate studies) chronobiological effects of mutated alleles 1, 2, or 3; but these disco mutations are caused by identical nucleotide substitution, so phenogenetic distinctions are meaningless; in one study only (Hardin et al., 1992a), B effects of an independently isolated disconnected mutation "$disco^{1656}$" shown to be same as those of $disco^{1=2=3}$
dunce	Learning/memory-defective; female-sterile; cAMP phospho diesterase depleted or null (depending on dnc allele)	F-r B τ 23; accentuated phase delay induced by light in early night
ebony	Dark body color; β-alanine and dopamine accumulate to anomalously high levels because of putative deficit in β-alanyl dopamine synthetase	Ca. one-fourth to two-thirds of individuals B arrh.; (depending on mutant allele); E normal
no-retinal-potential-A	Physiologically blind; olfactory-defective; phospholipase C-null	Early phase of B peak in LD; f-r τ 23; reduced sensitivity to light-induced shifts of B entrainment, including extremely attenuated re-entrainability when combined with cry^b

Table 2. _continued_

Mutant	Basic defect(s)	Rhythm defect(s) remarks
Pka-CI, a.k.a. _Dco_	Depleted cAMP-dependent protein kinase A (gene encodes catalyic subunit); many alleles recessive lethal	Certain (lethal) alleles, over +, cause ca. one-half to three-quarters of individuals to be B arrh.; viable homozygous or trans-heterozygous types: ca. three-fourths B arrh.; one viable homozygous type tested for E: normal
sine-oculis	Most so^1 individuals eyeless; all are ocelliless; connection between bilalerally symmetrical LNs (Figure 12) frequently defective or absent; _so_ gene encodes transcription factor	Ca. twice as many individuals as normal exhibit complex f-r rhythmicity, including splitting into components with separate τ's; reduced sensitivity to light-induced B phase shifts, with shift of spectral-sensitivity peak from (normal) green toward blue range

Mutants with erratic rhythm defects and those minimally analyzed

Mutant	Basic defect(s)	Rhythm defect(s) remarks
Drop-Miopthalmia	Extreme compound-eye reduction; homozygous lethal	Half of Dr^{Mio}/+ individuals arrh. for f-r B; (only 10 flies tested)
glass	Compound-eye and ocellar photoreceptor cells absent; H–B eyelet (Figure 16) absent; _gl_ gene encodes a transcription factor	Vosshall and Young (1995) listed no appreciable B effects of 4 separate _gl_ mutations and noted that arrhythmia induction reported previously for one such (Sehgal et al., 1991) was due to second-site mutation; subsequent, more intensive analysis of one mutant (gl^{60j}) indicated mild increase in arrh. (compared with wild-type) and somewhat sloppy B entrainment (Helfrich-Förster et al., 2001); when $gl^{60\,j}$ combined with cry^b, the flies are nearly blind to re-entraining light stimuli
mushroom-body defect	Projections from adult-brain MB calyces missing; calyces themselves enlarged and anomalous lobes present outside main neuropil of brain	Female sterile and exhibits poor viability; initially, not reported whether males or females B-tested, such that 3/14 arrhythmic (Vosshall and Young, 1995): later, shown that 14/20 14/20 males exhibited weak, complex, or no B rhythms; and 1/8 females arrhythmic; overall locomotor levels slightly higher than normal in DD
mushroom-bodies miniature	Mushroom bodies (MBs) in brain of adult female have abnormally small calyces; MB peduncles and lobes thin or missing; male MBs normal	F-r B rhythms of inbred mutant females quickly dampen to to arrh. (and exhibited certain other B anomalies in DD); 3/16 males weakly rhythmic; 13/18 females weak or complex B; however, outcrossed _mbm_ flies were largely normal; yet this outcrossing (replacement of one wild-type

continues

Table 2. *continued*

Mutant	Basic defect(s)	Rhythm defect(s) remarks
		genetic background with another) also caused the anatomical defect to go away
Pka-RII	Near or total loss of cAMP-dependent protein kinase A type II regulatory subunit; mutant is homozygous viable and fertile, but exhibits a variety of ovarian follicle abnormalities; also, reduced sensitivity to ethanol and cocaine as well as anomalous lack of sensitization to repeated cocaine exposure	After exhibiting normal B in LD, ca. 60% of tested individuals, exhibited rapid dampening to arrh. in DD
	Mutants shown to be rhythm-normal or largely so	
almondex	Roughened eye and eye-shape defect; progeny of mutant females frequently die during development	Dying embryonic progeny of mutant mothers exhibit CNS hypertrophy
central-brain deranged	Extra lobe present in adult brain near MB calyx; peduncles and lobes reduced	Learning slightly subnormal
central-complex broad	Portions of central complex in adult brain abnormally flat and broad	Abnormal learning
chaoptic	Rhabdomeres of compound-eye, photoreceptors deranged or absent, owing to depletion of membrane protein at extracellular surface of such cells	Five mutant alleles tested for B effects
drop-dead	Early adult death and progressive neurodegeneration; glia have abnormally short processes, causing neurons to lack complete glial sheath	B testing initiated with very young adults, which exhibited normal rhythmicity during abbreviated *drd* lifespan
Ellipse	Compound-eye shape and texture abnormal	A.k.a. *Egfr*, owing to encoded epidermal growth factor homolog; *Elp* mutant tested (for possible dominant effects on B) is a hypermorphic allele
ellipsoid-body open	Ellipsoid body with central complex of adult brain abnormally flat and broad	—
eyeless	Compound eye variably reduced a given (eye mutation); antennal or maxillary structures variably abnormal	Three mutant alleles tested for B effects
eyes-absent	Compound eyes do not form; ocelli normal	By itself or when combined with ocelliless: reduced sensitivity to light-induced B phase shifts
giant-cell defect	In adult brain near base of each MB calyx, anomalous large cell body present in each hemisphere	—
Glued	Eyes rough, small, oblong; optic lobes abnormal	*Gl/+* tested for B because mutation is recessive lethal

Table 2. *continued*

Mutant	Basic defect(s)	Rhythm defect(s) remarks
Irregular facets	Roughened eye	
lobula-plateless	Most of lobula-plate optic-lobe neuropile absent, although "giant-fibers" retained	Optomotor roll and pitch responses weak or absent
lozenge	Compound eye reduced and ovoid with roughened, glistening appearance	Seven mutant alleles tested for B effects; 3 such mutants gave only 50–60% rhythmic individuals, but only 5–6 flies tested in these cases; among *lz* types tested are those exhibiting various features of the mutants' pleiotropy (e.g., female sterility, reduced tarsal claws)
Microcephalus	Eyes small or absent	Dominant mutation, including that homozygote not reliably distinguishable from *Mc/+*
minibrain	Adult brain ca. half-normal in volume	Lamina optic lope and MB peduncle: normal sizes; mild leg-shaking under ether and may be allelic to *Shaker* mutations
mushroom-bodies-reduced	MBs reduced in size	Most individuals strongly B rhythmic in DD; τ ca. 0.5 hr shorter than normal
neither-inactivation-nor-after-potential-[A]	Visual physiological defect; prolyl *cis–trans* isomerase-depleted (opsin-processing enzyme)	—
neither-inactivation-nor-after-potential-E	Visual physiological defect, including reduced light-responsiveness, owing to severe depletion of rhodopsin from R1–6 compound-eye photoreceptors	Reduced sensitivity to light-induced B phase shifts
no-ocelli/eyes-narrow	Corneal facets in compound eye fused; photoreceptor rhabdomeres and optic-lobe cartridges absent, glial-cell mass present in peripheral retina	External eye appearance like that of *Gl/+*
ocelliless	No ocelli; compound eyes and overall adult body size somewhat smaller than normal	*oc*[1] mutant (the one B-tested) also female sterile owing to two breakpoints on the relevant chromosome (one affecting ocellar formation, the other egg development)
optomotor-blind	Absence of yaw-torque optomotor responses; giant-fiber movement detector in lobula plate absent	Other mutations at this transcription-factor-encoding locus cause externally visible abnormalities or lethality; these mutants not tested for rhythms
pebbled	Roughened eye	—
reduced ocelli	Ocelli small, often missing	—
reduced optic lobes	Proximal portions of optic-lobe complex (excluding lamina) reduced to ca. 50% normal volume	Not as severe volumetric reduction as caused by *sol* (see below); when combined with *sol*, certain responses to

continues

Table 2. *continued*

Mutant	Basic defect(s)	Rhythm defect(s) remarks
		complex moving visual stimuli are aberrant (deduced as "operant" behavioral deficits)
rough	Roughened eye	—
roughex	Roughened eye	
scabrous	Roughened eye	—
sevenless	Missing R7 photoreceptor in compound eye; possible effects on visually mediated adult behavior caused by other than R7-less feature	—
small-mushroom-bodies	MBs reduced in size	Reduced overall locomotion; all individuals B rhythmic in DD, but ca. half were weakly so; also; τ ca. 0.5 hr shorter than normal
small optic lobes	Medulla, lobula, and lobula-plate optic lobes reduced in volume and cell number, owing to depletion of certain category of (columnar) neurons	Certain responses to moving visual stimuli (e.g., fixation behavior, landing responses) abnormal, but yaw-torque optomotor responses normal; when *sol* combined with *so* mutation (see above), proportion of flies exhibiting complex f-r B rhythmicity increased by >3 times
swiss cheese	Adult neurodegeneration: many holes appear in CNS tissue sections; hyperwrapping of neurons by glia	Age-dependent degeneration somewhat like that of *drd*, but *sws* does not cause as early adult death
Vacuolar medulla	Vacuoles in distal region of medulla optic lobe; laminar monopolar neurons degenerate near blind in tests of movement detection	—

[a] The three categories of mutant types are essentially self-explanatory. Mutations in the first category (and *gl* from the second) have their chromosomal positions depicted in Figure 18. The abbreviations [e.g., B for the results of testing for behavioral rhythmicity; arrhythmic (arrh.) for an arrhythmic B or E (eclosion) outcome] are as in Table 1. For mutants in the second and third categories, all rhythmicity assessments involved adult locomotion (B). The noneffects on such cyclical behavior of mutations in the third category are such that the great majority of mutant individuals tested exhibited B rhythmicity with average free-running (f-r) periods in the normal range; or, in rare instances, ca. 1-hr different from controls. Such potentially anomalous mutant values were deemed normal, mainly because of the small numbers of individuals tested; these were in the range of ca. 15–30 for some mutant types, but often the n's were only ca. 5–10 [e.g., 7 of 8 *oc* individuals tested by Vosshall and Young, (1995) were rhythmic and gave an average τ of 22.7 ± 0.8 hr]. If additional information is desired about the "Basic defects" exhibited by a given mutant, the *Drosophila* pheno-(and molecular-) genetic database can be consulted at http://flybase.bio.indiana.edu/.

2. Miscellaneous biochemical variants

Having said this, one wonders why mutants of the types described in the subsection that now begins were not recovered in the course of such screenings. For these components of the chronogenetic list (Table 2), a wide array of functions originally known by criteria unrelated to rhythms was tested genetically for connections to circadian rhythmicity.

Cyclic AMP metabolism and other related functions have long been suspected to contribute to circadian control in other organisms. Most such results have stemmed from manipulating cAMP levels, or the knock-on consequences of this molecule's actions, nongenetically, although augmented by assessing the chronobiological roles of related macromolecules (e.g., Gillette, 1996; Whitmore and Block, 1996; Zatz, 1996; Li *et al.*, 1998). Findings of this sort seemed to prompt tests of certain *Drosophila* mutants for rhythm abnormalities. Thus, it was found that *dunce* learning mutants, which are deficient in cAMP phosphodiesterase (PDE), exhibit accentuated behavioral phase shifts after light pulses delivered in the early night, and *dnc* flies showed a 1-hr shortening of free-running periods (Levine *et al.*, 1994). Moreover, certain *Pka-CI* variants that have subnormal levels of a cAMP-dependent protein kinase (A kinase) catalytic subunit were shown to be frequently arrhythmic in DD when series of mutant individuals were tested behaviorally (Levine *et al.*, 1994; Majercak *et al.*, 1997). A similar degree of free-running arrhythmicity was observed in a *Pka-RII* mutant, which is severely depleted of A kinase regulatory-subunit activity (Park *et al.*, 2000b). This lack of full penetrance for arrhythmicity may be due to the fact that these A kinase mutants are only hypomorphs, not full nulls. Mutations of that type at the *Pka-CI* locus are recessive lethals. [This gene is also called *Dco* (Table 2), which should not be confused with *dco* (a.k.a. *double-time*) because designating the genes of *D. melanogaster* with upper-case vs. lower-case initial letters implies separate, unrelated loci.]

Periodic eclosion was found to be normal in *Pka-CI* cultures, and biochemical assays of mutant adults showed them to exhibit normal expression of *period* gene products (Majercak *et al.*, 1997). Thus *Pka-CI*, and probably *Pka-RII* as well, are rhythm-related genes that seem not to be clock factors. This kinase activity (and its stimulation by cAMP; Park *et al.*, 2000b) may participate in regulation of outputs mediated by central-pacemaking functions. This would be consistent with the *Pka-CI* mutations affecting only one kind of circadian rhythm. For *dnc*'s part, the PDE enzyme's regulation of cAMP levels could be concerned in part with the clock's responses to environmental inputs (Gillette, 1996; Whitmore and Block, 1996; Zatz, 1996). The additional phenotype of a mild period change in DD, however, correlated with that fact that cAMP levels show circadian fluctuations (Levine *et al.*, 1994), suggests an influence on the clock's free-running operation.

Potentially in the context of mammalian rhythms and involvements therein of cAMP-response-element binding or modulator proteins—CREB and a CREM isoform called ICER (Li *et al.*, 1998)—a *Drosophila* relative of these factors called dCREB2 was interrogated chronobiologically. A *dCREB2* mutation, which is nearly lethal developmentally, shortened free-running behavioral periods of the rare survivors and affected expression of the *period* gene (Belvin *et al.*, 1999). *per*'s products normally oscillate circadianly (see Section V), and similar fluctuations were observed for temporally monitored action of dCREB2. This molecular rhythmicity was inferred from the activity of transgene sequences encoding a reporter enzyme (see Section V and IX), fused downstream of cAMP-response elements (CREs); however, dCREB2 protein levels themselves exhibited no daily cycling (Belvin *et al.*, 1999). Inferences about regulatory circuitries accompanying these phenomena came from demonstrations that the fusion gene's expression level was severely attenuated by the *dCBEB2* mutation, and that per^{01}, per^{S}, and per^{L} affected cyclical activity of the CRE *cis*-acting regulatory elements in ways that paralleled the effects of these mutations on biological circadian rhythms (Belvin *et al.*, 1999).

The *Drosophila* relative of a human gene called *FMR1* (mutations of which cause "fragile-X" mental retardation) was identified (Wan *et al.*, 2000). Reverse genetics devoted to this fly factor—a putative RNA-binding protein encoded by the *dfmr1* gene—led to imprecise excisions of a transposon inserted at the relevant third-chromosomal locus (Dockendorff *et al.*, 2002). The resulting null alleles allowed for normal viability, consistent with an earlier study of such mutations that focused on neurophysiological analyses of such mutants (Y. Q. Zhang *et al.*, 2001). A third set of investigators, who call the fly's fragile-X-like mutations *dfxr*, report that other loss-of-function forms of the gene exhibit extremely low percentages of adult emergence (Morales *et al.*, 2002), but it could be that the third chromosomes used in this study carry another factor that affects viability.

One of Dockendorff *et al.*'s mutations (called allele 3, which is deleted of *dfmr1*, but does not disrupt adjacent genes) was tested for effects on circadian rhythms. Periodic eclosion was found to be anomalous: $dfmr1^{3}$ cultures yielded delayed, abnormally broad morning maxima with subnormal amplitudes (Dockendorff *et al.*, 2002). Analogous eclosion analysis of this mutant type—although in this case involving rare "escapers" from the near lethality reported to result from *dfxr* null mutations—also revealed substantial phase delays of adult-emergence maxima (Morales *et al.*, 2002). In contrast, a *fourth* research group, which generated partial deletions of *dfmr1* in a manner similar to the tactic taken by Dockendorff *et al.*, reported that one of their mutations allowed eclosion to be the same as displayed by their control cultures (Inoue *et al.*, 2002).

In behavioral tests, both Dockendorff *et al.* (2002) and Inoue *et al.* (2002) showed that approximately 85% of *dfmr1*-null individuals were arrhythmic in

DD; in the former study *dfmr1³*-mutated flies were observed also to exhibit odd bursts of locomotion in this condition. The companion study showed that about 70% of *dfxr* null-mutant escapers were behaviorally arrrhythmic, with most of the remainder exhibiting "erratic" rhythmicity accompanied by multiple free-running components; moreover, this version of a fragile-X-like mutant was found to be notably sluggish in these locomotor monitorings (Morales *et al.*, 2002).

The DFMR1/DFXR gene was independently identified by Dockendorff *et al.* (2002) as rhythm-related in an "EP overexpression" screen (see Section III.A.1). Two of the relevant insertions at the locus (UAS-containing, *tim-gal4*-driven) did cause apparently elevated levels of DFMR1 protein in "clock neurons" within the adult brain (see Section VI) and concomitantly increased free-running behavioral periods by 2–3 hr. LD monitorings of two types of *dfmr1*-null mutant revealed normal patterns of entrained locomotion (Dockendorff *et al.*, 2002; Inoue *et al.*, 2002), including anticipation of the environmental transitions (Section IV.C.1). This suggests that the mutants are not clockless, a notion that was supported by demonstrations of normal PER and TIM protein expressions in *dfmr1³* (which must be considered in the context of lengthy descriptions of the molecular chronogenetic findings that are described in Sections V.B.1 and V.B.2). Morales *et al.* (2002) reported similar findings: noneffects of a *dfxr* mutation on PER expression. Inoue *et al.* (2002), however, say that their monitorings of PER and TIM temporal dynamics revealed phase differences involving the quantities and qualities of these proteins (which, I say again, will be appreciable when we arrive at Section V). One chronomolecular effect of this mutation was found by Dockendorff *et al.* (2002): reduced amplitude of cyclical expression exhibited by the reporter fused to CRE (cf. Belvin *et al.*, 1999, discussed a short distance earlier). This result, along with the possibility that elimination of *dfmr1* affects PER and TIM expression (Inoue *et al.*, 2002), leaves the question open as to whether this gene is "clock-involved" or instead affects output pathways that feed into the regulation of circadian behavior (three of the studies under discussion) and eclosion (two of them). This issue is revisited in Section X.

Any neurochemical mutation could be thought of as a candidate for causing defects in circadian rhythms. In this regard, *ebony* (*e*) mutants were so tested. These flies have darkened external pigmentation (by definition) and aberrant levels of dopamine and β-alanine (e.g., Hodgetts and Konopka, 1973). Perhaps as a result of these biochemical anomalies, *e* mutants exhibit defects in visually mediated responses and courtship behavior (Hotta and Benzer, 1969; Kyriacou *et al.*, 1978). *ebony* mutants entered the rhythm arena long ago when it was demonstrated that a daily variation of tryosine aminotransferase activity was shifted to a 6-hr-later-than-normal phase by an *e* mutation (Tauber and Hardeland, 1977). When such mutants were tested for behavioral rhythmicity, one- to two-thirds of the individuals were aperiodic in DD, depending on which *e* mutant allele was homozygous (Newby and Jackson, 1991). These mutants also exhibited poor synchronization to LD cycles, a defect that was partly ameliorated by introducing a blinding norpA mutation (Newby and Jackson, 1990). It is as if visual inputs to the rhythm system are not subnormal, but deranged in e mutants; the result could be mediocre synchrony among individual pacemaker structures. This putative effect is not global, however, for e mutant cultures exhibited normal eclosion rhythms (Newby and Jackson, 1990), which requires synchrony among flies and probably indicates that the cellular substrates of this emergence rhythm are synchronized *within* a given developing animal.

The mutants other than those that arguably identified clock genes are difficult to interpret. Some of those resulting from mutageneses, such as *psi-2* or *Toki* (Table 1), have no known molecular correlate. Others, which were understood biochemically to some degree before being applied in rhythm tests (Table 2), range from cases that seem significant chronobiologically to those that are mere puzzles. Thus, the factors involved in cAMP function may materially participate in *Drosophila*'s rhythm system, based on precedents from other organisms. However, only one of these functions, dCREB2, is known to affect expression of putative clock genes. The substances whose levels are affected by *ebony* mutations are not interpretable as to how they may be involved in pacemaker functions or (more likely) clock-input processes and outputs dedicated to a particular feature of overt rhythmicity.

B. Principles and puzzles stemming from the candidate-mutant approach to chronogenetics

The foregoing passages may have served to drive home an appreciation of genetic *pleiotropy*. As introduced in Section II, this phenomenon is defined by multiple effects of mutations—in chronobiological cases, phenotypes that include, but are far from limited to, disturbances of daily rhythms. This main theme of the current section is to contemplate pleiotropic issues. For example, we may agree that genetic variations of this ilk need to be catalogued (Table 2), but we will come to wonder whether they affect biological rhythmicity, among other processes, by indirect routes. In other words, these genes *can* affect the process of interest when mutated. However, are the normal alleles at these genetic loci closely involved in regulating the core rhythmic functions? A negative answer seems to suggest itself by apprehension of a routinely observed feature of these pleiotropic mutants: Such putatively peripheral genetic factors tend not to lead to dramatically defective phenotypes. Thus, several of the mutations in question (Table 2) cause relatively small period changes or erratic arrhythmicity. The same needs to be said for the mutants toward the bottom of Table 1, with regard to these "minor," often mildly abnormal rhythm variants that have not been studied well enough to know whether they are pleiotropic. These phenogenetic attributes seem to discourage further investigations. This explanation, however, needs to be viewed partly in a historical context: What if the original *period* mutant had been *per*^{Clk} (with at best a 2-hr change from the normal cycle durations), and bad luck had militated against inducing *per*^{S} and the like? The case of the *period* gene might have lain as fallow as has that of *Toki*. Furthermore, what if the only *double-time* variants induced in rhythm-mutant screening had been null mutations? These have no effect on rhythmicity when heterozygous with *dbt*$^+$ and cause pleiotropically lethal developmental defects when they are homozygous. Equally disturbing, perhaps, would be the possibility that hypomorphic (leaky) *dbt* alleles could have resulted from mutant hunting. Note in this regard that a mutation of this type (derived from the lethal, nearly null *dbt*^{P} variant) causes only erratic arrhythmicity [referring to *dbt*^{PrevIX} in Table 1, which was generated by Kloss *et al.* (1998)].

For all we know, there simply are not enough mutated alleles at certain of the loci in the subsidiary lists (within Table 1 and filling Table 2) fully to appreciate what these genes can do to the clock. A probable exception to this problem is the aforementioned case of *Andante*, a gene which so far has not been altered other than to cause a mild period change. For other genes, ones that can vary such that more substantial rhythm abnormalities occur, induction of rhythm mutants will not necessarily result in the most beguiling variants at the outset. A corollary is that some (perhaps most) induced mutants might be barely recoverable in searches for rhythm variants based on phenotypes that require healthy cultures or solidly viable flies. Even if the first set of *dbt* mutants had not met this requirement, presumably the gene would sooner or later have been hit to result in heterozygous mutant flies with period changes of 2 or more hr or nonnull *dbt* homozygous types that are alive and thoroughly arrhythmic (namely the *g* and *ar* alleles of *dbt* in Table 1 and Figure 3). However, it could be that all *disco* or *dCREB2* mutants, which may have been unwittingly induced during all the years of screening, fell by the wayside because all such mutations lead to such sick flies that they were not even recovered for testing. Alternatively, more viable *disco* and *dCREB2* mutants could have been isolated unknowingly, but it might have been that only around half of the individual adults were arrhythmic. Thus, these variants would have gone unrecorded or been abandoned. The latter routinely occurred for the impenetrant mutants listed in Table 1, several which did not even have their genetic etiologies determined (cf. Figure 18; color insert). A further consideration is that some genetic loci in *Drosophila* may be more mutable than others: *per*, *tim*, and *dbt* could be relative genetic hotspots, whereas loci such as *Toki* or *dCREB2* might have suffered significantly fewer primary mutations.

There is no bottom line to this part of the rhythm-genetic story as it now stands, other than to say that the full chronobiological capacity of the *Drosophila* genome is likely to be underappreciated. Further mutageneses are warranted. Additional candidate genes with no known relationship to rhythms (at the moment) should be interrogated. Mutants involving these genetic loci could be tested, and the full gene-manipulative power of the organism should be brought to bear to

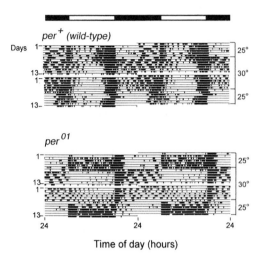

A   **Thermal and photic cycles interacting to influence *Drosophila*'s locomotor rhythmicity**

Figure 5. Locomotor cycles of *D. melanogaster* in changing temperature conditions. (A) A wild-type (*per+*) fly and a *per^01* one were each monitored for almost 1 month in LD cycles (signified by the black/night and white/day bars at the top); during the middle ca. 2 weeks, a 5°C temperature step-up was imposed as indicated on the right. These raster double plots of locomotor activity (like those in Figure 1) show that daytime activity decreased and nighttime activity increased in the warmer condition. The two behavioral records are from Tomioka *et al.* (1998), who described the *per^01* fly as switching from a diurnal to a nocturnally active pattern at the higher temperature; the *per+* example appears not to involve such a marked heat-induced change. (B) Locomotion influenced by daily temperature cycles. Twenty-one wild-type (*per+*) males were monitored in constant darkness (DD) for about 1 week in 12-hr:12-hr high:low cycles; the higher temperature was 28°C (white vertical bars) and the lower one was 25°C (black bars). Twenty males hemizygous for *per^01* were similarly monitored. The data are average-activity plots analogous to the per-day averages shown in Figure 1, but here a given histogram bar represents a per-*fly* mean (±SEM, as indicated by the dots). For this, the mean numbers of locomotor events for each day's worth of 30-min time bins were computed from each individual record; interfly means for each bin were then computed, after normalizing the data among the separate records (as in Hamblen-Coyle *et al.*, 1989), to cope with the fact that maximal levels of locomotion vary from fly to fly; thus the ordinate is dimensionless (see below), compared with the average-activity plots in Figures 1 and 4, which are based on single-animal records from LD (constant-temperature) monitoring. Interfly average-activity plots reflecting entrainment to such LD cycles are inset into the two temperature-cycling results, in which the ordinate values resulting from the normalization procedure are included with respect to the 111 *per+* and 113 *per^01* flies that were monitored in light:dark/constant-temperature cycles. The wild-type flies entrained in both the high:low/DD conditions and in the LD/constant-temperature ones, although the trough levels were relatively high and the "late-day" peaks (toward the end of the warm periods) were relatively early in the temperature cycles. The *per^0* flies, which are essentially arrhythmic in completely constant conditions (although see Figure 6A), can be inferred to

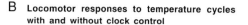

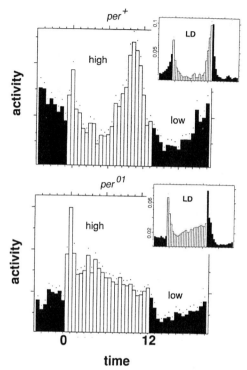

time

"respond only" to the onsets of warmth (main plot at the bottom) or to the beginnings of the light period (inset) in that there were little or no anticipations of these environmental transitions. The latter (LD) plot shows that the *per⁰* flies also exhibited sharp, transient increases in locomotion as a response to lights-off. These four sets of results are from Wheeler *et al.* (1993). Tomioka *et al.* (1998) carried out similar DD monitorings of *per⁺* and *per⁰¹* flies in 12-hr:12-hr high:low temperature cycles, although they used 30°C for the warmer condition and genetically eyeless flies as *per⁺* controls. Results from the latter investigators were similar, but not identical to those of Wheeler *et al.* (1993): For *per⁺*, Tomioka *et al.* (1998) observed ramping-up of activity that *anticipated* the cooler-to-warmer transition time and an end-of-thermophase hump *centered* near the warm:cool transition times (later than the locomotor maximum shown here). Tomioka and co-workers' *per⁰* records—average activity plotted in a manner similar to this panel of the current figure—showed relatively high levels of locomotion during the warm period and a "spike" that followed its onset (as depicted here); but an increase in activity was also induced by the onset of the cooler portion of the daily 25°C:30°C cycles, in contrast to the 25°C:28°C experiment of Wheeler *et al.* (1993) shown in the present plot.

alter the expression of these factors (as exemplified by the studies of Martinek *et al.*, 2001, and Dockendorff *et al.*, 2002) in the context of what is now known about the rhythm system in *Drosophila*.

C. Rhythm mutants applied to analysis of phenotypes other than basic circadian locomotor cycles

The value of testing a rhythmically defective behavioral variant for another circadian phenotype, eclosion, was stated earlier, and of course some of the locomotor-rhythm mutants started out as eclosion ones. This section deals with further time-based phenotypes involving the fly's development or (more commonly) its adult functions. Investigators have seemed relentlessly curious as to whether a given temporal phenomenon would come under the sway of a gene identified initially by effects on the principal circadian rhythms of *Drosophila*.

1. Abnormal entrainment and responses to phase-shifting stimuli

The first phenotype to be considered is actually a kissing cousin of free-running behavioral rhythmicity—periodic locomotor changes in fluctuating environmental conditions. (Also note that certain mutants, such as the original *per*'s and the *psi*'s, were isolated in LD tests.) Rhythm mutants that are arrhythmic in constant conditions are typically periodic in LD or when the temperature is made to oscillate (e.g., Dushay *et al.*, 1989; Sehgal *et al.*, 1991; Wheeler *et al.*, 1993; Tomioka *et al.*, 1998). In the former condition, *Drosophila* exhibits two locomotor peaks per day, one near the time of dawn, the other near dusk (Figures 1A and 5). The late-day maximum involves more locomotor events (per time bin) than occur around the time of lights-on. Anticipating both of these transitions, the flies ramp up toward their behavioral peaks during the last one-fourth to one-third of the (12-hr) daytime and nighttime (e.g., Hamblen-Coyle *et al.*, 1992). Anticipation of the L-to-D transition is typically more prominent and sustained than observed in advance of the D-to-L change (Figures 1A and 5B). However, these patterns are dependent on the ambient temperature: Relatively more *nighttime* activity was observed in overheated conditions by Tomioka *et al.* (1998), who monitored flies in LD for a few days at 25°C, shifted them to 30°C, then returned them to 25°C (Figure 5A).

In another kind of diel behavioral test, flies were not monitored automatically in their containers, but instead were moved into chambers at three separate times after dawn; locomotor events were observed by eye, but essentially no changes were counted between Zeitgeber Time 1 (ZT1) and ZT9 (ZT0 defining the time of lights-on) for five *Drosophila* species (Cobb *et al.*, 1987). This pattern of partly *evoked* behavior is different from the results of measuring spontaneous locomotor changes over the course of the daytime, during which a distinct midday siesta is routinely observed (Figures 1A and 5B).

In temperature cycles, wild-type *Drosophila* exhibit an early "morning" peak of locomotion which, in one study, seemed not to anticipate a 3°C temperature increase at this transition time (Figure 5B), in contrast to the anticipation of lights-on that is exhibited in LD cycles (exemplified in Figure 1A). However, Tomioka *et al.* (1998) found relatively high levels of activity surrounding the cool-to-warmer transition they applied (a 5°C stepup), that is, anticipation of the thermophase. Wheeler *et al.* (1993) observed a substantial peak during the second half of the warm period (in 25°C:28°C cycles), which subsided to a lower level of behavior well before the high-to-low temperature change (Figure 5B), whereas the end-of-thermophase peak observed by Tomioka *et al.* (1998) in 25°C:30°C cycles was later. Assessments of entrainment to temperature cycles varying by only 1–2°C gave poor or mediocre synchronization to such fluctuations (Helfrich, 1986; Wheeler *et al.*, 1993). However, 12:12 high:low cycles defined by 5°C changes not only entrained the flies' behavior in DD, but also overrode the arrhythmia-inducing effects of constant light (Tomioka *et al.*, 1998; cf. Section IX.B.1).

Against this background, arrhythmic per^0 mutants or flies deleted of the *period* gene seem to exhibit "forced" periodic behavior in LD or temperature-cycling conditions (e.g., Petersen *et al.*, 1988; Hamblen-Coyle *et al.*, 1989; Ewer *et al.*, 1990; Wheeler *et al.*, 1993; Tomioka *et al.*, 1998). Thus, these clockless flies become rather abruptly more active right after D-to-L and L-to-D changes (Figure 5B). The DD-arrhythmic *timeless* mutant (née *aj42*) behaves analogously, although recall that when the original strain was tested in LD most of the activity occurred during the night (Sehgal *et al.*, 1991). In contrast, *per*-null mutants (and apparently flies in tim^{01} strains derived from *aj42*) carry out the majority of their locomotion during the L period, although the plots (as exemplified in Figure 5B) do not look like simple square waves. Another clockless mutant, Clk^{Jrk}, does not behave in a *per*- or *tim*-null-like manner in light:dark cycles: Locomotor monitoring in this condition showed Clk^{Jrk} not to exhibit abrupt increases in locomotion after the lights come on (Allada *et al.*, 1998); this result presaged certain further considerations of *Clock*'s mediation of light effects on locomotor activity (E. Y. Kim *et al.*, 2002; see Section V.C.2).

When per^{01} was tested in temperature cycles (Wheeler *et al.*, 1993), increased locomotion was observed after a warm-to-hotter change (Figure 5B), but there was no peak at the opposite transition 12 hr later (let alone an evening maximum, as occurs in per^+). Tomioka *et al.* (1998) obtained similar results, but observed an "extra" spike of accentuated per^{01} activity following the warm-to-cooler transition (see the legend to Figure 5). per^S or per^L flies exhibited largely free-running behavior in 12:12 high:low cycles (DD), although there were relatively higher locomotor levels of activity during the warm period; the increased activity during this phase was interpreted to represent "masking" (Tomioka *et al.*, 1998), that is, a direct influence of the higher temperature on behavior, thus an effect that bypasses the clock's ongoing functions or those involved in resetting it.

Drosophila expressing *per*-null mutations exhibit changes in their behavior *subsequent* to the times when light or temperature changes trigger the rather abrupt increases in locomotion depicted in Figure 5. Thus, locomotor rates subside in a rather gradual manner during the ensuing 12 hr of light, darkness, or warmth (e.g., Wheeler *et al.*, 1993). So clockless flies do not modulate their behavior solely in response to environmental changes. However, the apparently forced nature of an initial light-induced increase in locomotion, for example, was suggested by the fact that *per*⁰ flies exhibit clearly periodic behavior in all kinds of shortened LD cycles (8-hr L:8-hr, D, down to 4:4 cycles), and their periodicities tracked whatever regime was imposed (Wheeler *et al.*, 1993). This null mutant was active mainly during the light phase (whatever its duration) in these 25°C tests (Wheeler *et al.*, 1993). A switch of *per*⁰¹'s activity from a diurnal mode to a nocturnal one was observed in an experiment for which the temperature was raised from 25°C to 30°C during 12:12 LD monitoring (Figure 5A). When *per*⁺ controls were monitored in abbreviated LD cycles by Wheeler *et al.* (1993), they free-ran with ca. 24-hr periodicities in cycles shorter than 8:8 ones (see later for more about these "limits of entrainment").

Imposing a shifted (12:12) LD cycle on wild-type—or genetically blind, but clock-enabled—flies re-entrained them, but this required 1–2 days of transient, not-yet-synchronized behavior; in contrast, *per*⁰¹ went immediately in synch with the new regime (Wheeler *et al.*, 1993). [See Emery *et al.* (2000b), discussed in Section IX.B.3, for more about this from the perspective of additional kinds of visual impairments.] Shifting a 12:12 temperature cycle to earlier or later phases re-entrained wild-type flies (after ca. 5 days of transients), whereas *per*⁰¹ shifted almost immediately to synchronize with the new thermophase (Tomioka *et al.*, 1998)

Thus, the responses of *per*-null *Drosophila* indicate that the mutant's locomotor modifications are basically *driven* by environmental changes. However, certain features of the mutant's behavior in these fluctuating conditions indicate that its varying locomotor levels are not entirely imposed from exogenous cues. Nevertheless, the diminishings of locomotor rates that commence a few hours after an environmental transition do not necessarily imply regulation by an endogenous oscillator. Thus, imagining that *per*-null mutants are thoroughly clockless, one can consider that their periodic behavioral fluctuations could be mediated by an "hourglass" effect: When the biological analog of such a device is flipped (say, at the time of D-to-L), increased locomotion would be induced by the exogenous agent (which does the flipping); then, as the physiological or biochemical sand runs out, this causes an underlying regulator of behavior to be less efficacious. When the top of the hourglass is depleted of its time-measuring material, this could allow levels of locomotion to increase once more.

Now comes the point of putting forth this labored metaphor for a putatively imposed influence on *per*⁰'s periodic behavior: Imagine that the sand is gone from the top of the device at about the time of the D-to-L transition. Little or no anticipation of that time marker will be apparent. However, if cycles longer than 24-hr are imposed on such mutant individuals, this time-giving material could be depleted well before the lights come on; increased locomotion that anticipates that transition could occur. This is what has been observed when per^0 flies are monitored in "long-T cycles" (28–36 hr). The mutant anticipated the D-to-L changes, and the longer the T duration, the longer and more pronounced was the locomotor increase in advance of lights-on (Helfrich and Engelmann, 1987; Helfrich-Föster, 2001). These results were interpreted to be consistent with *Drosophila*'s "behavioral clock" consisting of a *two oscillators*, a morning and an evening one, and that the former is relatively insensitive to the effects of *period* gene changes. In this regard, the morning peak in LD typically disappears when wild-type flies proceed from LD into DD, so there would be naturally nothing left for the (intact) morning oscillator to control in constant conditions; thus, per^0 would be arrhythmic in DD, which it is.

Whether there is a "*per*-insensitive second clock" that influences adult behavior in LD (Helfrich-Föster, 2001) or whether a less endogenous hourglass-type phenomenon is operative, the effects of clock-functional *per* mutations on diel behavior are at issue. Here, it is notable that in 12:12 LD the evening peak of locomotion is moved substantially by the effects of period-shortening or period-lengthening mutations. However, the morning peak is less affected (Hamblen-Coyle *et al.*, 1992), consistent with a notion that its driving oscillator is minimally influenced by the action of this gene. Yet the putative evening oscillator would be *per*-controlled because LD behavioral maxima occur during the daytime in per^S, per^{Clk}, or per^T, instead of at about the time of lights-off (Dushay *et al.*, 1990; Hamblen-Coyle *et al.*, 1992; Konopka *et al.*, 1994). [However, per^S exhibited relatively more nocturnal activity at 30°C compared with 25°C (Tomioka *et al.*, 1998).] The evening peak is shoved well into the nighttime when per^L or per^{SLIH} is monitored in LD (Hamblen-Coyle *et al.*, 1992; Hamblen *et al.*, 1998; Tomioka *et al.*, 1998). Similar results were found for long-period *doubletime* mutants (Suri *et al.*, 2000), which entrained to 24-hr T cycles and exhibited nighttime locomotor maxima. Such results are readily rationalizable in the context of light-at-dusk eliciting daily phase delays (thus, per^S's "internal daytime" would end well before the lights go off), and that dawn light causes phase advances (so dbt^h would still be operating within its internal night when the lights come on).

Application of the early-phase per^S mutant in tests of the arrhythmia-inducing *disco* mutations was useful. If *disco*'s free-running arrhythmicity is caused by clocklessness, such that its periodic diel behavior (Dushay *et al.*, 1989) is merely forced, then the locomotion in LD of such mutant flies would presumably be insensitive to the effects of per^S. However, a per^S *disco* double mutant exhibited an earlier-than-normal evening peak (Hardin *et al.*, 1992a), as if a functional pacemaker is retained in *disco* and subject to the clock-altering effects of this *period* mutation.

In other pacemaker-enabled mutants (that is, other than per^0, tim^0, and the like), the magnitudes of daily phase shift that have been observed in entrainment tests are impressive. Consider per^T, which effects a 7- to 8-hr delay every day, such that all mutant individuals behave in synchrony with 24-hr T cycles (Konopka et al., 1994). This does not necessarily mean that a mutant of this sort is hypershiftable, because the limits of entrainment for wild-type are wide (as introduced above): genotypically normal flies can behave in synchrony with LD cycles ranging down to T = 16 and up to T = 36 hr (Wheeler et al., 1993; Helfrich-Förster, 2001).

Phase-shifting experiments involving relatively brief light pulses are potentially instructive in this regard: per^{Clk} and per^L exhibited modest advances and delays, as did wild-type adults, to 10- or 15-min pulses (Dushay et al., 1990; Rutila et al., 1998a). However, per^S and per^L were found to be hypershiftable to 1-hr pulses (Saunders et al., 1994), the shortest such stimuli used in these experiments. In another set of PRC tests, both per^S and dbt^S flies had larger-than-normal phase changes induced by 2-hr light pulses (Bao et al., 2001). Six-hour pulses caused high-amplitude advances and delays for flies of all three per genotypes, including per^+ (Saunders et al., 1994). With regard to per^S's eclosion rhythm, rather short light pulses (40 min) led to high-magnitude phase shifts compared with the responses of wild-type cultures (Konopka, 1979). An embellishment of this kind of test also suggested heightened responsiveness of the mutant's eclosion-controlling clock system. here, relatively low intensity light pulses delivered to per^S cultures were able to locate a point of "singularity" during the middle of the subjective night, which was not exhibited by wild-type flies, but drove the eclosing per^S flies into arrhythmicity (Winfree and Gordon, 1977). Thus, as noted by these authors, this mutant responded as genetically normal cultures of D. pseudoobscura do in their phase responses to low-energy light.

Returning to considerations of locomotor rhythmicity, mutations at loci other than per have been tested for effects on the behavioral phase response curve (PRC). These different types of fly gave varying results depending on the mutant, but all were based on delivery of short light pulses. Modest (wild-type-like) phase shifts were observed for Toki, tim^{SL}, and dbt^L, but shifts of higher-than-normal magnitude occurred in tests of tim^{L1} (Matsumota et al., 1994; Rutila et al., 1998a; Rothenfluh et al., 2000a).

Accompanying several of these PRC experiments were demonstrations of differential effects of a given mutation on certain portions of the daily cycle (e.g., Konopka and Orr, 1980; Rutila et al., 1998a). In some cases altered activity (α)-to-rest (ρ) ratios were revealed (e.g., Saunders et al., 1994; Matsumoto et al., 1994). Certain of these results are in effect congruent, notably a 5-hr shortening of the PRC's subjective day (Konopka, 1979; Konopka and Orr, 1980), the same degree of free-running period shortening observed for per^S (Table 1). It is difficult to say at the moment how such light-response and other behavioral anomalies might prompt certain understandings of how these hypothetical clock mutations alter these features of the circadian system's functioning. However, some of the relevant molecular issues may be illuminated later (Section IX.B), but not in the case of *Toki*, with its increased α/ρ ratio (Matsumoto *et al.*, 1994). This genetic locus (Figure 18, color insert) remains unknown as to what kind of clock or other rhythm-related function it may encode.

Short light pulses have been delivered in another circumstance: Flies monitored for many days in DD were so stimulated, and the "split" activity components frequently observed in such tests could be induced to coalesce back to a unimodal pattern of daily locomotor segments (Helfrich-Förster, 2000). Incidentally, that the pre-pulse pattern of behavior could indeed split into free-running (subjective) morning and evening components was taken to buttress the notion that they are controlled by two different oscillators (Helfrich-Förster, 2001). However, the main purpose of this study (Helfrich-Förster, 2000) as it is considered currently was to assess possible sex differences in locomotor rhythmicity. If observable, these would be influenced by the different chromosomal genotypes. Indeed, for the three wild-type strains analyzed, males exhibited an earlier morning peak of activity in LD compared with females. In DD, the morning peak merged with the evening one in about half the individuals tested (males exhibited a stronger tendency to retain bimodality), and the sex-influenced "phase-angle" difference disappeared in the unimodal cases (in LD, the timespan between morning and evening peaks was longer in males, in the context of the latter peak being unaffected by sex genotype). In DD, males exhibited slightly shorter free-running periods compared with those of females, and the latter were more active than males in two of the strains (Helfrich-Förster, 2000). A previous test of locomotor levels in LD (general fly movements that, as noted above, were stimulated by moving the flies into observation chambers at various times of day) showed *males* to be the more active sex (Cobb *et al.*, 1987). Whether or not these sex-linked behavioral differences are consistent between studies, they hark (way) back to male versus female variations in a rhythm of oxygen consumption: The morning maximum was smaller than the evening one in females, and such *D. melanogaster* expressing certain cuticular mutations exhibited no morning peak; males (whether mutant or not) never exhibited an evening peak only and typically gave equal maxima at these two cycle times (Rensing *et al.*, 1968). In a subsequent study, an oxygen-consumption rhythm was also observed in pupae and pharate adults of this species (Belcher and Brett, 1973).

2. Rest-activity fluctuations as sleep-wake cycles

Subtracting the α component just referred to from the overall daily performance of flies leaves ρ. Such rest periods could be called sleep, but until recently such an appellation would have seemed a labored metaphor. To look into the possibility that *Drosophila* do sleep, the flies' active versus resting states were monitored at higher resolution than by merely recording whether the flies are grossly moving back and forth within glass tubes (Hendricks *et al.*, 2000; Shaw *et al.*, 2000; cf. Figure 1A). It was found that *Drosophila* are nearly quiescent for many minutes in a row during the ρ portion of a daily cycle (exhibiting only slight movements of appendages or "respiratory pumping"). Comfortingly, a very high correlation was found between grossly monitored locomotion and what can be construed as non-somnambulant activity (Shaw *et al.*, 2000). When the fly's putative sleep state was examined experimentally, it was concluded that ρ periods are "homeostatically" regulated (Hendricks *et al.*, 2000; Shaw *et al.*, 2000), as occurs in mammals and other supposedly higher forms (reviewed by Kilduff, 2000; Greenspan *et al.*, 2001). Thus, *Drosophila* deprived of rest during the normal ρ segment of a given cycle exhibit severalfold more rest during the subsequent α period. In tests of the per^{01} mutant, this homeostatic "rest rebound" was found to be like that exhibited by wild-type flies (Shaw *et al.*, 2000). Curiously, tim^0, which can be thought of as arrhythmia-inducing in a manner equivalent to the effects of per^0 (see Section V), was reported to eliminate the normal rebound after a rest deprivation (Hendricks *et al.*, 2000).

The effects on sleep of no *timeless* function were further examined by Shaw *et al.* (2002), who also retested a per^0-null and rolled in two additional clockless types. The tim^{01} mutant was found to be reboundless, but only after 3 or 6 hr of prior sleep deprivation; these flies exhibited quite robust rebounds after 7, 9, or 13 hr of being sleep-deprived. In fact, this was a supranormal response, similar to those observed in parallel for per^{01} or Clk^{Jrk} flies, which "reclaimed" ca. 2–3 times more sleep than did wild-type flies (within a half-day after the various deprivation sessions). More striking were the effects of a *cycle*-null mutation: cyc^{01} *Drosophila* could be said to reclaim way too much sleep ("3 min over baseline for each minute ... lost"), which persisted for as long as the flies were recorded (more than 2 weeks). Worse, when cyc^{01} was deprived of sleep for greater than 10 hr, this mutant (unlike any of the other clockless types tested) exhibited dramatic mortality (about one-third of them dying). Molecular-genetic correlates were obtained for this *cyc*-null effect (Shaw *et al.*, 2002), in that these flies exhibited reduced expression of heat-shock genes after they were made to lose sleep; however, activation of such factors before sleep deprivation rescued cyc^{01} from the morality just described, and a heat-shock-protein mutant created a genocopy of this clock mutation (exaggerated rest-rebound and sleep-induced death).

A different kind of gene—*Dat*, whose encoded arylalklylamine acetyltransferase is involved in monoamine metabolism—was also connected to regulation of the fly's sleep–wake cycles. This was among four genes identified in an differential-display protocol: mRNA was extracted from heads of flies that had been resting during the dark period, sleep-deprived during that time, or spontaneously awake during the light period; this material was reverse-transcribed and the cDNA copies amplified by polymerase chain reaction (PCR) using many primer combinations. *Dat* expression was thus found to be at relatively high levels during waking and after rest deprivation. This prompted tests of the already extant Dat^{lo} mutant, which has 15% of normal enzyme activity when the mutation is heterozygous with a deletion of the gene; a rest rebound was observed that amounted to about twice the magnitude of that exhibited by rest-deprived wild-type flies (Shaw *et al.*, 2000). The molecular version of *Dat* will reappear in another chronobiological context (Section X.C).

3. Neurochemical connections to rhythm mutants along with their temporally related and other phenotypes

Dopamine and octopamine are among the neuromodulators likely to be metabolized by DAT. The following pieces of related biochemistry are unknown as to their biological significance (let alone any connection to sleep regulation), but nonetheless come under the heading of chronogenetic phenomena. Thus, accumulation of newly synthesized octopamine and tyramine were found to be threefold reduced in brains dissected out of per^{01} adults; concomitantly, tryosine decarboxylase activity (TDC, which is required for production of the octopamine precursor tyramine) was reduced to about one-third normal by per^{01}, and TDC was briefly noted to be at about half-normal levels in per^L and per^S (Livingstone and Tempel, 1983). Elements of these seemingly meddlesome results were unwittingly examined further in another context: Andretic *et al.* (1999) showed that TDC is induced by repeated exposure to cocaine, that is, in experiments involving application of the drug at 6-hr intervals. In this context, tyramine was shown to be required for such behavioral sensitization to cocaine administration (McClung and Hirsh, 1999). per^{01}, Clk^{Jrk}, and cyc^{01} flies did not exhibit drug-induced increases in TDC activity, whereas tim^{01} showed normal induction after exposure to cocaine (Andretic *et al.*, 1999). Correlated effects of these mutations were found in behavioral assays: Wild-type and tim^{01} flies sensitized to the same kind of repeated drug exposures (by enhanced jumping and "twirling," for example), but various types of *per* mutants (per^S, per^T, per^L, and per^{01}) did not, nor did flies homozygous for Clk^{Jrk}, cyc^{01}, dbt^S, or dbt^L (Andretic *et al.*, 1999).

Perhaps because of an involvement of dopamine-receptor responsiveness in cocaine sensitization in vertebrates, time-dependent responses of *Drosophila* to applied dopamine were also examined by these investigators; they found that application of a "D2-like" dopamine-receptor agonist induced locomotor actions in a rhythmic manner (Andretic and Hirsh, 2000). The flies in question were only partly that, as they were decapitated, facilitating quinpirole application to the resulting hole at the neck. Responsiveness peaked during the middle of the night in LD and exhibited a mild free-running rhythm in LL. The curve for the latter was completely flattened in tests of headless per^{01} flies, although this mutant showed

A Ultradian locomotor cycles of a *per⁰* mutant

TIME OF DAY (HRS)

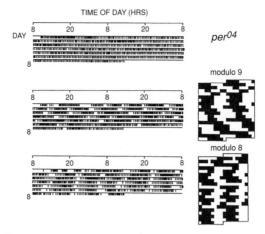

modulo 9

modulo 8

B High-frequency courtship-song rhythms

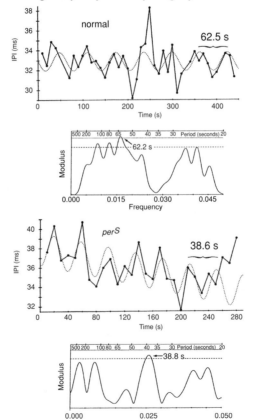

an early-nighttime peaklet in LD. A *per*-influenced "body oscillator" was invoked to explain the circadian modulation of quinpirole-induced locomotion (Andretic and Hirsh, 2000), but it is not known how the cocaine-response abnormalities tie into the fly's clock system, even though these experiments involved a temporal variable.

Figure 6. Ultradian rhythms of *Drosophila* locomotor behavior and of the male's courtship song. (A) Flies each expressing effects of the arrhythmia-inducing per^{04} mutation (Figure 3) monitored for DD activity. This mutation causes somewhat accentuated levels of locomotion compared with those recorded for the wild-type fly or the per^{01} mutant (Table 1). Thus, the activity markings in the raster-plotted actogram at the top are denser compared with, for example, the *disco* actograms in Figure 1B (which are per^{01}-like). This per^{04} individual was thoroughly arrhythmic. However, the middle and the bottom actograms suggest regular "beats" of relatively accentuated activity occurring (in these particular cases) three times per day, reflecting the ultradian locomotor cycles (with periods in the range of 5–15 hr) that are routinely observable for this particular *per* mutant and for other arrhythmic genetic variants in this species (Table 2). Filtering the digitally collected activity data (cf. Figure 1) for the per^{04} flies whose behavior led to the bottom two actograms and replotting these results by way of 9- or 8-hr horizontal time scales highlights the ≪24-hr cycles that define the locomotor fluctuations for these mutant individuals. These plots were adapted from Hamblen-Coyle *et al.* (1989). (B) Song rhythms of *D. melanogaster* flies controlled by the normal per^{+} allele (top pair of plots) and as affected by the fast-clock per^{S} mutation (bottom pair). Courting males of this species generate pulses of tone by vibrating their wings and "singing" to females. The intervals of silence between pulses (interpulse intervals, IPIs) fluctuate through a range of ca. 30–40 msec. The IPI fluctuations are systematic in clock-enabled flies, as revealed by the changing values (averages, for a series of ca. 45 10-sec time bins) that define the normal song rhythm exemplified at the top. The period value determined from curve-fitting (dotted line) was 62.5 sec, as shown. This fly was not a wild-type male (whose song-rhythm periods are about 1 min), but instead a genetic mosaic. It was part per^{+} (single *X*) and part per^{S}/per^{+} (*XX*) in different body regions and tissues—all of which were turned into genetic maleness by the effects of a sex-determination mutation known as *transformer* (*tra*), present in homozygous condition on an autosome within the mosaic's genetic background. This normally singing mosaic had heterozygous mutant tissue throughout most of its brain (and exhibited shorter-than-normal locomotor cycles; see Table 1 and Figure 4), but its thoracic ganglia (a.k.a. VNC) were largely per^{+}, which was inferred to be the reason that the song-rhythm period is in the normal range. That this is so is further indicated by the results of a spectral analysis of the IPI timecourse, which, independent of the curve-fitting shown in the top plot, led to a very similar best estimate of the period value. The bottom pair of plots resulted from a mosaic whose tissue distribution of *per* alleles was essentially opposite to that of the normally singing case; thus, this fast-cycling mosaic had a largely per^{+} brain, but a mostly per^{S}/per^{+} VNC—believed to be responsible for the 38.8-sec (and spectrally determined 38.6-sec) periodicity that defined this individual's IPI fluctuations. This shorter-than-normal cycle duration is well within the range exhibited by fully mutant per^{S} males or by pseudo-males that carry per^{S}/per^{+} and are homozygous for the aforementioned *tra* mutation (Kyriacou and Hall, 1980). These plots were adapted from Konopka *et al.* (1996).

Moving to another apparently random piece of *Drosophila* biochemistry, a weak daily rhythm of protein carbonyl content was observed and studied as an index of free-radical damage. The inferred "protein damage" showed a ca. 50% higher level for a mid-nighttime point compared with the baseline values at other times (Coto-Montes and Hardeland, 1999). *per*^{S} flies exhibited a more convincing maximum during the night and relatively high values during the day; the *per*^{01} timecourse was flattened. A different type of mutant was also measured for its protein carbonyl concentrations during a 12:12 LD cycle. This was *rosy*, which is null for xanthene dehydrogenase and thus lacks the free-radical scavenger urate. Homogenates of *ry*^{506} flies gave (roughly) two daily peaks during the late night and midday, associated with higher values than the maxima determined for the other three genotypes (Coto-Montes and Hardeland, 1999).

4. Ultradian fluctuations associated with rhythm mutants

We now return to considerations of the spontaneous locomotor behavior of *per*-null mutants and, as it turned out, other arrhythmic variants. All of these essentially, or supposedly, aperiodic mutants exhibit a kind of fluctuating locomotion in constant conditions: As a rule of thumb, approximately half the individuals tested behave such that high-frequency *ultradian* rhythms accompany their activity. These periods tend to be in the range of 5–15 hr (Dowse *et al.*, 1987). Occasionally, such rhythmicity is overt (Figure 6A), but usually it must be revealed by spectral analysis of the locomotor data (reviewed by Dowse and Ringo, 1993). In one study, however, application of merely a periodogram to locomotor data revealed high-frequency components accompanying about 30% of the *per*^{01} records that resulted from monitoring the mutant's behavior either in DD or in LD (Tomioka *et al.*, 1998).

When this ultradian rhythmicity was uncovered it was suggested that normal *per* gene function is involved in coupling high-frequency oscillators, which naturally talk to each other as a fundamental feature of building the circadian pacemaker (discussed by Dowse and Ringo, 1993). Therefore, the action of *per* could promote such coupling among the relevant cells or within a given individual cell, and output from the uncoupled multi- or intracellular entities in *per*^{0} mutants would underlie the ultradian rhythms. In either case, if these pacemaker structures are gone, then the ultradian periodicities should not manifest themselves; each individual would be profoundly arrhythmic such that sensitive spectral analysis would be unable to tease out any periodic behavioral components because there are none. However, *disco* flies, the great majority of whom seem devoid of certain pacemaker neurons (Zerr *et al.*, 1990; Helfrich-Förster, 1998), exhibit ultradian behavioral rhythmicity no less frequently than do *per*-null variants (Dowse *et al.*, 1989). If these brain cells are *the* key neural substrates of locomotor rhythms in *Drosophila* (see Section VI.A.2), the coupling hypothesis would predict _disco_ to be completely arrhythmic. That a set of behaviorally mon-itored flies expressing this mutation was not (Dowse _et al._, 1989) reflects on _per_^{0}'s etiology of ultradian rhythmicity and reflects badly on the supposition that the _period_ gene is involved in pacemaker couplings. A potentially better conjecture would go like this: No matter how one eliminates solid circadian rhythmicity—by clock-functional or neuroanatomical means—one uncovers the hidden capacity of the fly to exhibit ultradian rhythms by way of brain–behavioral processes that do not connect with the circadian system. It follows that any root cause of circadian arrhythmicity—well beyond the effects of _per_ and _disco_ mutations—would auto-matically lead to ultradianly rhythmic behavior. This is what was found by meta-analysis of a medium-sized list of such arrhythmia-inducing factors (Hall, 1998a). Moreover, monitoring adult locomotion in LL led to ultradian patterns within the grossly appearing behavioral arrhythmicity (Power _et al._, 1995b; although Tomioka _et al._, 1998, were not able to pull ultradians out of such records by peri-odogram analysis). Constant-light effects on the molecular clockworks—and how they come close to mimicking the effects of _per_-null mutations—are considered in Section IX.B.

Before the decoupling hypothesis formulated to explain the grossly ar-rhythmic, but ultradianly periodic behavior of _per_-null mutants got undermined by the further behavioral analyses just referred to, this presumption seemed to be correlated with certain physiological effects of _period_ mutations. For this, we first consider that a circadian rhythm of membrane-potential fluctuations was re-ported for individual cells monitored in the larval salivary gland (Weitzel and Rensing, 1981). Moreover, fairly high proportions of such cells in _per_^{01} glands were arrhythmic for this piece of physiology (measured indirectly by application of a voltage-sensitive dye). This erratic effect of the arrhythmia-inducing muta-tion on a cell-by-cell basis seemed incongruous to other investigators, for it has long been known that cells within the salivary gland are naturally coupled. Thus, intercellular communication between the rhythmic cells in a _per_^{01} gland could well have promoted periodic membrane-potential fluctuations in the other ones within the same tissue—unless intercellular coupling itself were affected by _per_ mutations. These defects in fact got reported for _per_^{01} and _per_^{S}, which were said, respectively, to exhibit almost no such communication among salivary-gland cells or substantially enhanced such coupling (Bargiello _et al._, 1987). If these physiolog-ical phenotypes were also manifested in—actually among—behaviorally relevant brain cells, this would permit rationalization of the various locomotor pheno-types: gross arrthythmicity of _per_^{01} intertwined with ultradian periodicities, and a fast pace caused by the overall neural pacemaker structure in _per_^{S}, mediated by stronger-than-normal coupling among cells, which, as individual structures, would naturally be running pacemakers with substantially less than 24-hr periodicities.

In terms of extending the salivary-gland phenotype to other tissues, studies involving communication among CNS neurons were never performed in *Drosophila* expressing a given *per* allele (not that that would be possible within such tiny insect brains). Potential effects of one such mutation—in *per*[L] salivaries, which should have exhibited weak coupling within the larval gland, but better than that observed among *per*[01] cells—were never tested physiologically. At all events, analyses of further *per* mutants or additional physiological analyses of *per*[01] and *per*[S] became moot. As descriptions of the presence of PER within CNS cells emerged, there was a growing sense that the consequences of this protein's functions are limited to what is going on *within* a given cell, including certain brain neurons (Sections V and VI). The accompanying *intracellular pacemaking* hypothesis prompted a reexamination of intercellular-communication abnormalities in larval salivary glands of *per* mutants. The seminal observations (involving injections of dye and its rate of spread among the glandular cells) proved irreproducible (Siwicki *et al.*, 1992a; Saez *et al.*, 1992; Flint *et al.*, 1993).

Another controversial aspect of *period* mutational effects involves an ultradian behavioral rhythm whose time constants are much shorter than days or hours. The normal phenotype concerns a certain feature of *Drosophila* courtship: Males that are interacting with females generate a courtship song by their wing vibrations; the most salient song component is a series of tone pulses. The rate at which these clicking sounds are generated, expressed by way of interpulse-interval (IPI) measurements made over the course of 5-min courtships (Figure 6B), speeds up and slows down systematically (Kyriacou and Hall, 1980). The cycle durations are normally ca. 55 sec in the songs of *D. melanogaster* (about 1 min elapses between the times of longest IPIs, i.e., slowest pulse rates). Such periodicity is a species-specific courtship parameter, for the song-rhythm periods are ca. 40 sec in *D. simulans* (Kyriacou and Hall, 1980, 1982) and ca. 70 sec in *D. yakuba* (Demitriades *et al.*, 1999). Within *D. melanogaster*, the *per*[S] (Figure 6B) and *per*[L] mutations were found to shorten and lengthen the IPI periodicities, respectively, and *per*[01] males exhibited essentially no regular IPI oscillations (Kyriacou and Hall, 1980). *per*[Clk] males exhibited a normal ca. 1-min rhythm (Dushay *et al.*, 1990), but perhaps the mere 6% change in clock pace with regard to this mutant's circadian period (Table 1) was not connected with a noticeable-enough alteration of singing rhythmicity. Additional mutants not appreciable as clock-defective (per se) were tested as well: *psi-2* and *Andante* males were briefly noted to exhibit a mild lengthening of song-cycle durations (Jackson, 1982; Kyriacou *et al.*, 1993), and the *gat* mutation caused singing behavior to be ostensibly aperiodic (Jackson, 1982).

The genetic etiology underlying one the behavioral difference between *D. melanogaster* and *D. simulans* was shown, by song recordings of reciprocally hybrid males, to map to the *X* chromosome (Kyriacou and Hall, 1986). Given that the *per* gene is on that chromosome in both such species, transgene-mediated transfers of *per* "alleles" that were all or partly derived from *D. simulans* clones were performed. It was shown that, in *D. melanogaster* males whose genetic background includes *per*[01], the partially hybrid males generated *D. simulans*-like (ca. 40-sec) rhythms, even if only the middle region of the *period* gene was derived from that species (Wheeler *et al.*, 1991). A harbinger of this kind of interspecific transfer experiment involved daily locomotor-rhythm differences that are apparent in the activity records of D. *melanogaster* compared with D. *pseudoobscura* flies (a lower α/ρ ratio in the latter): Transferring the *pseudoobscura* form of *per*, fused to upstream-regulatory sequences cloned from D. *melanogaster* (Figure 14, see color insert) made the partially hybrid D. *melanogaster* hosts behave rather more like the species that was the source of the donor DNA (Petersen *et al.*, 1988). This finding will be revisited when temperature compensation of *Drosophila*'s circadian rhythms is taken up in Section VIII.

Meanwhile, we consider intraspecific *period*-gene variations in conjunction with song-rhythm analyses: These phenotypes, including the *per*-mutational effects, were reported to be irreproducible (Crossley, 1988; Ewing, 1988). Enhancements of the analytical procedures, applied to the original song records and additional recordings made from males expressing the various *per* alleles, were asserted to confirm the initial findings (Kyriacou and Hall, 1989; Kyriacou *et al.*, 1990b). Accompanying these tests were demonstrations that, after all, ca. 1-min rhythms existed in the key per^+ records reported in one of the supposedly iconoclastic studies (Crossley, 1988). More important, the bona fide nature of such normal song rhythmicity and of the *per*-mutational effects were reproduced in a thoroughgoing independent investigation (Alt *et al.*, 1998). Having said this, there is nothing about *period* expression, or the nature of the gene's product (Sections V and VI), that permits rationalization of the manner by which this clock factor regulates a rhythm operating over such short time scales.

Nevertheless, the behavioral significance of song rhythmicity has been examined in experiments that have in part drawn upon *per* variants or electronic mimics of them. Thus, stimulation of a female's mating receptivity was shown to require that the normal species-specific rhythmic component accompanies the pulses being produced; the underlying (average) pulse rate (ca. 30/sec in D. *melanogaster* and ca. 20/sec in D. *simulans*) must also be operating if a female of the species in question is to exhibit quasi-normal mating receptivity (Kyriacou and Hall, 1982, 1984, 1986). The key elements of these behavioral effects were also reproduced in independent tests of the song-rhythm effects (Ritchie *et al.*, 1999). An acoustic priming effect of playing rhythmic pulse song to reproductively naive females (Kyriacou and Hall, 1984) also proved reproducible, although not by way of a completely independent investigation (Griffith *et al.*, 1993).

However, the efficacy of the electronic mimics of D. *melanogaster* or D. *simulans* songs was not reflected in the mate choice of females being serenaded by live males. The latter were set up to generate fully normal D. *melanogaster* songs, on the one hand (per^{01} carrying a "*mel*–per^+" transgene), or average IPIs characteristic of that species, but a short-period D. *simulans* rhythm on the other (per^{01}

transformed by $sim-per^+$): $D.$ $melanogaster$ females did not discriminate between these two types in terms of their mating receptivity to one or the other male, both of which were simultaneously in her presence (Ritchie and Kyriacou, 1994). A further genetically related mating effect that suggested itself in conjunction with these sorts of various behavioral inquiries was that per^S females might respond best to the short-cycling songs of per^S males, and that the same kind of "sender–receiver coupling" would be operative in pairings of per^L males and females. Yet both mutant female types responded optimally to IPIs oscillating with normal (ca. 1-min) periods, compared with stimulations by mutant-like 40- or 80-sec cycles (Greenacre et al., 1993).

5. Reproductive biology

While we are on the subject of courtship phenomena, let the following reproductivity related findings be registered. Gross counts of courtship events for male–female pairs of 16 Drosophila species in LD cycles yield one fairly sharp maximum or plateau per day; these occurred during the L phase or overlapped the L-to-D transition time for most species, but $D.$ $melanogaster$ pairs gave a distinct nighttime peak (Hardeland, 1972). Interpretation of these courtship-bout quantifications was undermined by the lack of control for whether a female had previously mated [this causes her to elicit diminished male courtship (e.g., Griffith et al., 1993)]. In subsequent studies, testing for time-of-day influences on courtship activities was based on introducing naive males and females to each other at a series of times throughout most of the L phase (Cobb et al., 1987); no temporal changes in latencies to initiate courtship or mating were found (between ZTI and 9) for pairs of four separate Drosophila species [including $D.$ $melanogaster$ and $D.$ $yakuba$, which had also been monitored by Hardeland (1972)].

Further questions in this ballpark (or boudoir) brought intraspecific genotypic variation to bear on courtship variables in $D.$ $melanogaster$. When it was asked whether the durations of courtship bouts vary rhythmically (Roche et al., 1998), the answer was "no" for wild-type male–female pairs. The average durations of such episodes (during which the male orients toward the female, follows her when she moves, and may or may not be singing) were not different when comparing the courtships of per^+, per^S, and per^L males. Turning to mating receptivity of females, a fairly distinct trough was observed for wild-type flies in LD cycles, at about the time of lights-off; during the remainder of such cycles, female receptivity either plateaued or exhibited a peak in the early morning; this behavioral oscillation was shown to run reasonably well in DD and perhaps also in LL (Sakai and Ishida, 2001). Male–female pairings in which individuals were either both per^{01} or both tim^{01} retained morning mating maxima in LD, but the periodic feature of these matings disappeared in free-running conditions (notably, in DD, female receptivity stayed near the LD trough level). Pairings of mutant with normal individuals showed that these genetic effects apparently operate on the female (Sakai and Ishida, 2001).

Once a *Drosophila* female has mated, her egg-laying rate increases dramatically. Such oviposition rates vary periodically (e.g., Rensing and Hardeland, 1967). There is a relatively high daytime peak or plateau in LD and a more consistent, robust peak around the time of lights-off. This pattern is dependent on the chemical conditions, in that food high in acetic acid caused the rather broad L peak to disappear (Fluegel, 1981). These fluctuating rates of egg laying are also dependent on genotype. First, McCabe and Birley (1998) showed that the evening peak persists fairly well in both DD and LL, although free-running periods for wild-type females were somewhat shorter than the ca. 24- to 25-hr ones in LD. In this regard, per^S and per^L females exhibited rather systematic shortenings and lengthenings of their free-running oviposition cycles, and per^{Ol} females were all over the place (21 hr in LD, 14 hr in LL, 19 hr in DD). However, when the LD phase of this rhythm was compared among the *per* types with regard to the evening maxima (McCabe and Birley, 1998), they were all essentially the same, that is, uncorrelated with the earlier- and later-than-normal evening peaks of locomotion in per^S and per^L discussed earlier (Section IV.C.1). Moreover, in the study being currently considered, per^S's "standard" late-afternoon locomotor peak was curiously absent (McCabe and Birley, 1998).

Egg-laying and other reproductive-fitness components have also been examined from a different clock-mutant perspective. Null mutations at the *per*, *tim*, *Clk*, and *cyc* loci were found to cause decreased progeny production in pair matings involving like-mutant types; the causes were determined to be subnormal numbers of eggs laid and higher-than-normal percentages of unfertilized eggs (Beaver *et al.*, 2002). This study went on to reveal "low-sperm" phenotypes in *per*-null and *tim*-null males mated with wild-type females (Table 3).

Female *Drosophila* exhibit another clock-controlled reproductive attribute: Provided that one assesses the following phenotype in very low temperature conditions (Saunders and Gilbert, 1990; Williams and Sokolowski, 1993), these flies exhibit ovarian diapause once the photoperiod drops below a critical day length (CDL; Saunders *et al.*, 1989). This diapause phenomenon is underpinned by a photoperiodic clock, in the sense that high proportions of females do not shut down their egg maturation—in particular, a block in yolk uptake by oocytes (Saunders *et al.*, 1990)—when subjected in "Nanda–Hamner" experiments to T cycles other than those of either 24-hr duration or some multiple of that value (Saunders, 1990; cf. Nanda and Hamner, 1958). Interestingly, the female's photoperiodic clock so inferred still runs with its usual daily cycle duration when she is homozygous for per^{Ol} or deleted of the *per* gene (Saunders *et al.*, 1989; Saunders, 1990). The Nanda–Hamner experiments also uncovered 24-hr photoperiodic cycles in per^S and per^L females (Saunders, 1990). Those two mutations did not otherwise affect this time-based phenomenon, in that the usual

Table 3. Phenotypes Other Than Daily Locomotor Cycles or Periodic Eclosion Affected by Rhythm Mutations in *Drosophila*[a]

Phenotypes	Mutations affecting them	Remarks
Rest rebound after "sleep" deprivation	cyc^{01}, tim^{01}, per^{01}, Clk^{Jrk}	Dramatically accentuated rebound for the *cycle*-null mutant and mortality following lengthy deprivations; the *timeless*-null mutant exhibited subnormal rebound, but only after relatively brief deprivations (in one study); the *period*- and *Clock*-null mutants, along with the last one listed (a neurochemical variant): accentuated rebounds, but nowhere near the supranormality exhibited by cyc^{01}
Octopamine and tyramine synthetic rate; tyrosine decarboxylase (TDC) activity	per^{01}, per^S, per^1	Monoamine synthesis rate lowered by per^0, (the other two: not tested); TDC (which acts in the pathway synthesizing these substances) lowered in all three mutants
Melatonin (MT) concentration; hydroxy-indole-O-methyl transferase (HIOMT) activity	per^{01}, per^{04}	No MT detectable in either mutant; HIOMT (which acts in the MT-synthetic pathway) at extremely low levels in both
Tyrosine amino transferase (TAT)	e	TAT half normal; only an early-nighttime peak of such activity (compared with late-day + late-night peaks in wild type)
Sensitization to cocaine; drug-induced TDC increase	per^{01}, per^S, per^T, per^L; Clk^{Jrk}; cyc^{01}; dbt^S, dbt^L	All mutants exhibit sensitization absence or decrement; the three arrhythmic types were tested for cocaine-induced TDC increase (seen in wild type) and did not exhibit it
Quinpirole-induced increases in movements of decapitated flies	per^{01}	This dopamine-receptor agonist induces such B activity in a daily-rhythmic manner, but the mutant timecourse was flat
Protein carbonyl concentration	per^{01}, per^S	This index of free-radical damage peaks during the night; no such rhythm in per^{01}; accentuate nighttime peak and daytime values in per^S
Ultradian locomotor rhythms	per^{01}, per^{04}, Df-per, per^L, tim^{01}; *disco*; *restless*; $AC72$; $E64$; $EJ12$; cl-7; cl-8	With reference to Tables 1 and 2, these "basically arrhythmic" mutants (except per^L) are such that ca. 10–80% of individuals (depending on the mutant), exhibit cryptically periodic behavior in DD, with ca. 5- to 15-hr-cycle durations; cl-7 and 8 are *D. pseudoobscura* arrhythmics
Membrane-potential oscillation in larval salivary gland	per^{01}	Higher-than-normal proportion of individually monitored cells arrhythmic in the mutant; attempt to replicate the normal phenomenon (which had been assessed via voltage-sensitive dye applications) by direct measurements of membrane potentials failed to reproduce it

Table 3. *continued*

Phenotypes	Mutations affecting them	Remarks
Coupling among larval salivary-gland cells	per^{01}, per^S	Weak or no coupling (and associated physiological parameters) in per^{01}; increased coupling strength in per^S; the basic phenotype (intercellular dye spread) irreproducible in two subsequent studies
Larval time-memory	per^S	Altered timing of phase-shifting light pulse, as delivered to larvae and tested days later for effects on adult B rhythm
Courtship and general locomotor vigor	$dfmr1^3$, a.k.a. $dfxr^-$	$dfmr1^3$ males exhibit severely subnormal courtship; sluggishness reported to accompany the arrhythmic and weekly rhythmic locomotion of $dfxr^-$ flies, which could explain the courtship subnormalities, although bursts of *high*-level activity said to be a feature of locomotor-monitoring data for $dfmr1^3$; both kinds of mutations at this locus (see middle column) shown to cause neuroanatomical defects, including ones associated with "brain-clock" neurons, but not extending to all neural structures examined
Courtship song rhythm	per^{01}, per^S, per^L; *D. simulans* per^+; dy^{And}; $para^{ts1}$; nap^{ts}; psi-2, gat	The normal ca. 60-sec cycles of interpulse-interval oscillations (Figure 6B) largely eliminated in per^{01} males; shortened cycle durations in per^S or *D. melanogaster* (per^{01}) carrying *D. simulans* per^+; lengthened cycles in per^L and dy^{And}; "song clock" stopped by heat treatments of the (latter two) temperature-sensitive action-potential mutants; psi-2 and gat effects: weak, tentative (see Figure 18)
Mating receptivity	per^{01}; tim^{01}	F-r cycle of female receptivity (peak in early subjective day; trough in late subjective day/early subjective night) was such that mating willingness stayed at trough levels in the two mutants
Oviposition rhythm	per^{01}, per^S, per^L	With regard to f-r subjective dusk peak of egg laying, shortened and lengthened cycle durations in the two rhythmic types; per^{01} females cyclical but with noncircadian periods in constant conditions
Fertility and fecundity	per^{01}, per^{04}; tim^{01}, tim^{03}; Clk^{Jrk}; cyc^{01}	Numbers of eggs laid and percentages of unfertilized eggs lower and higher than normal (respectively) in all these mutants; "sperm release" subnormal in young adult males expressing the *per* or *tim* mutations
Ovarian diapause	per^{01}, Df-per/Df-per	Critical day length for diapause induction shortened from 14 to ca. 11 and 9 hr in these two arrhythmic types, but "photoperiodic clock" underlying the phenomenon still runs in females expressing either of these *per* variants

continues

Table 3. *continued*

Phenotypes	Mutations affecting them	Remarks
Phototaxis	per^{01}, per^S, per^L	In a certain maze test, *per* photonegative compared with "neutral" behavior of wild type; per^S showed slight photonegativity; *per* (females only) were anomalously photopositive; these findings (Palmer *et al.*, 1985) not mentioned in text, and different kind of phototaxis testing (performed subsequently by Dushay *et al.*, 1989) revealed no effects of these mutations
Geotaxis	cry^b, *Pdf* dosage effects	Accentuated geotaxis-maze scores for the former; for the latter variants, males with no PDF (cf. Table 6) gave higher-than-wild-type scores; mutant females: similar, but a less marked behavioral accentuation; adding one or two copies of Pdf^+ allele caused decrease of the *Pdf*-null-induced increase, a gene-dosage effect observed in males only
Learning	per^L, dy^{And}; *psi-2*; *psi-3*	Courtship conditioning of male behavior attenuated in these long-period mutants (*psi-3* has this attribute only for E); the effects were largely irreproducible in two subsequent studies, one of which also showed per^L (as well as per^{01} and per^S) to be normal for "classical" (shock-odor) conditioning
Habituation to giant-fiber pathway stimulation	per^{01}, per^{04}	Earlier-than-normal onset of response decrement (to the relevant electrophysiological stimuli) observed in constant light (LL) for both mutants
Anatomical rhythm in optic lobe	per^{01}; tim^{01}	Daily fluctuations of axonal diameters within "cartridges" of lamina neuropil attenuated or eliminated by either mutation
Neurosecretory (n-s) cell positioning	per^{01}; *cl-7*, $cl\text{-}7^{10}$	Anomalously high proportion of such cells in dorsal brain of adults scattered in ectopic locations (even more dorsal than usual); the *cl-7*'s are arrhythmia-inducing mutations in *D. pseudoobscura*, which cause anatomical anomalies in males (barely for $cl\text{-}7^{10}$) and in females, against background of genetically normal females showing more "extradorsal" n-s cells than wild-type males
Heart morphology	per^{01}	Pericardial cells of pupal heart shrunken and wrinkled
Duration of development	per^{01}, per^S, per^L	Shortened and lengthened durations in the two period-altered mutants (including in eclosion-rhythm-eliminating LL); per^{01} showed varying differences from normal cultures depending on photic conditions

Table 3. *continued*

Phenotypes	Mutations affecting them	Remarks
Adult lifespan	per^{01}, per^{T}, per^{L}	Reduced survivorship for these mutants in LD cycles; other studies led to negative results for per^{L} (in analogous but not identical conditions), but one of them reported a lifespan decrement for per^{01}

[a] These miscellaneous phenotypes are listed in no particular sequence (although roughly adhering to the order in which these phenomena are described in the main text). The mutational effects usually were reported as the result of a given one-time study. In some cases (noted under Remarks), further investigation of the phenomenon failed to reproduce the mutant phenotype(s) or at least produced different kinds of results. Nearly all the mutants listed were introduced in Tables 1 and 2, and the abbreviations used here (e.g., *Df*, E, f-r) are as in those tables.

CDL of ca. 14 hr was measured in analyses of these females (Saunders *et al.*, 1989). However, *period* gene functioning could not be said to be without effect on this feature of insect photoperiodism, because both per^{01} and *per*-deleted females exhibited significant shortenings of the CDLs, which were 3 and 5 hr shorter than normal, respectively (Saunders *et al.*, 1989; Saunders, 1990).

Tauber and Kyriacou (2001) attempted to rationalize the "*per*-independent" photoperiodic clock subserving diapause in *D. melanogaster,* partly in the context of the *per*-null mutants exhibiting shorter-than-normal CDLs. Inasmuch as these authors' model is self-stated to be "fanciful" and because its key features involve temporal dynamics of specific clock molecules other than PER, the reader is referred to this theoretical paper if he or she desires "to explore different ways in which identified molecular components of the circadian pacemaker [see Section V and later ones] may play a role in photoperiodism" (Tauber and Kyriacou, 2001). In any event, here we have a circadian clock function that does not depend on some action or the other of the *period* gene in order to function robustly. It is pathetic that the other circadian clock variant in this species have not been tested for effects on this female phenotype—as Tauber and Kyriacou (2001) reveal by tacitly speculating on what might be the effects of, for example, *timeless* mutations on diapause induction.

6. Miscellaneous phenotypes: physiology, behavior, development, and morphology

The fragile-X-like mutations in *Drosophila* (Section IV.A.2) lead to quite a bit of pleiotropy, including a phenotype that could have been dealt with in the just-previous section. But these *dfmr1* variants (or *dfxr* ones if you like) have other problems as well. Fortunately, the effects of the mutations fall short of causing mindlessly ubiquitous abnormalities. In this respect, $dfmr1^{3}$ mutant flies exhibited normal odor- and light-mediated behaviors (Dockendorff *et al.*, 2002). The latter kind of sensory responsiveness included normal visual-system electrophys-iology for an enigmatically synonymous *dfxr*-null mutant (Morales *et al.*, 2002; these investigators uniquely report their fragile-X-like mutants to exhibit ex-tremely low viability). However, courtship tests of *dfmr1*[3] males showed them to be severely impaired (Dockendorff *et al.*, 2002). A majority of mutant individuals did not proceed past early courtship stages, and bout durations were subnormal. These behavioral mutants are also neuroanatomical ones: Certain *dfmr1*[3] brain neurons, which contain the products of clock genes and other rhythm-related ones (Sections VI and X), exhibited abnormal numbers of "collateral branches" (Dockendorff *et al.*, 2002) extending from the axons of these cells, and *dfxr* mu-tants were shown to display "overextension" of such axons (Morales *et al.*, 2002). These morphological abnormalities could contribute to the circadian defects dis-played by *dfmr1*[3] or *dfxr*⁻ cultures and flies (Table 1), but would not necessarily connect with the courtship deficits. Yet other neural structures could be altered by the mutations, especially because the product of this gene seems to be expressed in "most, if not all, neuronal cell bodies" (Morales *et al.*, 2002). Indeed, a category of neurons whose axons arborize in a relatively distal optic lobe showed subnor-mal numbers of neurites entering this ganglion (Morales *et al.*, 2002). However, examination of additional neural structures (the mushroom bodies, cf. Section IV.A.1; and compound-eye photoreceptors) turned up no anatomical problems there (Dockendorff *et al.*, 2002; Morales *et al.*, 2002). These results in effect fit with the fact that absences of the fragile-X-related protein do not cause subnor-malities in "every" kind of behavior.

There are other time-sensitive attributes related to the functions of cer-tain *Drosophila* tissues and to whole-fly behaviors. These have been considered candidates for establishing genetic connections to the control of circadian func-tions. For example, the animal's heart-beating (at a late developmental stage) was said to be erratic in *per*[01] (Livingstone, 1981; Dowse *et al.*, 1988). However, this proved almost certainly not so when the pertinent parameters (associated with the basic heartbeat rate of ca. 3 Hz) were subjected to scrutiny and doc-umented (Dowse *et al.*, 1995). Incidentally, causing pericardial cells to be logy with heavy water makes the heart rate slow down accordingly (White *et al.*, 1992a). This is mentioned because circadian pacemakers are also well known to give long-period circadian rhythms under the influence of deuterium oxide. This effect extends to *Drosophila*'s locomotor rhythmicity, and it is worth noting (also in passing) that *per* mutants do not seem to respond in particularly interest-ing ways to D_2O feeding: *per*[S], *per*[L], and *per*[T] all exhibited period lengthenings to similar extents as *per*⁺, if not always in the same proportions (vis à vis the τ value associated with a given genotype). At least there was no override of the heavy-water effect in a short-period mutant, and D_2O-treated *per*[L] did not exhibit synergistically lengthened periods (White *et al.*, 1992b; Konopka *et al.*, 1994).

A flyer analogous to asking the heartbeat question was taken when learning-like phenomena associated with male courtship were interrogated chronogenetically. Several mutations that cause long-period circadian rhythms were reported to cause decrements in the extent to which a courtship-trained male retained the effects of such experiences (Jackson *et al.*, 1983). The subnormally performing mutant males were *per*^{L}, *And*, and the two *psi* variants (only one of the latter pair affects locomotor rhythmicity). However, *per*^{S} and *per*^{0} males were normal for conditioned courtship (Jackson *et al.*, 1983). Nevertheless, it was as if a slow clock might actively interfere with the normal rate of information storage, whereas no such problems would occur if time-keeping were sped up; in its absence, putting the requisite information into memory could somehow bypass this particular etiology of clocklessness (in *per*^{01}). These conjectures became moot when *per*^{L} males and those expressing other clock-slowing forms of the gene were again tested for courtship-related learning and newly examined for classical conditioning; essentially all experience-dependent behaviors were found to be normal (Gailey *et al.*, 1991). Moreover, *Andante* males, which had exhibited ostensibly the most severe decrement in courtship conditioning (Jackson *et al.*, 1983), were reexamined by tests of the original (ca. 25-hr) mutant and of the newer *And*-like *dusky* mutants that had proved to be similarly long period (Newby *et al.*, 1991). All such males learned normally, and the notion that the small-wing-cell phenotype of *And* flies might be reflected in their brains—somehow to cause both rhythm and learning defects (Newby and Jackson, 1995)—led to no such detectability in anatomical examinations of mutant head specimens (van Swinderen and Hall, 1995). Therefore, the genetic connections in *Drosophila* between regulation of rhythms and "higher" (associative) learning are limited (so far) to the cases of the *dunce*, *Pka-CI*, and *dCreb2* genes [Section IV.C.3; and see Dubnau and Tully (1998) for a review of cAMP-responding elements and their involvement in long-term associative memory].

Fruit flies also exhibit experience-dependent changes that are nonassociative. One such phenotype involves habituation to stimulation of the giant-fiber pathway, which originates in the anterior visual system and brain. Electrically mediated brain stimulations lead to a response decrement, measured at one of the pathway's endpoints (thoracic flight muscles). Curiously, flies expressing either of two *per*^{0} mutations exhibited an earlier onset of such habituation when maintained in LL compared with LD (Megighian *et al.*, 2001). The habituation values for wild-type flies tended to be the same as the baseline metrics determined for these mutants in LD, but exposing *per*$^+$ flies to constant light did not lead to shortened timecourses for habituation to the stimuli.

The experience-dependent, courtship-song, and heart-beating phenomena operate with respect to time scales much shorter than 1 day. Potential influences of the *period* gene on phenomena defined by much longer timespans have also been examined. For this, it was considered whether the overall duration of development might involve a pleiotropic effect of *per* mutations. Indeed, per^L cultures required longer-than-normal numbers of days to go from eggs to adults; per^S ones developed faster than normal (although there was less of a temporal effect of this mutation compared with that of per^L); and per^{01} cultures exhibited erratic differences from the normal developmental timecourse, depending on the culturing conditions (Kyriacou *et al.*, 1990a). In this regard, the per^L and per^S effects were observed in LL, a situation in which the circadian clock does not run insofar as eclosion and adult-locomotor rhythmicity are concerned. Constant light creates a near phenocopy of the per^{01} effect on the protein encoded by this gene (Section IX.B.1). It is therefore notable that developmental times of wild-type cultures of this species in LL were shorter than those in LD (albeit only in one of the two such studies performed: Sheeba *et al.*, 1999a; contra Kyriacou *et al.*, 1990a). The *further* PER reduction caused by per^{01} compared with the protein decrement that would be induced by subjecting developing per^+ animals to LL (Price *et al.*, 1995) might therefore be expected to shorten the mutant's developmental timespan relative to the wild-type fly. This is what was observed in LL-rearing conditions (Kyriacou *et al.*, 1990a). We should keep these *per* mutant phenotypes in mind when descriptions of the gene's expression at various stages of the life cycle come into play (Section VI.A.1).

Production of *per* mRNA and protein persists into adulthood, naturally, and in this regard a possible influence of the gene on postdevelopmental lifespan has been considered. For example, could a "live-fast, die-young" scenario play out for per^S or per^T flies? The first set of lifespan assessments found an apparent effect of one of the original *per* mutations (Konopka, 1987): per^S lived *longer* than wild-type flies (but per^L did not) in four of the five tests performed, including and especially one that applied 28-hr LD cycles. [This kind of environmental manipulation was imposed in the spirit of a previous study in which wild-type *D. melanogaster* was reported to exhibit subnormal survivorship in 21- or 27-hr T cycles (Pittendrigh and Minis, 1972); in Konopka's (1987) hands, the earlier findings were irreproducible.] Subsequent examinations of this temporally dependent phenotype showed reduced lifespans for per^T and per^L flies, although in one such experiment a comparison between per^T and per^+ females gave no effect of the fast-clock mutation (Klarsfeld and Rouyer, 1998). Yet in most of these tests, including survival-curve determinations in 24- and 16-hr T cycles, per^T flies exhibited 10–15% shorter-than-normal lifespans, and the reduction in per^L was 5–10% [the T-cycle effects of Pittendrigh and Minis (1972) once again were not observed by Klarsfeld and Rouyer (1998)]. Perhaps these phenotypes reflect a generalized health decrement of the *period* mutants. This notion was buttressed by the shorter-than-normal lifespan that was found for per^{01} in a test of this allele's effect (Ewer *et al.*, 1990). Note that living fast (per^S, per^T) or slowly (per^L) in terms of the putative internal temporal states for these mutants did not consistently truncate or extend survival [at least in terms of comparing the results of Konopka (1987) to those of Klarsfeld and Rouyer (1998)]. None of these tests involved lifespan assessments in constant darkness, wherein, for example, adding up a series of ca. 30-hr days for per^L might have maximized the chance of revealing a stretched-out survival curve. Recall in this regard that three of these *per* mutant types can be said to live 24-hr days in 12:12 LD because all those expressing a functional allele entrain to such *T* cycles (e.g., Hamblen-Coyle *et al.*, 1992; Konopka *et al.*, 1994).

A last-gasp case involving the effects of rhythm mutations on what seems to be "anything" stemmed from a molecular screen for genes whose apparent expression levels differed in lines of *D. melanogaster* selected (long ago) for relatively high versus low geotaxis scores (Toma *et al.*, 2002). The *cryptochrome* and *Pdf* genes were implicated by microarry screening (see Table 1 and Sections X.B.1 and X.C). This led to retrospective geotaxis testing of the cry^b mutant and of *Pdf* variants (including transgenics for the latter that come into play in Section X.B.1; also see Figure 13, color insert). Toma *et al.* (2002) found that former to give elevated geotactic scores over various baseline values, although not reaching the level of the "high" line; *cry* was tapped into in this study because its mRNA level was *lower* than expected in the high-geotaxis line, correlated with a similar behavioral phenotype caused by the decrement-of-function mutation in this gene. Toma *et al.* (2002) found *Pdf* mRA levels also to be down in the high-geotaxis line, and these authors attempted to describe the effects on geotaxis of a *Pdf* mutation and/or gene-dosage alterations for a normal form of this gene (Renn *et al.*, 1999), as they interacted with sex [unfortunately, not mentioning sex-influenced *Pdf* mRNA levels reported by Park and Hall (1998)]. Whether or not such geotactic results are comprehensible in the narrow sense (including problems encountered in by Toma and co-workers in describing their own *Pdf*-related geotaxis data), they make little or no overall sense in light of the "main effects" of these rhythm factors (Sections IX and IX).

So, we have a long, idiosyncratic list of pleiotropic possibilities for the biological effects of rhythm variants—mostly *per* mutations (Table 3). The various temporally related phenotypes, if altered in *per* or other mutants, range from the reasonable—altered features of locomotion beyond the basic abnormalities of such daily cycles—to the rather occult. At least some of these "other phenotypes" involve time-based phenomena, although some of them operate well *within* a 24-hr period or over the course of *several* days. Most of the, dare one say, unreasonable phenotypes associated with these mutants have been of little heuristic value. Maybe this will get better for certain of the recently uncovered connections between *Drosophila*'s chronobiology and "all the rest of it." For example, at least it is the case that the *cry* and *Pdf* genes overlap in part of their brain-expression patterns (compare Figures 11 and 12 discussed in a later section with Figure 16; see color insert). The fly's rhythm-related attributes did not come into play in the re-identifications of *cry* and *Pdf* (Toma *et al.*, 2002). In contrast, phenotypes in the temporal ballpark, those different from the norm in one mutant or the other (Table 3), are seemingly more comprehensible. They are arguably of little explanatory value, however, in terms of sharply defined genotype–phenotype connections on the one hand or, on the other, for furthering the potential understanding of clock properties, either in terms of a widely appreciated rhythm such as adult locomotor cycles or an ill-understood pacemaker such as the photoperiodic clock. Mercifully, a few tests of rhythm mutants for defects in phenotypes other than behavioral rhythmicity proved negative—as in retesting the effects of *per* mutation on intercellular communication, heartbeating, learning, and phototaxis (Table 3).

The current section closes with one further negative result. This particular chronogenetic question was asked in the context of a clock gene's expression at the molecular level and thus poises this treatment to move on to that part of the story. Thus, as we will soon see, the *Drosophila* eye is a tissue in which clock genes are prominently expressed, both in terms of elementary descriptive findings and tacit reliances on eye expression to ask more analytical questions about the quantities and qualities of molecules encoded by these genes.

There is a circadian rhythm running in the *Drosophlia* eye: Daily decreases are observed in the amount of visual pigment, indirectly measured by microspectrophotometry (MSP) and in the sensitivity of photoreceptor cells [via electroretinogram (ERG) measurements]; reduced values for these parameters are measured in the early morning and recover to baseline levels 2–4 hr later (Stark *et al.*, 1988; Chen *et al.*, 1992). This gated nature of apparent rhodopsin turnover persists in DD, and thus the photoreceptor cycling was a prime candidate for being affected by *per* mutations. Yet, only *per*^{S} had one comprehensible effect—shortening the free-running rhythm of visual-pigment turnover, but not the sensitivity cycle; *per*^{01} was *rhythmic* in terms of pigment measurements (this mutant did not get analyzed for ERG), and *per*^{L} ranged from only 5% longer than normal in its MSP rhythms to 8% *shorter* than normal for its ERG cycle (Chen *et al.*, 1992). Therefore, the biological significance of all this eye expression, as exhibited by material encoded within the clock genes discussed in Section VI, remains unknown.

V. MOLECULAR GENETICS OF CENTRAL-PACEMAKING FUNCTIONS

A. Cloning clock genes

1. *period* and its first-stage gene product

Monitoring the presence of putative clock-gene products in adult photoreceptors turned out to be of paramount importance for deducing the manner by which these molecules seem to function. However, first the transcription units defined by certain rhythm mutations had to be defined. The *period* gene was the initial putative clock factor that got cloned. This was accomplished by rhythm assays of DNA fragments isolated from at least the close vicinity of the *per* locus. This required high-resolution mapping of it, which had been achieved by Young and Judd (1978) and especially by Smith and Konopka (1981, 1982). Molecular identification of this gene was accomplished by correlating stretches of putative *per* DNA with mRNAs complementary to pieces of this *X*-chromosomal genomic material (Reddy *et al.*, 1984)—notably a 4.5-kb species (Bargiello and Young, 1984). More definitively, transgenic strains carrying a series of adjacent and overlapping such DNA fragments were generated in a *per*^{01} genetic background. Certain such fragments, all of which included the genomic source of at least part of the 4.5-kb transcript, "rescued" the behavioral and eclosion arrhyhmicity caused by that mutation (Bargiello *et al.*, 1984; Zehring *et al.*, 1984; Hamblen *et al.*, 1986; Citri *et al.*, 1987; Hall and Kyriacou, 1990).

However, the first generation of germ-line transformants effected mediocre coverage of *per*^{01}-induced arrhythmias: The free-running periods were anomalously long, only a fraction of the behaviorally tested transgenic individuals were rhythmic at all, or both. This was because the rescuing DNA fragments did not include the entire *period* gene (in the studies reported in 1984–1986), such that certain of the transgene-encoded mRNAs were a tad shorter than the 4.5 kb they need to be for behavioral normality at either the 5′ or the 3′ end of the transcript in question. This point is made for reasons that go deeper than a historical curiosity, because certain of the partially rescued transgenic types exhibited *per*$^{+}$ expression in an informative subset of the gene's normal spatial pattern (as discussed in Sections V.C.2 and VI.A.2).

2. Molecular identification of six additional clock genes

The clock-gene candidates identified subsequently, mainly by way of the rhythm-defective mutants described above, were in the main cloned throughout the second half of the 1990s (although in one instance, the gene had been precloned by developmentalists earlier in that decade). These accomplishments involved:

(i) Isolation of *X*-chromosomal DNA at the *shaggy* locus that was tagged by the EP transposon inserted there (Table 1 and Figure 3). This genomic material was used as a probe to recover complementary cDNAs (from an adult-head library) that were candidates for encoding portions of the protein misexpressed in the short-period EP line; sequencing these cDNAs showed that they were derived from the previously cloned *shaggy* locus (Martinek *et al.*, 2001).

(iia) Chromosomal walking from second-chromosome-derived clones that had to be near the *tim* locus (by virtue of mapping it at high resolution), followed by identification of a partly deleted transcription unit in the *tim*^{01} strain (Figure 3, color insert) that almost certainly defined the gene (Myers *et al.*, 1995).

(iib) Identification of a cDNA-encoded polypeptide (engineered to be produced in yeast cells), which physically interacted with a *per*-encoded protein fragment in the same cells (Gekakis *et al.*, 1995)—a "yeast-two-hybrid" approach that identified *tim* sequences independently of the gene's chromosome-positional cloning (see ia) and presaged the PER–TIM interaction that was in the process of being inferred genetically (Sehgal *et al.*, 1994; Price *et al.*, 1995; Rutila *et al.*, 1996).

(iiia) Mapping the $Clock^{Jrk}$ and $cycle^{01}$ mutations with a fair degree of precision to separate third-chromosomal loci (Figure 18, color insert), followed by making a judicious guess that an "expressed sequence tag" (EST) database—which includes nucleotide sequences from a host of cDNAs—would contain archived material that might be mutated in Clk^{Jrk} and cyc^{01} and whose informational content would imply that this material encoded transcription factors. The ESTs in question (Section V.B.3) were identified, and isolation of additional cDNAs led to identification of the full open-reading frames for these two genes; moreover, the two sets of candidate clones mapped to the *Clk* and *cyc* loci, and the stretches of DNA in question (taken from the Clk^{Jrk} and cyc^{01} strains) harbored nucleotide (Figure 3) that allowed rationalizations of the mutant phenotypes (Allada *et al.*, 1998; Rutila *et al.*, 1998b; Park *et al.*, 2000a).

(iiib) Isolating *Drosophila* genomic DNA that hybridized to sequences previously identified as corresponding to the *Clock* gene (*mClk*) of mouse (Darlington *et al.*, 1998); it was originally defined by a chemically induced rhythm mutant (Vitaterna *et al.*, 1994), which was followed by positional cloning of the corresponding murine DNA (Antoch *et al.*, 1997; King *et al.*, 1997).

(iiic) Effecting yet another identification of *Clk* sequences with the aid of the aforementioned EST database (Bae *et al.*, 1998), much as Allada *et al.* (1998) had done, but without the mutant connection.

(iv) Isolation of the *double-time* gene by transposon tagging (Kloss *et al.*, 1998), facilitated by the aforementioned P-element mutation at the locus (Table 1 and Figure 3), which is a recessive lethal variant, but is allelic to the period-altering *dbt* mutations (Price *et al.*, 1998). *dbt* cloning was accomplished by Kloss *et al.* (1998) in much the same manner as the $EP(X)$ 1576-tagged material at the *sgg* locus was molecularly identified.

(v) Getting to the molecular biology of *Timekeeper* was once again aided by *Drosophila* genomics. Given the kinds of functions that turned out to be identified (or reidentified) by *sgg* and *dbt* (discussed In section V.B.4), a gene located in the vicinity of *Tik*'s genetic map position on the third chromosome (Figure 18, color insert) was a candidate for having been hit by this mutation. [Not to flog the fly's genome database too fervently, the nucleic acid sequence in question had been determined years before by relatively primitive cloning procedures and localized by *in situ* annealing to the same chromosome region to which *Tik* was genetically mapped 15 years later (Saxena *et al.*, 1987; cf. Lin *et al.*, 2002a).] Sequencing this material, obtained from *Tik*-mutant flies, showed that it causes two amino-acid substitutions (Figure 3) compared with the polypeptide encoded by the normal form of the gene (Lin *et al.*, 2002a). Remarkably, a spontaneous "revertant" (*R*) mutation was identified not long after *Tik* was isolated (Lin *et al.*, 2002a). In this derivative strain, the longer-than-normal circadian period determined by behavioral analysis of the original mutant (Table 1) was partly ameliorated (1 hr of lengthening instead of 3). The *TikR* variant still harbors the two amino-acid changes present in the original mutant (and remains a recessive-lethal mutation), but there are further intragenic changes in the revertant; one results in a net deletion of seven residues from the polypeptide, and the other causes a third amino-acid substitution some 80 residues C-terminal to the clustered location of the originally mutated sites (Figure 3). These connections between the circadian behavioral and molecular findings essentially demand the conclusion that the genetic locus named *Timekeeper* (Lin *et al.*, 2002a) and the molecularly defined function (Saxena *et al.*, 1987) are one and the same.

The elementary molecular biology proceeded slowly at first for the seminal case of *per*. Standard characterizations of the cloned DNA and products stemming from the gene seemed, on the one hand, not to nail *per* as necessarily a clock factor, and, on the other, provided few insights as to how the final product—PER protein—might be functioning chronobiologically. However, as the following substory will discuss, the case was eventually cracked, at least to a first approximation. This created a certain momentum that promoted understandings of the manner by which the TIM, CLK, CYC, and DBT proteins are likely to act, along with how they interact with PER functions. For its part, SGG's function was revealed to interact with TIM.

B. Basic actions of and interactions among clock-gene functions

1. Daily oscillations of *per* products in their normal and their mutated forms

The informational content of PER—inferred from sequencing fragments of this gene, then all of it and cDNAs complementary ot it—led initially to apprehension of a largely featureless amino-acid sequence (Shin *et al.*, 1985; Jackson *et al.*, 1986; Reddy *et al.*, 1986). In retrospect, that is good, for one could infer the discovery of a potentially *unique category of protein*, compared with, for example, some known enzyme involved in general features of metabolism. The only hint provided from subsets of the PER sequence was that a ca. 40-residue threonine–glycine (T–G) repeat located in the middle of the polypeptide (Figure 3) meant that the protein might be a proteoglycan located at the surfaces of cells or the spaces between them (Jackson *et al.*, 1986; Reddy *et al.*, 1986). This potential lead was followed up by biochemical studies of PER immunoreactivity in tissue extracts (Reddy *et al.*, 1986; Bargiello *et al.*, 1987). However, the suggestion turned out to be a red herring, even though PER-as-proteoglycan was for a time thought to connect with the aforementioned defects in intercellular communication that were once believed to be caused by certain *per* mutations (Bargiello *et al.*, 1987). Ultimately PER itself was shown, by subsequent protein characterizations, almost certainly to possess nothing in the way of proteoglycan-like properties (Edery *et al.*, 1994b).

Another PER connection to something biochemically tangible involved its similarity—of a ca. 260 amino-acid (aa) stretch, N-terminal to the T–G repeat—to a sequence located within a *Drosophila* gene product called SIM. This protein is encoded by the *single-minded* (*sim*) gene, which was identified originally by embryonic neural-lethal mutations (Crews *et al.*, 1988). The SIM sequence as initially established was inadequate to reveal what became clear later: This protein is apparently a transcription factor, similar in turn to another such protein called ARNT (Nambu *et al.*, 1991); the latter is the mammalian aryl hydrocarbon receptor nuclear translocator (Hoffman *et al.*, 1991). Both it and SIM contain a basic helix–loop–helix (bHLH) domain (Figure 3), signifying DNA-binding activity. C-terminal to the bHLH segment of these proteins is the aforementioned 260-aa region. This came to be called the PAS domain (Figure 3). The acronym is based on the first letters of the seminal trio of proteins found to contain this domain. Later, PAS-containing proteins were found to make up a large family, and their functions extend well beyond chronobiology, or neural development, or cellular responses to aryl hydrocarbons like dioxin (reviewed by Taylor and Zhulin, 1999; Gu *et al.*, 2000).

Other kinds of *per*-related studies, contemporary to these conceptual-protein analyses and protein-biochemical assessments (such as they were), seemed to move the gene and its product toward comprehensibility. For the studies in question, an antibody was generated against a synthetic peptide, corresponding to a 14-aa stretch that happened to be located between the PAS domain and the T–G repeat (Siwicki *et al.*, 1988). This reagent was produced simply to assess the tissue distribution of PER expression; perhaps the structures labeled in adult *Drosophila* would include neural substrates of the fly's daily biological rhythms. Indeed, marking of certain brain neurons was mediated by this antibody, but other tissues were labeled as well, such as photoreceptors in the compound eye and ocelli and cells throughout the gut (Siwicki *et al.*, 1988). This was the tip of the *per*-expression iceberg, as was indicated at the same time by generation of a *per-lacZ* fusion-gene transformant: β-galactosidase (β-GAL)-reported *per* expression was observed in the locations just mentioned (Liu *et al.*, 1988), along with the detection of the (bacterial) reporter-enzyme activity in various appendages and abdominal organs (see Table 4).

Tracking the expression of antibody-labeled PER included a crucial temporal component: Flies were taken from the middle of the day and of the night for sectioning and staining. The notion that this might be informative came from earlier mRNA assessments, although in those cases, the *per* transcript was shown *not* to exhibit abundance fluctuations in tests that were based on whole-fly extracts (Reddy *et al.*, 1984; Young *et al.*, 1985). However, the immunohistochemical inspections quickly showed very low signals in compound-eye photoreceptors in the day-sectioned specimens, compared with robust staining in the eyes of flies sacrificed during the nighttime (Siwicki *et al.*, 1988). Subsequent immunohistochemical tests, performed at higher temporal resolution, showed that the anti-PER-stained brain neurons also "cycled" in terms of signal intensities, although the trough and peak times for the neurons were a bit later than in the case of photoreceptor stainings (Zerr *et al.*, 1990). In the later cell type (at least), PER was inferred (Siwicki *et al.*, 1988) and later shown at high resolution (Liu *et al.*, 1992) to be primarily a nuclear antigen.

These protein-staining timecourses prompted a reexamination of *per* mRNA in the temporal domain of expression. RNA extracts from adult heads led to easily demonstrable cycling of the transcript's abundance (Hardin *et al.*, 1990). In experimental retrospect, the failure to discern such cycling in whole-fly, mixed-sex extracts was explained by the fact that *per* mRNA in the ovaries does not cycle (Hardin, 1994), but that material forms a high enough fraction of the head-plus-body gene product that it effectively swamps the *per* oscillations occurring in all other tissues known. Whole-male extracts contain *per* mRNA whose daily oscillations were readily revealed (Hardin, 1994).

The peak time for *per* head-RNA (Hardin *et al.*, 1990) is a few hours earlier than the late-night maximum histochemically inferred (Zerr *et al.*, 1990) or biochemically demonstrated (Zeng *et al.*, 1994) for the abundance of PER protein. Crucially, these macromolecular oscillations were found to be affected by *per* mutations: *per*[S] was shown to cause earlier-than-normal peaks or troughs of PER immunoreactivity in LD cycles and a ca. 20-hr molecular cycle in DD (Zerr *et al.*, 1990). Parallel effects of *per*[S] were found in terms of transcript cycling; moreover, *per*[01] caused complete flattening of the RNA oscillation (Hardin *et al.*, 1990). This question was moot immunohistochemically because *per*[01] eliminates the relevant antigenicity (Siwicki *et al.*, 1988). This was as expected because, meanwhile, that mutation had been found to be caused by an intragenic, premature stop codon (Figure 3), which is located upstream of the 14-aa immunogen (Siwicki *et al.*, 1988; cf. Baylies *et al.*, 1987; Yu *et al.*, 1987b). This molecular etiology of *per*[01], identical ones for per^{02} and per^{03} (Hamblen-Coyle et al., 1989), and the fact that heterozygosity for certain overlapping X-chromosomal deletions remove the gene, but allow for viable flies (Smith and Konopka, 1981; Bargiello and Young, 1984; Reddy et al., 1984), showed that period is a nonvital gene. These results are consistent with an earlier supposition, which was based on per^{01} being nonallelic to a set of lethal mutations in its immediate vicinity on the X chromosome (Young and Judd, 1978), a genomic region that had been seemingly "saturated" for mutationally identified vital genes.

The site changes in per^{S}, per^{L}, and the other, more recently isolated per alleles were also determined by nucleic-acid sequencing (Baylies et al., 1987; Yu et al., 1987; Hamblen et al., 1998). Most such alterations (except for per^{04}) involve amino-acid substitutions (Figure 3), although it is fair to say that almost none of these changes in the functional per mutants is understandable in terms of how an altered protein can cause a particular kind of period-altered phenotype. While we are on this subtopic, let it be registered that a whole host of additional amino-acid substitutions have been made in PER by in vitro mutagenesis. Most of these were made at or in the vicinity of the serine defined by per^{S} by virtue of an asparagine missense mutation between the PAS domain and the T–G repeat (Figure 3), and many, but not all such engineered changes led to shorter-than-normal behavioral periodicities (Baylies et al., 1992; Rutila et al., 1992). However, and as is the case for the in vivo-created period mutants (Table 1, Figure 3), the connection between a particular form of PER protein and the resulting chronobiological phenotype remains largely obscure.

These mysteries aside, the principal points to consider at the moment involve a circadian-clock gene, which per now seems almost certainly to be, that is expressed in a manner defining a daily rhythm of its own. Moreover, this temporal regulation of the molecule would appear to include feedback. The latter inference comes from the facts that altered PER protein leads to a changed mRNA-abundance rhythm and that the absence of this protein abolishes molecular rhythmicity of the transcript that encodes it. PER feedback effects on the mRNA that encodes it might seem a priori to be negative—for an oscillation, what goes up needs to be brought down—and this was demonstrated empirically: Transgene-induced overexpression of PER in the eye caused a substantial lowering of mRNA as encoded by the endogenous per^{+} allele (Zeng et al., 1994). However, there seems to be a lack of full negativity of PER's influence on "its" mRNA because the PER-null state does not cause transcript levels to stay constitutively at the (normal) peak mRNA level (Hardin et al., 1990; Qui and Hardin, 1996a).

At whatever level per products might be expected to settle in the absence of PER, fluctuations of the encoding mRNA in wild-type fly are clear, and this oscillation involves more than temporally varying transcript abundances: per-fusion transgenics in which 5′-flanking sequences from the gene (Figure 9, color insert) are fused to gratuitous reporter factors showed that mRNAs encoded by the latter cycle (Zwiebel *et al.*, 1991; Hardin *et al.*, 1992b). Also, nuclear run-on experiments demonstrated that *per* exhibits circadian oscillations in *transcription rates* per se (So and Rosbash, 1997).

More generally, circadian rhythmicities of clock-gene transcripts, associated with negative feedback effects of translated proteins on the mRNAs that encode them, formed the Zeitgeist of contemporary molecular chronobiology (Dunlap, 1999). The current such scenarios are beyond the scope of this review. Also, it is not and need not be the case that all clock-gene products within a given species cycle with affiliated feedback effects of the proteins. However, daily oscillations of macromolecules are an important component of the ever-expanding view that is being directed at the functions mediated by several of *Drosophila*'s rhythm-related genes.

2. *timeless* as producer of oscillating molecules and their interactions with *per* functions

The *tim* gene is the next case in point. Incidentally, *timeless* in *Drosophila* is one of several clock factors for which mouse ("*m*") relatives were found (reviewed by Reppert and Weaver, 2001). However, the inferred protein similarities between the insect and the mammalian forms is not great. This prompted a search for another *tim*-like gene in *D. melanogaster*; one was found and variously dubbed *tim2* (Benna *et al.*, 2000) or *timeout* (Gotter *et al.*, 2000). This fly gene is appreciable as an actual ortholog of *mTim*. However, the chronobiological significance of *tim2/timeout*, and for that matter that of murine *mTim*, is unknown (Gotter *et al.*, 2000). Yet the original *timeless* mutant in *Drosophila* is now well established with regard to circadian arrhythmicities that go deeper than the biological phenotypes on which criteria tim^{01} was isolated. Thus, this mutation eliminates *per*'s mRNA rhythm in addition to causing a periodic eclosion and adult locomotion (Sehgal *et al.*, 1994). Not long afterward, the positional cloning of *tim* itself was accompanied by demonstrations that its products cycle with similar parameters to those established for *per* ones (Sehgal *et al.*, 1995). The mRNA produced by *tim* goes up and down by virtue of fluctuating transcription rates (So and Rosbash, 1997) This is consistent with the daily oscillations that were demonstrated for a reporter enzyme activity, encoded by sequences fused to upstream regulatory sequences (Figure 9, color insert) cloned from this clock locus (Stanewsky *et al.*, 1998).

The peak times for *tim* mRNA and TIM are slightly earlier compared with the *per*-product maxima (Zeng *et al.*, 1996; Marrus *et al.*, 1996; Myers *et al.*, 1996). However, even though such temporal changes do not occur in lock-step, these two genes' worth of function are mutually interacting ones (Sehgal *et al.*, 1995): per^{01} flattens the tim^+ mRNA timecourse (and vice versa; see above), which is the same effect as tim^{01} exerts on its own (internally deleted) transcript (cf. Figure 3), and per^{S} leads to the same shortening of free-running tim RNA cycling as caused by this mutation for oscillations of the transcript that encodes the altered PER^{S} protein (Hardin et al., 1990). The molecular features of this interdependence do not extend fully to the protein level: Whereas loss of tim function leads to a drop in PER levels (Price et al., 1995), those of TIM (in the dark) remain robust (but temporally constant) in flies expressing a per^{0} mutant (Myers et al., 1996; Zeng et al., 1996).

Before interactions between $period$ and $timeless$ gene products were examined directly, it was demonstrated that PER can interact with other proteins: PER's PAS domain mediates binding to other polypeptides—in fact, with the PAS regions present in certain other members of this protein family (Huang et al., 1993). Additional empirical threads were pulled together to produce a more expanded interaction scenario, which will be implicit as the following points are listed:

(i-a) tim^{01}-induced absence of TIM (Figure 3) leads to a substantial lowering of PER abundance (Vosshall et al., 1994; Price et al., 1995), an effect that happens to be similar to the result of exposing flies to constant light (Zerr et al., 1990; Price et al., 1995).

(i-b) Ostensibly consistent with the negative effects of tim^{01} on PER levels is the fact that turning on tim expression, using a transgene in which that gene's coding sequences were fused to an inducible promoter, led to PER accumulation (at cycle times when it is usually low), although this effect is partly due to a posttranscriptionally regulated increase in per mRNA (Suri et al., 1999).

(ii-a) PER and TIM were shown to associate with one another in fly-head extracts (Lee et al., 1996; Myers et al., 1996; Zeng et al., 1996).

(ii-b) The tim product and sequences encoding it were discoverable without mutation—by virtue of a yeast-two-hybrid screen in which a PER–PAS fragment fished out a stretch of cDNA that encodes part of $Drosophila$ TIM (Gekakis et al., 1995).

Further biochemical experiments, also involving polypeptide fragments, showed that a region of PER including part of its PAS can interact with TIM, in particular, a region containing the latter's "nuclear-localization signal" (NLS; discussed later), although this is in the context of TIM being in the main a featureless protein according to its amino-acid sequence (Figure 3). For example, subsequent characterization of the full-length tim open-reading frame (ORF) showed that it encodes no PAS domain, so TIM's interaction with PER cannot involve PAS-mediated heterodimerization (Huang et al., 1993).

(ii-c) There are other stretches of amino acids that mediate interactions between these two proteins: the "cytoplasmic localization domain" (CLD) of PER (discussed later), binding to a relatively C-terminal region of TIM (Saez and Young, 1996).

(iii) The interactions between PER and TIM seem to be involved in the nuclear entry of both (full) proteins, at least in the sense that tim^{01} and tim^{rit} each lead to weak and anomalous accumulation of PER in the cytoplasm (Vosshall *et al.*, 1994; Matsumoto *et al.*, 1999); the same occurs for TIM in per^{01} flies (Hunter-Ensor *et al.*, 1996; Myers *et al.*, 1996).

(iv) The aforementioned interaction regions that promote PER's formation of heterodimers with TIM have features that were determined separately from the biochemical tests of polypeptide associations (Saez and Young, 1996); thus, PER by itself goes into the nucleus of transfected, cultured cells when a polypeptide fragment lacks a CLD located within the N-terminal half of the protein (a result presaged by findings in actual flies that carried *per* transgenes encoding certain portions of the protein; Vosshall *et al.*, 1994); an analogous CLD for TIM was mapped in the cell-culture experiments to a region near its C-terminus, and an NLS for this protein was similarly localized to a relatively central region of TIM (Figure 3).

(v) Nuclear localization of PER involves an N-terminal region (Vosshall *et al.*, 1994); and entry of this protein seems to be a gated event, in the sense that that protein does not automatically go into the nucleus (of actual fly cells) immediately it is made (Curtin *et al.*, 1995). This stretching out of the time between *per*'s mRNA upswing and appearance of the encoded protein in a subcellular compartment where PER arguably exerts its biochemical effects can be thought of as lengthening the oft-discussed "delay" of the protein peak with reference to the mRNA one (e.g., Leloup and Goldbeter, 2000).

Among further implications of the results just summarized, and others described earlier, are those that point to interpretability of a certain τ-altering *period* mutation. This would be per^{L}. The amino-acid change in this mutant is a valine-to-aspartate substitution (Figure 3) within the relatively N-terminal part of PAS known as PAS-A, a stretch of some 50 aa (mostly hydrophobic ones) which is repeated (PAS-B) farther downstream within this domain. Recall that lower-than-normal levels of per^{+}-encoded products lead to period lengthening (Smith and Konopka, 1982; Cooper *et al.*, 1994). Moreover, per^{+}-derived transgenes that are inadequate in terms of the gene's regulatory sequences can mediate basic circadian rhythmicity (see Section V.C.2), but the free-running periods are typically longer than normal; correlating grossly determined *per* mRNA levels in these transgenics (without collecting the flies at specific timepoints) showed that the lower the average transcript abundance, the longer is the period (Baylies *et al.*, 1987; cf. Coté and Brody, 1986). This expression level/clock pace correlation was ratified by application of a reverse-genetic strategy, based on the fact that double-stranded RNA corresponding to a given gene (a.k.a. RNAi) can interfere with its function. Therefore, inverted-repeat sequences based on portions of the *per* gene were transformed into *Drosophila* strains and put under the control of *tim*-gene regulatory material (Martinek and Young, 2000). The double-stranded ("fold-back") *per* RNA that was thus designed to be produced in putative pacemaker cells lengthened behavioral periodicities by about 2 hr in per^+ flies (Martinek and Young, 2000), indicating that this RNAi setup reduced the level of functional *per* mRNA but did not eliminate it.

In yeast cells, a PER fragment substituted as the PER^L protein led to relatively weak interactions with TIM, and they were weaker still as the temperature was raised (Gekakis *et al.*, 1995). This would seem to be correlated with the following two sets of findings: (i) Putting wild-type *Drosophila* in constant dim light leads to longer-than-normal free-running periods of the locomotor rhythm, and per^L is somewhat hypersensitive to the effects of such LL exposures (Konopka *et al.*, 1989). (ii) per^L flies have their free-running periods lengthened even further when they are subjected to increasing temperatures (Konopka *et al.*, 1989; Ewer *et al.*, 1990); the same is true for per^L genocopies in which the pertinent valine residue (Figure 3) was substituted by asparagine or arginine (Curtin *et al.*, 1995; Huang *et al.*, 1995). Curtin *et al.* (1995) also showed that nuclear entry of PER is delayed in the three per^L types compared with the already stalled entry time in wild-type flies (also see Lee *et al.*, 1996), that the extra time required in the mutants corresponded with their lengthened τ's, and that there was even more of a delay when the temperature was elevated.

Now, if PER^L exhibits weak interactions with (native) TIM *in vivo*, and if dim light drives PER levels downward, a pleasingly similar relationship would seem to exist between these two kinds of molecular impairment and the long-period biological rhythms that are observed in both cases. These suppositions are consistent with what happens in a situation where PER cannot interact with TIM at all (in tim^{01}) or when wild-type flies are exposed to constant and relatively bright light: PER levels are substantially lowered (Zerr *et al.*, 1990; Price *et al.*, 1995). Pulling these various strands together leads to the notion that a mutated TIM-interaction domain in per^L results in lower-than-normal PER (at a key stage of the protein cycle), with the result being a slow clock of the same sort (overall) that is caused by subnormal concentrations of the normal protein.

This scenario is valuable principally as a device for trying to remember some of the phenotypic and genotypic complexities revolving around *period* and *timeless*, which have been more extensively studied than other clock genes in insects. The problem with the device is that elements of the empiricism that created it are puzzling in light of certain further findings. Thus, the overall abundance of

PER^L does not appear to be appreciably lower compared with normal PER (Huang *et al.*, 1995), although the amplitude of the mutant's protein cycling is subnormal (Rutila *et al.*, 1996). The latter data were collected in conjunction with analyzing the *tim*^{SL} mutation, which partly suppresses *per*^{L}'s lengthened period. This *timeless* allele also blocks the heat-induced further lengthening of τ that is exhibited by *per*^{L} alone (Rutila *et al.*, 1996). However, in the same kind of yeast two-hybrid assays that showed heat-lability of PER^L-with-TIM interactions (Gekakis *et al.*, 1995), a TIM^{SL} fragment failed to ameliorate this defect (Rutila *et al.*, 1996). This does not obviously fit with the *in vivo* effect of *tim*^{SL} on free-running behavioral rhythmicity. Moreover, this *timeless* mutant bears an amino-acid substitution (Figure 3) outside the aforementioned intra-TIM regions that were determined to interact with portions of PER (Rutila *et al.*, 1996; cf. Saez and Young, 1996). What made the most sense in terms of thinking about *per*- and *tim*-encoded functions interacting with one another was that *tim*^{SL} advanced the nuclear entry of PER^L protein, eliminating most of the "extra" delay that is observed in *per*^{L} flies (Rutila *et al.*, 1996). None of these findings belies the conclusion that PER and TIM physically interact in the wild-type flies. However, full comprehension of the mutants' biological phenotypes in terms of how the normal and mutated proteins (or fragments thereof) behave in biochemical assays or heterologous cell types is not yet possible.

There is one further part of the nuclear-entry story to be considered, which has to do with the relative times that PER and TIM go into that subcellular compartment in actual fly cells whose circadian function is interpretable. The question was whether the two proteins enter together, as could be inferred from studies of the cultured cells transfected with *per*- and *tim*-expressing constructs (Saez and Young, 1996). Yet, immunohistochemical scrutiny of certain brain neurons (see Section VI) at different times of night in LD showed that PER became predominantly nuclear approximately 2 hr before TIM did, at ZT19 and 21, respectively; both proteins together were observed to be almost exclusively in the nucleus by the end of the night (Shafer *et al.*, 2002). Given that the turndown of *per* or *tim* mRNA levels with respect to the daily cycles of these transcripts' abundance fluctuations begins before the night is half over, it would appear that PER alone is capable of mediating the onset of this transcriptional repression. Such suppositions dovetail with the results of transfecting a *per*-expressing clone into cultured cells (Rothenfluh *et al.*, 2000c); the transgene was designed to produce a constitutively nuclear form of PER; it was sufficient to repress *Clk* gene expression (discussed below in Section V.B.3).

Rothenfluh *et al.* (2000c) went on to transform *Drosophila* with a *per* construct deleted of sequences encoding PER's cytoplasmic-localization domain (Figure 3) and found that the resulting "TIM-independent nuclear entry" caused dominant quasi-arrhythmicity behaviorally. It was as if the level of native PER produced from the *per*$^+$ allele carried by these flies was quashed by the repressive effect of the recombinant PER protein. This set of results was taken to correlate with protracted "depression" of *per* and *tim* expression observed in flies expressing an ultralong *tim* mutation (Rothenfluh *et al.*, 2000a). It seemed as if the prolonged nuclear localization of PER:TIM^{UL} that was also observed by Rothenfluh *et al.* (2000c) meant that PER was tied up by the mutated TIM protein. Getting rid of the latter by exposing *tim*^{UL} flies to light (which gets ahead of the story) led to a downturn in *per* and *tim* expression. Rothenfluh and co-workers' results jibe with elements of those obtained by Shafer *et al.* (2002). However, the former investigators concluded that TIM-independent repression of transcription is restricted to the late-night/early-morning times, when both TIM and PER have moved into the nucleus (and when TIM is degraded in response to light; Section IX.B.2). The histochemical inferences of Shafer *et al.* (2002) part company with Rothenfluh *et al.* (2000c), who clinged to the supposition that PER:TIM heterodimers are the emissary of nuclear entry (Saez and Young, 1996) and the initiator of repression in the middle of the night. Shafer and co-workers concluded that PER alone can do this job at that phase.

So the view of TIM's function may be in need of substantial modification. It might function primarily as a facilitator of *further* nuclear entry of PER, or perhaps serves only to protect PER from degradation before it goes in. Proteolysis of TIM—later on in the cycle, around dawn—is a prominent feature of the second clock-input subtopic (in Section IX). Thus, as a preview of the molecular-input story, TIM's interaction with PER at a cycle time after the former follows the latter into the nucleus (Shafer *et al.*, 2002) could be connected (mainly?) with what happens to TIM after light enters into the equation. A slightly broader view would be that nuclear PER is "rate-limiting" in terms of negative effects on its own production as well as those of *tim* mRNA and TIM protein. Perhaps this fits with the fact that cytogenetically mediated alterations of *per*$^+$ dosage affect biological rhythmicity: Even the 50% decrease or increase in *per* product levels that can be inferred from these manipulations has noticeable consequences, whereas similar alterations of *tim*$^+$ dosage do not cause changes in the pace of the clock (Section III.A.3).

How would nuclear PER, if not TIM, effect apparent turndowns in the transcription rates of the genes that encode them? Such time-based diminishing of gene activity was surmised owing to the fact that, as the protein abundances are on the rise, the mRNA levels are falling—hence, a further suggestion that the feedback effects are negative. This phenomenon was further presumed not to involve direct interactions of PER or TIM with the pertinent DNA sequences—the 5'-flanking regions of the respective gene—because neither protein possesses a known DNA-binding motif. For example, unlike the other founding family members SIM and ARNT, PER has no bHLH amino-acid sequence (Figure 3). A better presumption would be that the negatively acting factors would function by interacting with positive transcription-enhancing proteins. Precedents for this had been established in which negative transcription factors, devoid of DNA-binding sequences, acted accordingly by binding to positive ones and poisoning the resulting heteromultimers (see, for example, Ellis *et al.*, 1990; Van Doren *et al.*, 1991, 1992).

3. *Clock* and *cycle* functions regulating *per* and *tim* expressions

Candidate factors for filling in the gaps of the model just implied were provided by the Clk^{Jrk} and cyc^0 mutations. They lead to biological arrhythmicity by definition of their isolation phenotypes. The mutations were then quickly shown also to cause very low levels of *per* and *tim* gene products (Allada *et al.*, 1998; Rutila *et al.*, 1998b). Such transcripts and proteins trickle along with no apparent temporal variations. The almost certain explanation for these molecular phenotypes is that both CLK and CYC are bHLH, PAS-containing proteins (Figure 3). The supposition that this was so facilitated their cloning in the first place, as noted previously with regard to database searching and mapping the resulting bHLH/PAS-encoding candidate sequences to the loci defined by Clk^{Jrk} and cyc^{0l}. The CLK^{Jrk} protein is truncated by a nonsense mutation (Allada *et al.*, 1998), which would nevertheless allow the remaining protein fragment to bind to DNA, but not be activated because it lacks the relevant (relatively C-terminal) glutamine-rich sequences (Figure 3). This rationalizes the semidominance observed for $Clk^{Jrk}/+$ flies, about half of which individuals are behaviorally arrhythmic. That is, the mutation acts as a "dominant-negative," interfering with overall function of the inferred protein complex (see later), whereas flies carrying a *Clk* deletion over the + allele are fully rhythmic. Thus, the heterozygous effects of Clk^{Jrk} are worse than those of no gene at all.

 Both cyc^0 mutants are apparently complete "nulls," in that they each harbor nonsense mutations (Figure 3) relatively near the 5' end of *cycle*'s open-reading frame (Rutila *et al.*, 1998b; Park *et al.*, 2000a). So, the discoveries of *Clk* and *cyc*, and how the qualities of these gene products nicely fit with the chronobiological and molecular defects caused by mutations at these two loci, paint an emerging picture in which the mutually associated CLK:CYC polypeptides have their functioning combated by PER:TIM heterodimers, or at least that one of the latter two proteins would negatively interact with CLK or CYC (Figure 7, see color insert). These products of the *Clk* and *cyc* genes would not have to fluctuate in order for this crude version of a circadian pacemaker to work, in that it would be sufficient for a negatively acting factor to oscillate. For *cyc*'s part, neither its mRNA (Rutila *et al.*, 1998b) nor its protein (Bae *et al.*, 2000; Wang *et al.*, 2001) cycle in their gross concentrations. However, the *Clk* gene products do exhibit daily abundance fluctuations, as will be discussed in a later section (Section V.C) that attempts to flesh out details of the circadian pacemaker mechanism (Figure 8, color insert).

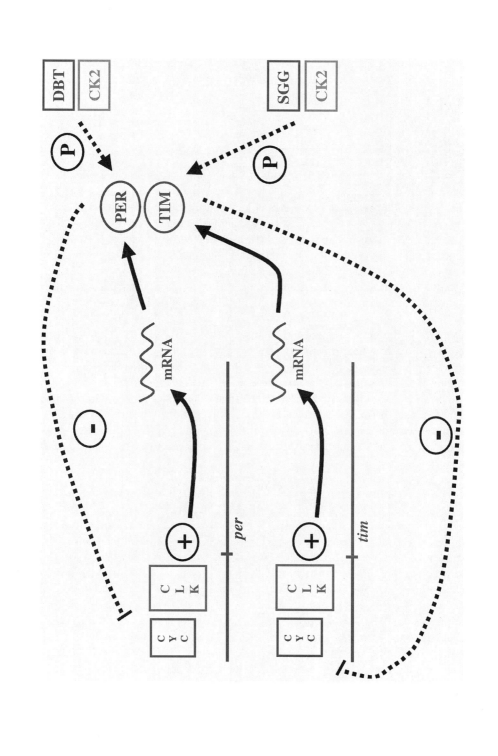

4. Roles of *double-time*, *shaggy*, and *Timekeeper* in posttranslational regulation of the negatively acting proteins

Before delving into these particulars, the basic function provided by two additional genes that are apparently clock factors will be described. One is *double-time*; prior to considering the nature of its product, certain posttranslational modifications of

---

Figure 7. Rudiments of clock-gene actions in *Drosophila* and interactions among their products. How six of the "major" *D. melanogaster* genes (Table 1) and the proteins they encode basically act to compose a circadian pacemaker is diagrammed. Those two genes are transcriptionally activated largely by the combined positive actions (circled plus signs) of CLK:CYC heterodimers—"largely," because mutational elimination of the two transcription factors leaves a low level of constitutive *per* and *tim* expression. As these two genes' worth of mRNA begin to accumulate in the late day- and early nighttime, PER and TIM (or at least their functionally meaningful actions) stay at low levels. After about 4–6 hrs of such lags, the concentration of PER has risen enough so that it can enter the nucleus of pacemaker cells, followed by TIM. These two proteins can heterodimerize, but it is not clear how much of this occurs in the cytoplasm, where TIM-bound PER may be protected from degradation and by which nuclear entry of PER:TIM may be accelerated. Within the nucleus, PER, TIM, and possibly the heterodimer, can interact with CLK:CYC; it is possible that either of the monomeric proteins or the heterodimer or some combination of the three entities can bind with CLK, CYC, or both. In any case, CLK:CYC's ability to bind *per* and *tim* is substantially attenuated (circled minus signs) by such binding(s), leading to a turndown in the transcription rates for these genes. Note that the lag between their transcriptional initiations and accumulation of the proteins they encode—or at least their negative actions on CLK:CYC—allows for *per* and *tim* mRNA cycling (wavy lines). However, *per* and *tim* seem to be transcriptionally inactivated in advance of PER and TIM entering the nucleus, as if the initial turn-off of these genes is due to CLK going toward its trough at this time (see Figure 8). With regard to the other protein oscillations, some number of hours after PER and TIM begin to rise (starting in the early nighttime), they start to fall, in part because the encoding mRNAs are diminishing in their abundance. However, these two proteins would exhibit relatively high steady-state levels (notwithstanding *per* and *tim* mRNA cyclings) if PER and TIM did not have rather short half-lives. The relatively rapid postsynthesis degradation of these proteins is apparently influenced by PER and TIM becoming progressively phosphorylated (circled P's) during their accumulation phases, as if P-PER and P-TIM are good substrates for the relevant (but unknown) proteases. PER is a substrate for the DBT kinase and TIM is one for the SGG kinase; both of these "negatively" acting proteins have their phosphorylations mediated in part by action of casein kinase 2 (CK2). These three enzymes—which can be thought of as mundane catalytic factors, whose functions are in no way limited to the circadian system (true)—are regarded as pacemaker components because mutations at the *dbt*, *sgg*, and *CK2α* loci (Figure 18) can change the pace at which the clock runs (Table 1, which includes the *And* mutant, emerging as having identified a CK2β-encoding gene whose kinase-regulatory subunit is also involved in mediating "P" additions to both PER and TIM). Certain *per* and *tim* mutations also cause circadian period alterations; these genes have, as well, been mutated to cause arrhythmia (as has *dbt*). *Clk* and *cyc* mutations lead only to arrhythmicity in terms of the effects of the mutations induced so far induced at these genetic loci.

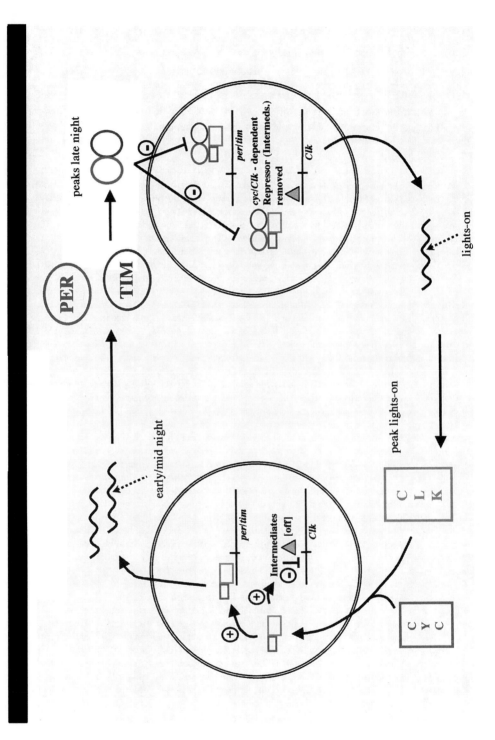

those encoded by other clock genes must be registered. Thus, PER was found to be increasingly phosphorylated in conjunction with its rising abundance during the second half of the night (Edery *et al.*, 1994b). When PER levels are lowered by the effects of a *tim*-null mutation or exposure to constant light, the protein not only is brought down to constitutively low levels, it also does not exhibit the late-night "mobility shifts" to higher molecular weight forms (Price *et al.*, 1995)—shifts that are not observed when protein extracts of wild-type flies are treated with phosphatases (Edery *et al.*, 1994b).

Attachment of phosphate groups to proteins well outside the confines of chronobiology can help target them for degradation. Bear in mind that cyclical production of a protein could not assure its circadian oscillations; the half-life of the polypeptide must be reasonably short for its abundance to go up and down over the course of a given cycle (Wood, 1995). Therefore, it was intriguing to contemplate the *dbt* product as a regulator of this process, insofar as PER is concerned, when the open-reading frame at this locus was inferred to encode a casein kinase, similar to the Iε subtype known in mammals (Kloss *et al.*, 1998). Recall that *dbt* had been defined in part by a recessively lethal transposon inserted at the locus. This *dbt*^{P} mutant develops through the late larval period. Then and there, putative circadian-pacemaker cells had been detected by observing *per* and *tim* expression in third-instar larval brains (Kaneko *et al.*, 1997), including circadian oscillations of apparent PER and TIM levels (inferred by immunohistochemically observed fluctuations) in various clusters of neurons (see Section VI.A.1). In *dbt*^{P} larvae, neither PER nor TIM protein fluctuated in terms of their immunohistochemically inferred levels, and further assessments of PER's level showed it to be at a constitutively high level (Price *et al.*, 1998).

The *double-time* kinase is involved in setting the pace of the clock, and immunochemical timecourses for PER and TIM in extracts of adult heads showed

---

Figure 8. Expansions of the basic pacemaker mechanism: the interlocked feedback loop hypothesis. This diagram is adapted from Glossop *et al.* (1999), who invoked it to explain certain effects of clock mutations on product levels (beyond the elementary such effects diagrammed in Figure 7, such as *Clk*^{Jrk}- and *cyc*^{0}-induced lowerings of *per* and *tim* products). Thus PER:TIM binding to CLK:CYC not only depresses *per/tim* activation, but it also releases CLK:CYC-dependent repression of *Clk*. Presumably, the "positive-looking" CLK:CYC factors would not clamp down on *Clk* transcriptional activation directly, but instead would regulate production of a hypothetical "intermediate" factor that possess such repressive action. In any event, CLK production goes into its rising phase, probably as mediated by the action of an unknown activator. As CLK accumulates, it promotes increases in *per* and *tim* transcription rates (Figure 7) and stimulates repression of "its own" (*Clk*) transcription (see above). The result of these interplays is that *per* and *tim* mRNA cyclings (wavy lines) are out of phase with that of *Clk* mRNA: mid-nighttime peaks for the former two; daytime peak for *Clk* and (according to the findings that support this scheme) CLK as well, in that a PER- or TIM-like accumulation delay does not occur. The other symbols (+, −) are as in Figure 7.

effects of dbt^S and dbt^L on the abundance oscillations of these proteins that paralleled the consequences of these mutations in terms of locomotor rhythmicity (Price et al., 1998). Therefore, among DBT's substrates is a clock protein, and in fact this enzyme molecule can physically interact with PER in tissue extract and transfected cultured cells (Kloss et al., 1998). Given the low (but non-null) expression of double-time in the dbt^P mutant (Kloss et al., 1998), arguably that mutation leads to hypophosphorylation of (at least) PER and causes that protein to be "too stable" such that it cruises along at inappropriately high levels. The DBT kinase could be imagined to vary in its activity over the course of a daily cycle. However, it is not necessary to assume this to apprehend progressive phosphorylation of PER, and indeed dbt mRNA and protein levels appear to be temporally flat, as assessed in extracts of adult heads (Kloss et al., 1998, 2001). Thus, the products of this gene do not define "state variables" subserving clock functions. That dbt is nonetheless a bona fide pacemaker component is reinforced by a gene-interaction phenomenon: dbt^{ar} alone causes locomotor arrhythmicity in DD and eliminates PER and TIM oscillations, but combining this mutation with per^S partly alleviates the behavioral effects of this double-time mutation (Rothenfluh et al., 2000b).

TIM is also a substrate of one or more kinases, although its normal form seems to get phosphorylated to a lesser extent than does PER (Zeng et al., 1996); also, mutated TIM^{SL} appears to be hyperphosphorylated during the late night compared with the normal form of the protein (Rutila et al., 1996). Pharmacological inhibition of kinases led to nondegradable TIM (Naidoo et al., 1999). That DBT is not a regulator of phospho-TIM production brings the case of shaggy to the fore. The protein encoded by this gene has long been known to be a glycogen-synthase kinase-3 (GSK-3) relative whose substrates include factors that regulate embryonic pattern formation and other features of Drosophila development (e.g., Siegfried et al., 1992; Blair, 1994; Peifer et al., 1994). Analysis of the intragenic location of the $sgg^{EP(X)1576}$ transposon insert (Table 1) and of SGG protein isoforms whose levels might be changed in this mutant (when driven by yeast GAL4) suggested that a certain form of the kinase (Figure 3) is responsible for short-period rhythmicity (Martinek et al., 2001). Thus one of the several SGG subtypes—which have different N- and C-termini, but the same kinase domain (Figure 3)—may target a clock protein as one of its substrates. This turned out to be TIMELESS: Transgene-mediated overexpression of sgg caused the TIM timecourse in DD to be shifted in the early direction (Martinek et al., 2001); that is, the progressively premature abundance peak and nuclear entry time were correlated with the shorter-than-normal period exhibited by this EP-induced mutant (Table 1). PER's nuclear entry was shown to be earlier than normal as well in the face of sgg overexpression. However, altered TIM quality was revealed to be the primary culprit. For this, sgg underexpression was nicely contrived by rescuing the lethal effects of classic mutations at this X-chromosomal locus (Martinek et al., 2001). The resulting hymormorphic expression of sgg in flies arising from animals in whom sgg^+ was induced during development led to two chronobiological phenotypes: long-period rhythms (Table 1) and migration of TIM in anomalously rapid manner (in Western blots) at all timepoints. This protein difference from the norm, which was not observed for PER, suggested that TIM is hypophosphorylated when *sgg* is underexpressed in the mutant/transgene-rescue combination (Table 1). Overexpression of *sgg* in the double-transgene situation led to hyperphosphorylated TIM, that is, unorthodox conversion of most of the protein to slowly migrating forms; these were decreased in their apparent molecular weights by phosphatase treatment (Martinek *et al.*, 2001). Finally, fragments of TIM protein were shown to have (radioactive) phosphate groups added when incubated with a mammalian form of the GSK-3 enzyme, which is a close protein relative of SGG kinases in *Drosophila*.

The chronobiological factor identified by the dominant, long-period *Timekeeper* (*Tik*) mutation (Table 1) kept getting alluded to or depicted as yet another kinase. That *Tik* identified the catalytic subunit of casein kinase 2α as playing a rhythm role was demonstrated by enzyme-activity assays (Lin *et al.*, 2002a) as well as molecular genetics (Figure 3, color insert). Thus, $CK2\alpha/+$ (synonymous with *Tik*/+) flies exhibit a significant reduction down to ca. 60% of wild-type levels, which exhibited no daily cycling (also true for the DBT kinase). The *Tik* revertant (for which *TikR*/+ flies give much closer to normal behavioral periods compared with *Tik*/+) showed ca. 85% of the normal activity level for this particular kinase (Lin *et al.*, 2002a). To get around the contribution to this kinase by the normal allele in heterozygous flies, $CK2\alpha$ as encoded by a recombinant, *Tik*-mutated form of the coding sequence was assessed for activity; the levels were only about 10% of normal, essentially the same value as obtained using a recombinant version of *TikR*. This implies that the reverting effects of this allele do not simply restore near-normal catalytic function. Lin *et al.* (2002a) suggest that the *TikR* polypeptide cannot be incorporated well into the holoenzyme, so that that factor would consist largely of normal catalytic subunits in *TikR*/+ flies. Given that *TikR* is still a recessive lethal mutation, the results and suppositions just stated indicate that this allele is a near-null variant. It would follow that the original *Tik* allele is another example of an antimorphic clock mutation, leading in this case to a catalytic subunit that can participate in holoenzyme formation and damage its overall function (once again, an effect of a *changed* form of a gene and its product, which is biologically more damaging in heterozygotes than mutating a factor to the null state).

C. How clock-gene actions can build a circadian pacemaker

1. Assembling feedback loops

Seven clock factors' worth of gene products is on the table, and certain features of the ways that they act and interact are in hand. It is now time to delve into the ever-deepening details about how these factors seem to function. However, before mining the several relevant investigations, here is a disclaimer: It will be neither possible nor warranted to discuss all the apparent gene actions and interactions. Stated another way, it is inappropriate to provide a minireview within this larger one that might do little more than update the reader in terms of the current pacemaker model. In fact, there is a continuing profusion of minireviews on this very subject, although most of them are newsy, uncritical treatments of the chronogenetic subjects. (By the way, the maxireview of the subject represented by this monograph was supported by literature searching that extended through August 2002.)

Having said this, there is not all that much definitive information to *review* about the cellular biochemistry of the manner in which a given clock-gene product actually functions. At least it is the case that we are missing a fair fraction of the necessary information, which, when it is ultimately obtained, may reconcile certain of the inconsistencies that have unfolded. However, it is unlikely that the inferred functions of the known pacemaker factors are misconstrued to the extent that the entire edifice will come tumbling down.

At all events, it is not necessary in a survey of the whole insect chrono-genetic subject to home in on hypercriticism of the pacemaker-mechanism com-ponent of the story. Thus, the reader can decide: Stick a pin into the pages of a scientific magazine or review organ, and you will encounter a minireview that purports to provide an entrée into the "current view" of what the pacemaker is all about. Nevertheless, beware of the possibility that the review diagram of "the circadian-pacemaker" mechanism is based on an overreliance on demonstrating that a given clock protein *can* do such-and-such—for instance, in heterologous cultured cells. Yet, *does* it do this in the actual organism? Even certain of the phenomena that can occur in extracts of *Drosophila* tissues, involving things like protein interactions or biochemical "parameters" associated with levels of such clock-gene products, may be problematic. Bear in mind as well that most such stud-ies are most commonly initiated by grinding up adult fly heads. This includes the eyes, which arguably contain the great majority of the molecules in question, that is, within ca. 13,000 photoreceptors, compared with about 2,000 *per*-expressing neurons and glial cells within the head (cf. Figure 11, color insert). [The estimate of 1,800 glia was obtained by counting anti-PER labeled cells of this apparent type in serial sections (B. Frisch and J. C. Hall, unpublished).]

However, eye tissue is uninterpretable in terms of the biological meaning of *period* or other clock-gene expressions in photoreceptors (Section IV.C.6). One wonders why chronobiochemical experiments in *Drosophila* have not taken advantage of eyeless mutants in order to boil the homogenate down (figuratively) to assessments of the pertinent parameters in terms of cells whose meaning is more firmly in hand (at least for some of the neurons). In particular, use down the years of the apparently rather cleanly rhythmic eyeless *eya* mutant (Table 2), compared with the chronobiologically more pleiotropic *so* one (Section II.A), would seem to have been warranted.

Could an unwitting preoccupation with eye molecules have contributed to the shifting inferences that have stemmed from these chronomolecular analyses? Consider in this regard a scenario in which the main clock-protein interaction in terms of "closing the negative feedback loop" could be argued to involve TIM alone binding to CLK. This was inferred from the behavior of these molecules in tissue extracts from per^{01} adult heads (Lee et al., 1998), such that TIM might be the inhibitor of transcription-enhancing actions mediated by CLK and its CYC partner. Thus, the PER:TIM interaction could be viewed mainly to serve the purpose of getting TIM into the nucleus. However, the potential primacy of PER as the actual, *in vivo* factor that closes the loop by breaking up the CLK:CYC heterodimer has surfaced more recently (Rothenfluh et al., 2000c; Shafer et al., 2002). The problem is that various kinds of molecular chronobiological experiments have prompted some feature of an early-stage model. Results of subsequent studies necessitated augmentations of the next iteration of the scheme, not surprisingly, but also might have turned certain features of the previous version on their head. Therefore, the conclusional force of the various clock-factor actions and gene-product interactions inferred from the studies about to be described should be regarded as provisional. This is not to say that the *mutational effects* of rhythm-related factors on *nonmolecular* phenomena (Sections II–IV) were devoid of their own investigatory inadequacies.

In any case, we plunge ahead molecularly. With regard to the basic actions of *Clock* and *cycle* several inquiries have been made following the establishment of the basic function of the gene products. That mutations at these loci lead to biological arrhythmicity as well as constitutively low levels of *per* and *tim* products caused by a substantial drop in the transcription rates of those two genes—along with the fact that positively acting transcription factors are readily inferred to be encoded by *Clk* and *cyc*—suggested the elementary scenario introduced earlier (see Figure 7): CLK and CYC would participate in activating *per* and *tim*; later in the daily cycle, PER, TIM, or both would interact with at least one member of a CLK:CYC pair, poisoning the functional heterodimers presumed to be formed by the latter two proteins in terms of their binding to regulatory sequences at the *per* and *tim* loci. Note that the "and/or" aspects of this scheme are based on the facts that PER and TIM can associate in part via the former's PAS domain (Saez and Young, 1996). Implicitly, the PASs of CLK and CYC (or possibly other stretches of amino acids within those proteins) could be bound by TIM, if the PAS-less feature of the latter permits it to interact with PER (true)—although it may be that interactions between PAS domains of two heterologous proteins (Huang et al., 1993) more easily allows an appreciation of PER interacting with CLK, CYC, or both.

A test of this model came from experiments reported contemporaneously with the discovery of the first *Clk* and *cyc* mutants. Thus, expressible *Clk* sequences were transfected into cultured *Drosophila* cells, which remarkably expressed *cyc* all by themselves (Darlington et al., 1998). Cotransfection of these cells with *per* or *tim* (so-called) promoter constructs (Figure 9, color insert) fused to luciferase encoding sequences (*luc*) led to severalfold induction of the reporter enzyme. The stimulations of *per* or *tim* transcription, presumably mediated by the cooperating action of (introduced) CLK and (endogenous) CYC, could occur when certain boiled-down subsets of the latter genes' 5'-flanking regions were fused to *luc* (Darlington *et al.*, 1998), as will be discussed later (Section V.C.2).

In a subsequent test of the model shown in Figure 7, this portion of the *per* gene (5' to the transcription-start site, which is shown in Figure 9) was found biochemically to bind CLK and CYC (produced recombinantly by *in vitro* translation), such that absence of either protein left the *per*-regulatory fragment unbound in the relevant "test-tube" assays (Lee *et al.*, 1999). Addition of PER, TIM, or both inhibited CLK:CYC *per* DNA binding, and neither of the former two negatively acting proteins could be shown to interact with this sequence, consistent with the lack of DNA-binding motifs in the PER or the TIM proteins. This *in vitro* approach was extended to show that CLK and bind to CYC in the relevant solutions, and that PER and/or TIM can join the partnered positive factors (Lee *et al.*, 1999). Moreover, CLK:CYC heterodimerization was not broken up by addition of the negative factors, as if the aforementioned poisoning of CLK:CYC's DNA binding involves inhibition of that activity per se, and not a molecular event such as TIM grabbing CLK away from CYC. Finally, PER or TIM were each shown in coimmunoprecipitation experiments performed on fly-head homogenates to bind with CLK (Lee *et al.*, 1998).

These experiments included a demonstration of the protein interaction possibilities that were previewed earlier: PER *can*, when present with TIM, interact with CLK; however, perhaps it does not have to do, because in PER-less homogenates TIM could still be found associated with CLK (Lee *et al.*, 1998). Alternatively, PER might really need to deal with CLK:CYC on its own, at least before the nighttime is nearly over (Shafer *et al.*, 2002). Thus, it could be misleading that PER is dispensable in biochemical experiments (Lee *et al.*, 1998). In fact, even the *per* and *tim* promoters were shown to be dispensable for a crude form of biological rhythmicity. Here, a complex genotype was assembled in which a "pan-neural" gene's promoter (Section V.C.2) simultaneously drove UAS-mediated expression of *per*$^+$ and *tim*$^+$ coding sequences in a *per*^{01};*tim*^{01} genetic background; with no meaningful *per* or *tim* promoters to be regulated by CLK and CYC, about half of the behaviorally monitored flies were nonetheless rhythmic (Yang and Sehgal, 2001). Thus, bypassing major elements of the feedback loop left components of the system functioning in a chronobiologically consequential manner, although the behavioral rhythms in this genetic situation really sucked (Yang and Sehgal, 2001).

Nevertheless, chronobiochemists had been grinding away further to analyze the dynamics of these molecular interactions, assuming all the while that they are significant for *normal* rhythmicity, if not necessary for *some* degree of biological cycling. Using extracts of fly heads, it was shown that CLK:CYC dimers are bound by PER:TIM during the late nighttime as well as early morning and that most of the CLK protein detectable interacts with CYC throughout the daily cycle (Bae *et al.*, 2000). Indeed, oscillating CLK protein appears to be present in "limiting amounts." The PER:TIM heterodimer seem to interact preferentially with the CLK part of a CLK:CYC pair, such that PER and TIM do not waste their time sopping up any of the CYC monomers; the latter are said to present in "excess" because the peak level CLK is way below that of the CYC plateau (Bae *et al.*, 2000). If these molecular inferences are correct, they could make the retrospective prediction that half-normal concentrations of CLK protein would alter the clock's pace, but the same kind of genetically mediated lowering of CYC's level would not. As was discussed in Section III.A.3, the effects on free-running behavioral periods of the relevant mutant and gene-deletion heterozygotes potentially allow for this connection to be made between formal phenogenetics (the long-period phenotype caused by $Clk^-/+$ vs. the normal periods observed for certain *cyc*-null heterozygotes) and concrete chronochemistry. However, one of the putatively null variants for *Clk*, when heterozygous with the normal allele, led to normal free-running periodicity (Table 1), and effects on this locomotor phenotype of *cyc*-null/+ heterozygotes are ambiguous (Rutila *et al.*, 1998b; Park *et al.*, 2000a; Table 1).

Notwithstanding these putative puzzles and others, the pacemaking mechanism as outlined in Figure 7 and tested both within and without the fly has not been dramatically undermined. None of the "bypass" experiments described earlier, in which, for example, *per* and *tim* RNA cyclings as controlled in part by their promoter regions are eliminated, led to normal biological rhythms. However, the results of these transgene manipulations, and implications of various remarks in the foregoing paragraphs, indicate that the mechanism has not been proved in terms of all the details and *in vivo* realities one wishes to perceive. Instead of some kind of final proof of the model, what we have now is an expanded view of the manner by which the implied actions of these four genes' worth of clock factors interact.

First, consider that *Clk* products were found to oscillate in their abundance, even though it is not necessary to envision changing levels of a positive factor to appreciate functioning of the feedback loop. As introduced earlier, circadian fluctuations of the negatively acting PER and TIM proteins are in principle sufficient (Figure 7). For *Clk*'s part, expression of the gene itself seems to be rhythmic, in that the mRNA was found in one set of studies to peak at a time essentially antiphase to the *per* and *tim* transcript maxima (Bae *et al.*, 1998; Lee *et al.*, 1998). However, Darlington *et al.* (1998) reported two separate daily maxima for *Clk* mRNA (one in the late nighttime, the other about one-quarter cycle later). Yet, as we continue to harrow ahead, CLK protein was also found to cycle (Lee *et al.*,

1998; Bae *et al.*, 2000), and its abundance maximum at the end of the night was indistinguishable from the one such peak for the encoding mRNA observed by these investigators.

CLK appears to be a phosphoprotein (Lee *et al.*, 1998) at the two time-points for which the relevant homogenates were assessed: end of the night and ca. 4 hr later (also see Kim *et al.*, 2002). CYC, too, may be posttranslationally modified (in an unknown manner). Mercifully, its abundance and that of *cyc* mRNA seem unchanging at different times of day or night (Rutila *et al.*, 1998a; Bae *et al.*, 2000; G. K. Wang *et al.*, 2001). Therefore, *Clk*-product cycling, but not that of *cyc*, may be related to clock regulation, beyond the matter of the basic involvement of these DNA-binding proteins in *per* and *tim* expression. In addi-tion, the temporal variation of *Clk* expression is itself under clock control, in that the transcripts produced by this gene are flat in per^{01} or tim^{01} mutants (Bae *et al.*, 1998). Curiously, however, the amount of *Clk* mRNA was also dramatically low-ered by the effects of either mutation such that the transcript abundance dribbled along at the trough level (Bae *et al.*, 1998).

These aspects of the gene interactions, inferred from the mutational effects just described, were explored by further application of genetic variants: Glossop *et al.* (1999) predicted that the Clk^{Jrk} or cyc^{01} mutations would result in low levels of *Clk* mRNA because either mutation leads to very low PER and TIM (Allada *et al.*, 1998; Rutila *et al.*, 1998b), and because *per* or *tim* mutations that eliminate those proteins bring the *Clk* transcript level down to its daily minimum (Bae *et al.*, 1998). However, in head extracts from Clk^{Jrk} or cyc^{01} homozygotes, *Clk* mRNA levels were constitutively *high*, staying at the wild-type peak level (Glossop *et al.*, 1999). This implies that another function of CLK:CYC in addition to *per* and *tim* activation involves repression of (at least) *Clk* gene expression (Figure 8). *per* and *tim* for their part would be required for depression of *Clk*, which restates the fact that *Clk* mRNA is low in *per*- or *tim*-null mutants. Clk^{Jrk} or cyc^{01} knock out the presence or function of the *Clk* repressor, so that PER:TIM-mediated derepressor function is not needed (and would be nearly absent, given the effects of these mutations on *per* and *tim* product levels). The "activator of *Clk*" would be some unknown function, certainly not those mediated by CLK or CYC themselves (*Clk* mRNA is high in mutants lacking the two proteins just named). PER and TIM become irrelevant for *Clk* derepression when CLK or CYC is nonfunctional: The null state for either of the latter genes results in the absence of the need for one feature of *Clk* and *cyc* functions, which is to activate a hypothetical repressor of *Clk* mRNA production (Figure 8). In the wild type, this repressor needs to be suppressed by the action of PER:TIM; but the irrelevance of (at least one of) those two proteins in the absence of *Clk*- or *cyc*-encoded function is underscored by the high levels of *Clk* mRNA observed in per^{01}; Clk^{Jrk} and per^{01}; cyc^{01} double mutants (Glossop *et al.*, 1999).

The ensemble of these results, even without identification of the *Clk* activator or its repressor, suggested the existence of "interlocked negative feedback loops" to explain something approaching the totality of interactions among these four clock genes. As diagrammed in Figure 8, PER:TIM inhibit the actions of CLK:CYC during the late night. This inhibition leads to repression of *per* and *tim* expression (Figure 7). The extra consequence, if you will, of the eventual fall in PER and TIM levels during the morning hours is that CLK:CYC dimers are released, such that not only can they reactivate *per* and *tim*, but they can also repress *Clk* expression (albeit not directly). Thus, *Clk* mRNA is brought down to its trough by the end of the day. As was just implied—remembering that *per* and *tim* mRNA levels are substantially rising at this time—the two feedback loops involve distinctly different peaks for the transcripts produced within the separate components of the scheme (Figure 8).

This is perhaps how the "overall" clock should operate: Different transcriptional regulators peak at separate times, not only in terms of how they regulate one another, but also such that various *output factors* can peak at different times of day or night. These factors—if it were the case that the genes encoding them are expressed cyclically—might want to be high or low at distinct times, depending on the biological parameters influenced by them. By way of example, the fly might have a component of its locomotion peak at the end of the day, but exhibit maximal sensitivity of a certain sensory function in the middle of the night. These possibilities, along with descriptions of the relevant biological and molecular phenomena, will be revisited within Section IX.

Meanwhile, let us consider further the antiphaseness of *period/timeless* versus *Clock* product cyclings (Figure 8). These dynamics have been manipulated by investigators who wondered what would happen to molecular rhythmicity and its biological consequences if *Clk*-coding sequences were brought under the control of *per* or *tim* 5′ regulatory ones (Figure 9). In the *per-Clk* transgenic type, *Clk* mRNA so manipulated cycled with similar dynamics to that of normal *per* mRNA, including a relatively high abundance peak characteristic of the latter, compared with the lower maximum of the *Clk* transcript in wild type (Kim *et al.*, 2002). Expression of the *per-Clk* transcript led to accentuated peak levels for *per*, *tim*, and *Clk* mRNAs encoded by the endogenous genes. The temporal profile for CLK produced by *per-Clk* roughly approximated the normal dynamics for this protein (although the peak levels of CLK were higher then normal, consistent with the transcript result). Given that the temporal profiles for *Clk* mRNA and CLK protein are rather superimposable in wild-type extracts, the implication is that posttranslational mechanism contribute significantly to CLK cycling (Kim *et al.*, 2002), or else the *per*-like cycling of transcripts encoded by *per-Clk* would result in more of a shift for the CLK curve. This conclusion was supported by the results of temporally monitoring macromolecules emanating from a *tim-Clk* fusion construct: *tim*-like *Clk* mRNA cycling, but a CLK-like protein profile (Kim *et al.*, 2002).

To recapitulate elements of these findings and to set up the behavioral assays that accompanied them: *per-Clk* allows for quasi-normal cycling of endogenous *per* and *tim* (with boosted mRNA peaks, but normal ones for the proteins), but drives higher-than-normal abundance of CLK. Yet, free-running locomotor activity rhythms of *per-Clk* transgenic individuals were largely normal (Kim *et al.*, 2002). Subjecting these flies to photic stimuli was necessary to reveal their behavioral defects. Thus, in LD cycles, they showed no D-to-L anticipatory behavior (Figure 5); instead, they exhibited accentuated activity following lights-on (more locomotor events than those exhibited by controls carrying no transgene and + alleles at all clock loci). Therefore, CLK could be involved in modulating the direct effects of light on behavioral activity. To look into this, Kim *et al.* (2002) exposed *per-Clk* flies in DD to 10-min light pulses. One-half hour later, the transgenic individuals and control ones showed enhanced locomotion, which lasted about 30 min in the controls, but the light-induced effect on *per-Clk* activity was much longer lasting. Intriguingly, there was a circadian component to these acute light effects: in controls, a larger increase of the locomotor spike in the relatively early subjective day compared with 4 hr later; but there was a rather weak effect on *per-Clk* at the latter time, and this transgenic type responded maximally in the middle of the night, when locomotion is normally low (Kim *et al.*, 2002).

Now we bring the clock-related kinases encoded by *double-time*, *shaggy*, and *Timekeeper* into play and how they concern themselves with phosphorylations of other pacemaking proteins (Edery *et al.*, 1994b; Zeng *et al.*, 1996; Naidoo *et al.*, 1999; Lee *et al.*, 1998). Analyses of DBT's functions indicated that the sole clock-related substrate for this kinase may be PER (Kloss *et al.*, 1998). In contrast, the SGG enzyme does not phosphorylate PER (Section V.B.4), so that its only chronobiological substrate may be TIM (Martinek *et al.*, 2001). The hedges just implied stem from the facts that (i) DBT and SGG almost certainly or surely (respectively) catalyze phosphorylation of proteins that function during *Drosophila* development, correlated with the lethal effects of null or nearly null *dbt* or *sgg* mutations; and (ii) the kinase for the third of the known phosphoproteins possessing pacemaker function—CLK—is unknown.

Speaking of substrates, one of those for SGG is a protein involved in pattern formation during *Drosophila* embryogenesis. This is ARMADILLO, a cytoskeletal anchor protein (β-catenin), which was pointed to initially by embryonic-lethal *arm* mutations (e.g., Peifer *et al.*, 1994). ARM seems to participate in the formation of intercellular adhesive junctions with respect to its functioning in the "*wingless/frizzled*-receptor signaling pathway." Here, factors mediating intercellular communication (such as the WINGLESS protein itself) and intracellular control of transcription participate in elaboration of the animal's body plan as well as influence several features of postembryonic development (e.g., Boutros and Mlodzik, 1999; Reifegerste and Moses, 1999). Such apparently stray information is mentioned because of some recent "protein modeling" analyses of the TIMELESS amino-acid sequence. This protein remains a "pioneer" one (with no overall primary amino-acid sequence similarity to any other polypeptide), yet TIM seems to contain three ARM-like dimerization domains (as discussed, but not demonstrated, by Kyriacou and Hastings, 2001). The meaning of such stretches of amino acids within the TIM molecule is unknown, although two of the ARM-like domains within the clock protein overlap with regions of TIM that can bind to PER in cultured *Drosophila* cells (Saez and Young, 1996). Standing back from such intrapolypeptide details, the putative similarities between portions of the TIM and ARM proteins allow one to rationalize that both molecules are phosphorylated by the SGG kinase in the context of the different biological processes influenced by this enzyme.

In biochemical timecourses performed to characterize *Timekeeper*, heads were taken flies heterozygous the original *Tik* mutation (Figure 3) in constant darkness. Extracts of this material were processed to monitor the levels of other clock-gene products, and the *Tik/+* genotype was found to cause increased PER levels (compared with the normal peak abundance of that protein), then a delay in its disappearance (Lin *et al.*, 2002a). A modest increase in hypophosphorylated forms of PER was also noticed in the mutant extracts. The same kind of material taken from *Tik/+* heads showed similar biochemical anomalies for TIM's temporal profile (Lin *et al.*, 2002a). These results are consistent with both PER and TIM being normal substrates for the CK2 holoenzyme (whose regulatory β subunit, as a reminder, is apparently encoded by the *Andante* gene discussed in Section III.B.1). Indeed, Zeng *et al.* (1996) had previously shown PER and TIM to be phosphorylated by CK2 *in vitro*. By focusing on the activity of recombinantly produced CK2α, it was shown that this activity mediated attachment of phosphate groups to PER, and less strongly phosphorylated TIM, in test-tube assays (Lin *et al.*, 2002a). Therefore one way to apprehend the effects of CK2α mutations *in vivo*, in light of the biochemical results, is that decrements in a phosphorylating function cause anomalously robust and persistent levels of the (core) clock proteins, reiterating a scenario in which PER and TIM want to turn over in a rather rapid manner. If such processes are impinged upon, the pace of the circadian clock slows down, as it does under the influence of hypomorphic *double-time* and *shaggy* mutations (Table 1). Further analyses of the circadian factor defined by *Timekeeper* (discussed at the very end of this subsection) led to an additional supposition about the cell-biological consequences of phosphorylation events mediated by casein kinase 2.

Most of the further information obtained about clock-related enzymatic activity revolves round the DBT kinase and phospho-PER (although the CK2α story started by the *Tik* mutation is catching up, as we will see shortly). Recall that a mutation causing subnormal levels of *dbt* mRNA leads to noncycling PER that also seems to be too stable (Price *et al.*, 1998; Kloss *et al.*, 1998). One imagines that a long-period *double-time* mutant would also exhibit lower-than-normal kinase levels. There is evidence for this (Suri *et al.*, 2000), but the phosphorylation of PER was minimally affected in the relevant dbt^h homozygotes and $dbt^g/+$ heterozygotes. In fact it seemed as if PER was hyperphosphorylated in extracts from these mutants (Suri *et al.*, 2000), suggesting that DBT kinase activity does not account for all or even much of the overall "PER phosphorylation program." Given what is being uncovered about the *Tik*-defined casein kinase 2, this conclusion is no longer disturbing.

However, another potentially distressing result from analyzing *dbt* mutants involved their effects on the temporal dynamics of other clock-gene products, considered irrespective of their phosphoprotein quality. First, when the gross mRNA and protein cycling of *per* and *tim* products were monitored in extracts from the two long-period *dbt* mutants (the then-new ones), the transcript-abundance peaks were later than normal (Suri *et al.*, 2000). This is as expected (Price *et al.*, 1998), although in an opposite type of *double-time* variant, the fast-clock dbt^S mutant, *per* mRNA also accumulated later (but declined sooner) than in wild type (Bao *et al.*, 2001). Returning to the slow-clock dbt^h mutant, mRNA/protein timecourse comparisons showed that the usual time lag between the *tim* transcript and the TIM peak (in an LD cycle) was essentially eliminated (Suri *et al.*, 2000). Given that this mutant is behaviorally periodic (with a 29-hr free-running cycle duration), perhaps the oft-discussed mRNA/protein delay is not needed for biological rhythmicity, at least for *timeless* gene products. Alternatively, the "no-delay" mRNA and protein cycles in dbt^h homogenates may reflect mainly what is going on molecularly in this mutant's eyes, knocking out a biological rhythm in that tissue (cf. Chen *et al.*, 1992). However, somehow a disparity between the *tim* transcript and the protein peaks would be maintained in brain neurons. For *per*-product cycles in these *dbt* mutants, the head-homogenate approach suggested that a temporal gap between the mRNA and the PER maxima is also largely eliminated, although the point was made in a more graphically explicit manner for *tim* mRNA and TIM protein (Suri *et al.*, 2000). An opportunity to address these issues by *in situ* observations is implied by the results of Bao *et al.* (2001). These investigators demonstrated that PER protein accumulates in compound-eye photoreceptor nuclei of dbt^S flies at a later cycle time compared with wild type. However, mRNA cycling was not monitored in this sectioned material (and has never been seriously so examined in a temporally controlled manner for *Drosophila* clock genes), so transcript versus protein dynamics could not be compared. Would the PER peak be delayed relative to that for encoding mRNA in a given tissue expressing a given clock-mutated genotype?

At all events, studies of additional mutants involving *double-time* and other clock genes provided further evidence for interactions among these factors insofar as biological rhythmicity, phosphorylation, and protein turnover are concerned. Thus, the fast-clock per^S and per^T mutations were shown to ameliorate the frequent arrhythmicity caused by heterozygosity for dbt^{ar} and either dbt^P or a deletion of the gene; the frequently rhythmic doubly mutant adults exhibited shorter free-running periods (31–37 hr) compared with the whoppingly long (ca. 40-hr) behavioral cycles exhibited by the residually rhythmic per^+ individuals heterozygous for dbt^{ar} and a null or nearly null variant (Rothenfluh et al., 2000b). Concomitantly, per^T was found to restore the PER and TIM oscillations that are eliminated by dbt^{ar}; acting alone, the latter mutation caused PER and TIM to be constitutively high and low, respectively (Rothenfluh et al., 2000b). It was suggested from these results that the DBT kinase and the PER clock regulator normally interact (Kloss et al., 1998) by way of a segment of the latter protein that includes locations of the per^S and per^T mutations; these two sites are near one another, within the so-called Short Domain (Baylies et al., 1987, 1992; Yu et al., 1987b; Hamblen et al., 1998; Rutila et al., 1992). The DBT–PER interaction in the question may "retard" PER phosphorylation in wild type, but mediocre heterodimerization in these short-period *per* mutants could partially restore the hypophosphorylation that is theoretically caused by dbt^{ar} (yet, see Suri et al., 2000).

Evidence related to these notions was generated by assessing the effects of an internal PER deletion, designed to be missing part of the Short Domain, and called ΔC2 (Schotland et al., 2000). This tactic for testing the phosphorylation-related significance of this PER region suggested that it cannot globally be viewed as (normally) inhibiting the action of DBT (Rothenfluh et al., 2000b). This is because $per^{\Delta C2}$ led to *hypo*phosphorylation of PER (Schotland et al., 2000). These transgenic flies were in a *per*-null genetic background, and behavioral tests of them revealed periods that were shorter than those of controls (the latter flies were hemizygous for per^{01} and carried a per^+-derived transgene that did not work very well because the periods were ca. 27 hr, a problem discussed in Section V.C.2). The biological effect of $per^{\Delta C2}$ is superficially similar to those caused by "minor damage" of the Short Domain, that is, resulting from *in vivo*-created or *in vitro*-produced amino-acid substitutions within this region (Baylies et al., 1987, 1992; Yu et al., 1987b; Hamblen et al., 1998; Rutila et al., 1992). Additional molecular assessments of $per^{\Delta C2}$ effects showed that the Short-Domain deletion leads to reduced amplitudes of PER and TIM oscillations (along with increased average "stability" of these proteins over the course of a daily cycle) and rather poor repression of *per* transcript levels at times when they are usually low (Schotland et al., 2000). These results may be at variance with those that suggested PER^T to be more easily degraded than the normal protein—in particular, when such turnover is attenuated by presumed reduction of casein kinase Iε-like activity in a dbt^{ar} genetic background (Rothenfluh et al., 2000b). However, it is dubious to compare these two kinds of PER changes because the intragenic deletion can change the overall characteristics of the protein structure, beyond just eliminating hypothetical phosphorylation sites.

On a more positive note, we now consider histological, biochemical, and genetic results that provide additional insights about interactions between the PER protein and the DBT kinase. Whereas dbt mRNA and its encoded enzyme do not exhibit daily abundance cycles grossly, oscillations associated with DBT were observed at the level of subcellular distribution. This protein was found to be nuclear in the late night/early day, but distributed throughout the cells that were observed (compound-eye photoreceptors and clock-gene-expressing brain neurons) during the early nighttime (Kloss et al., 2001). The cyclical movement of a portion of a given cell's worth of DBT into the cytoplasm was inferred to be PER-dependent because predominant nuclear staining (mediated by anti-DBT) was observed in per^{01} heads. The same relative lack of cytoplasmic signal was seen in tim^{01} as well, but this is no doubt due to the TIM-null state causing very low PER levels—here, in the context of TIM being insufficient to hold DBT in the cytoplasm (given the per^{01} result, against a backdrop of that mutation leaving TIM at a robust level). The primacy of DBT:PER interactions (insofar as this, the first "clock-kinase," is concerned) was demonstrated in a correlative manner by immunoprecipitation experiments (Kloss et al., 2001): The enzyme polypeptide was demonstrated to interact with PER, which in turn associated with TIM, although the latter protein was not required for formation of DBT:PER dimers, which could formed in extracts of tim^{01}.

What is the functional meaning of the interactions among these proteins in an (overall) ternary complex? The specific question asked in this regard took advantage of prolonged PER:TIM interactions that are observed in the tim^{UL} mutant (Table 1). PER was found to remain hypophosphorylated in this genetic condition, but when TIM^{UL} degradation was induced by light (Section IX.B.2), the release of "TIM-free PER" was found to be correlated with increased phosphorylation of the latter protein (Kloss et al., 2001). It was concluded that TIM rhythmically influences DBT-dependent phosphorylation of the PERIOD protein.

More about casein kinase 2, in light of the DBT cell biology just discussed: Unlike the latter kinase, the CK2α subunit appears constitutively to remain in the cyctoplasm of LN pacemaker cells (at different times of day and night). After these kinds of observations by Lin et al. (2002a), they went on to study the effects of *Timekeeper* mutations on the cellular chronobiology of this kinase type. For this, the investigators took advantage of the fact that the lethal *Tik* and *TikR* mutations allow survival of developing *Drosophila* into the third-instar larval stage. In such animals, homozygous for either mutation, nuclear entry of

PER appeared distinctly delayed; for example, antigenicity of that clock protein presented both cytoplasmic and nuclear signals in the early daytime, whereas most of the corresponding staining in wild-type neurons is in the nucleus 3 hr before lights-on (Lin *et al.*, 2002a). These results led to the conjecture that a key consequence of CK2-mediated phosphorylation of PER and TIM (as previously discussed) is to "promote their transition to the nucleus."

2. Regulatory details of clock-gene transcript productions

The foregoing section closed by wallowing in the details of posttranscriptional—in particular, posttranslational—features of clock-gene product regulation. We now consider the interplay between those levels of control and that which functions at an earlier stage of gene regulation. The following description of the ways that regulatory sequences of clock genes have been manipulated, in this section primarily from the perspective of temporally varying product levels, will also bring the story to the brink of considering altered spatial control of these genes' expression.

Recall that cycling of clock factors is controlled in part at the transcriptional level. Heterologous reporter sequences fused to 5′-flanking regions cloned from the *period* or the *timeless* locus can lead to products that exhibit daily oscillations (e.g., Hardin *et al.*, 1992b; Stanewsky *et al.*, 1997a,b, 1998). However, the reporter proteins or their activities do not necessarily cycle; if this is to occur, such factors must be turned over with reasonable rapidity. This does occur for firefly luciferase activity in *per-* or *tim-luc* transgenics, expressed either in flies (e.g., Plautz *et al.*, 1997b) or cultured *Drosophila* cells (Darlington *et al.*, 1998). However, β-GAL, translated *Escherichia coli lacZ* mRNA transcribed from *per* 5′-flanking fusion genes (Figure 10; see color insert) stays flat throughout the day and night (Dembinska *et al.*, 1997; Stanewsky *et al.*, 1997a). β-GAL-encoding mRNA cycles, even when the only *per* material fused to *lacZ* is the fly gene's 5′-flanking region (Zwiebel *et al.*, 1991). However, in transgenics designed to encode β-GAL downstream of the N-terminal half of PER, the antigenicity or catalytic activity of the reporter material did not exhibit daily fluctuations (contra Zwiebel *et al.*, 1991). Only when the bacterial enzyme was fused downstream of two-thirds of PER or very near the clock protein's C-terminus did this reporter show cycling (Dembinska *et al.*, 1997); the latter PER-fusion protein, but not the N-terminal two-thirds one, also allowed *per*[0] flies to exhibit weak behavioral rhythmicity (Stanewsky *et al.*, 1997a).

Whatever the fate of given protein, produced under the control of a fusion construct that is partly composed of clock-gene sequences, the mRNA cyclings observed when only the 5′-flanking regions of *per* or *tim* are present imply that transcriptional regulation is sufficient to mediate the molecular cycles insofar as *cis*-acting DNA is concerned. Direct measurements of temporally fluctuating transcription rates for *per* and *tim* mRNA confirmed these inferences from the fusion-gene transformants. However, the upswing and peaks of these rates occurred in advance of those for the steady-state mRNA levels by 2–4 hr (So and Rosbash, 1997). Consistent with these findings were those stemming from comparisons of reporter mRNA timecourses with those of native *per* mRNAs in "5'-flanking-only" transgenics: The half-risetime for the former transcripts was considerably earlier than for those encoded by the endogenous gene (Brandes *et al.*, 1996). A faithful representation of normal *per* cycling by oscillating steady-state levels of reporter mRNA required fusion of 5'-flanking sequences of this clock gene to two-thirds of the *per* coding sequences (Stanewsky *et al.*, 1997b); the latter were in turn fused to (downstream) reporter material. The same result (allegiance to the cycling dynamics of endogenous *per*) was obtained when the gene's 5'-flanking sequences along with its large first intron (Figure 9) were fused to *luc* (Stanewsky *et al.*, 2002). Thus, factors acting on a portion of the endogenous *per* mRNA influence the dynamics of its accumulation, once again indicating control of clock-gene product oscillation at a posttranscriptional level. It is as if transcription-rate cycling works together with regulation of product levels and their posttranslational qualities to tune the amounts of these *per*-promoted molecules at a given time of day or night. Only then would the varying amounts of the final products allow for normal biological rhythmicity in terms of phase and free-running period.

   This supposition is consistent with the chronobiological effects of truncated or otherwise *in vitro*-mutated *period* transgenes. It had long been known that incomplete forms of *per*, in terms of sequences flanking the coding region on either side, lead to mediocre rescue of per^0-induced arrhythmia: Proportions of rhythmic individuals, their free-running period, or both are abnormal (e.g., Bargiello and Young, 1984; Zehring *et al.*, 1984; Hamblen *et al.*, 1986). These transgenes were constructed in advance of much or any knowledge about the *per* transcription unit, simply to ask whether "any" DNA from the locus would possess biological activity. Remarkably, several sequences that were biologically active lacked *all* 5'-flanking material and a substantial proportion of *per*'s large first intron (the first exon is noncoding). Yet most of these transgenes, no matter where they got inserted in the genome, allowed for reasonable rhythmicity in a *per*-null genetic background, although only ca. 40–70% of the individual flies were periodic behaviorally, and their "weak" free-running periods tended to be 1–3 hr longer than normal (e.g., Liu *et al.*, 1991).

   These results implied the presence of positive regulatory sequences within the first intron or downstream of it. That such sequences can exist within the transcription unit itself, possibly in the midst of coding material, was revealed in a transgene-mediated dissection of *per* "dosage compensation," aimed at interrogating parts of the gene that influence its apparently higher level of "average" expression on a *per*-dose basis in males compared with females. Tentative answer: *Cis*-acting dosage-compensating sequences are present within intron 1 and within the 3' half of this _X_-chromosomal locus (Cooper _et al._, 1994). These tests of altered _per_ transgenes included application of a construct from which only the first intron was deleted. In this situation, meaning that the 5'–flanking region is intact, the transgene had been previously shown to function in a mediocre manner with regard to basic robustness of _per_ expression: Individual flies were more frequently and strongly rhythmic compared with cases involving removal of the 5'-flanking region along with most of if the first intron (see prior discussion and Figure 9), but their periods were 2–4 hr longer than normal (Liu _et al._, 1991). Largely similar results were obtained when all _per_ introns were missing, in cases of bioassaying cDNAs fused to 5'-flanking sequences (Citri _et al._, 1987; Cooper _et al._, 1994). These behavioral tests were performed in conjunction with assessing the significance of alternatively spliced forms of the primary _per_ transcript. (Note that the hypothetically different isoforms of _Clk_ mRNA, cited earlier and described in the legend to Figure 3, have not been examined as to their biological meaning.)

Analysis of _per_ cDNAs led to the inference that this gene produces three PER isoforms (Citri _et al._, 1987), although the corresponding mRNAs could not be detected in a later study (Cheng _et al._, 1998). Instead, two splice forms were found that differ only by way of an alternative intron in the 3' untranslated region (UTR) of the mRNAs (depicted in Figure 15; see color insert). Locomotor assays of transgenics expressing a 3'-unspliceable form of the gene or one in which this intron had been already removed (as their only source of PER) led to free-running periods that were 0.5–1 hr too long compared with the behavioral consequences of a transgene that produced both mRNAs (Cheng _et al._, 1998). Transcript cycling for each type (called A and B') seemed the same, although accumulation of A-encoded PER was slower throughout the normal risetime (at night) compared with that translated from the B' form.

Deleting must of _per_'s 3' UTR suggested that it was dispensable for "normal" rescue of per^0-induced arrhythmia (Chen _et al._, 1998), although these results are not thoroughly consistent with earlier ones (Bargiello _et al._, 1984; Hamblen _et al._, 1986). Moreover, the more recent study was hampered by usage of transgenes containing a rather short 5'-flanking region; this pseudo-rescues arrhythmicity by mediating ca. 27-hr behavioral cycles (Baylies _et al._, 1992), whereas a longer upstream region consistently yields periods just slightly longer than those of wild-type adults (e.g., Citri _et al._, 1987; Cooper _et al._, 1994; Cheng _et al._, 1998). In any case, replacement of the 3' UTR by mRNA sequences encoded by a _Drosophila_ tubulin gene led to periodicities ca. 3 hr shorter than control values; correspondingly, it seemed, PER accumulated during an earlier-than-normal segment of the night (Chen _et al._, 1998). This reinforces the notion that _per_'s 3' UTR influences translatability of the mRNAs, which normally might be rather delayed, and differentially so with reference to the naturally occurring type B' and type A transcript forms just described.

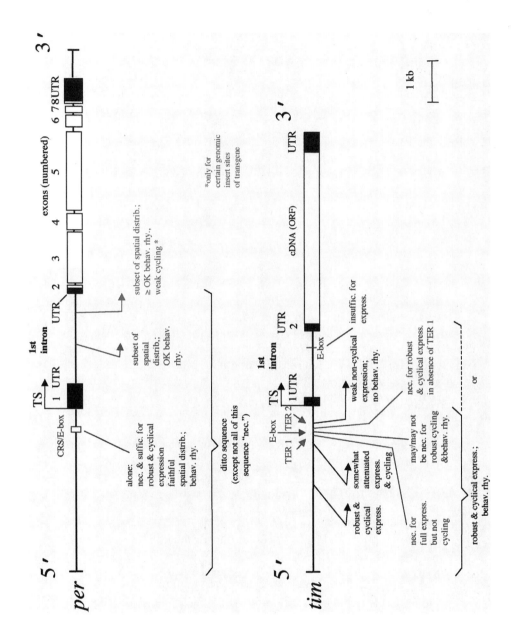

Figure 9. Dissection of gene-regulatory roles played by noncoding sequences within the *period* and *timeless* loci of *D. melanogaster*. *per*- and *tim*-derived transgenes are diagrammed at the top and bottom, respectively. For *period*, intragenic regulatory regions are shown with respect to a ca. 13.3-kb transgene, which, in a *per*[01] genetic background, restores near-normal behavioral rhythmicity (reviewed by Hall and Kyriacou, 1990, which refers to reports in which this arrhythmicity-rescuing DNA fragment is noted to be "13.2 kb" long, even though it contains about 100 more base pairs than that amount). Portions of this 13.3-kb DNA fragment that are not transcribed into mRNA are depicted by the thin segments of line, including, for example, the ca. 4.2 kb of DNA that flanks the transcription-start (TS) site on its 5' side, and the ca. 2.3-kb first intron, which is just downstream of (in the 3' direction from) exon 1 (the other introns, farther downstream, are much shorter). This and the other exons are represented by the numbered rectangles. Exon 1, the 5' (ca.) one-half of exon 2, and the 3' majority of exon 8 are "noncoding," that is, untranslated regions (UTRs) within the *per* mRNA. The TS site is indicated by the arrow pointing rightward (in the 5'-to-3' direction); and the 5'-flanking DNA upstream of this site typically contains gene-regulatory sequences (at least the "promoter," not far to the left of TS). For this clock gene, a 69-bp circadian-regulatory sequence (CRS) is located ca. 530 bp upstream of TS; CRS contains an 6-bp E-box sequence and is able (as well as required) to mediate proper temporal and spatial expressions of *per* coding sequences (the open-reading frame, or ORF, contained within exons 2–8), as is indicated below the diagram (also see text). These CRS/E-box manipulations are reviewed by Kyriacou and Rosato (2000). Given the results of such experiments, it should be the case (and is) that—when the entire 5'-flanking region, along with exon 1 and the first intron, are contained in a *per*-derived transgene—normal gene-regulation results; hence, "ditto," referring (for example) to the fact that the bracketed region fused to the ORF effects the *per*[01] behavioral rescue referred to above and drives daily cycling of molecular reporters (reviewed by Hall, 1998b). Curiously, however, neither the entire bracketed region nor any of it is required for partial such rescue (and crude mimics of the temporal/spatial patterns indicated); these results are denoted by the rightward-pointing arrows below the diagram, referring to an 8.0-kb *per*+-derived transgene that works as stated (e.g., "OK behavioral rhythmicity) and to an even smaller 7.2-kb DNA fragment that does as well (see Figure 4B), with the proviso that this transgene got inserted in a salutary chromosomal location, as described in the text and indicated here by the asterisked remark; the 7.2-kb *per* fragment (whose 5' end is at the rightmost, rightward-pointing arrow below the gene line) contains 338 bp of the first intron. For the *timeless* regulatory case, the conventions in the diagram at the bottom are like the *per* lines and rectangles. Here, however, a ca. 4-kb 5'-flanking region, noncoding exon 1 (leftmost black rectangle), the 2-kb first intron, and part of exon 2 are shown fused to *tim* cDNA sequences that are all ORF (none of the short downstream introns are present). This *tim* 5'-flanking/first intron construct was used to good effect behaviorally (rescue of *tim*[01]-induced arrhythmicity, when fused to the cDNA) and molecularly (mediation of reporter cycling, when fused to *luc*) by Rutila *et al.* (1998a) and Stanewsky *et al.* (1998), respectively. In another study (Wang *et al.*, 2001), 4.3 kb of 5'-flanking material (additional 300 bp added to the left end of the line shown) were fused either directly to the cDNA (i.e., first intron not included) or to *tim* genomic DNA (first intron present); both constructs rescued *tim*[01] in terms of behavior (Ousley *et al.*, 1998), but only the second one allowed for appreciable transcript cycling, compared with a situation in which the only putatively gene-regulating motifs present were in the 5'-flanking region. Above the line, specific such regulatory sequences are denoted: an E-box, flanked by 11-bp *tim*-E-box-like repeats (TERs); their roles are stated by the passages near the bottom (also see text). Unlike the case of *per* regulation, a portion of *tim*'s 5'-flanking region (including the TERs and the E-box) is required for at least quasi-normal temporal regulation and behavioral rescue of a *tim*-null mutation—the same rescue and daily cycles of molecular expression that occur when the entire 5'-flanking region (or it plus farther-downstream regions 5' to the ORF) are fused to the cDNA or to reporter sequences (see remark accompanying the horizontal bracket at the bottom). Elements of the results stemming from analyses in which only relatively 3' portions of the regulatory region were present in the relevant transgenes are stated in passages that accompany the rightward-pointing arrows below the line, and in a separate passage referring to the fact that an E-box within the first intron of *tim* mediated no expression of reporter sequences in transfected, cultured *Drosophila* cells (see text).

Returning to consideration of *per* transgenes truncated at the other end, a severely boiled down construct devoid of 5′-flanking DNA and containing only ca. 15% of the first intron (Figure 9) failed to rescue *per*[0]s arrhythmicity from most of the genomic locations were it got inserted (Hamblen *et al.*, 1986). It was as if transcriptional control had finally collapsed, such that this 7.2-kb transgene (Figures 4 and 9) had to land near a "random" promoter sequence, which should be present at some of the chromosomal insertion sites, in order to be expressed at all. This did occur for two such genomic locations and allowed the 7.2-kb of DNA, which includes all *per*-coding sequences, to mediate now-familiar long-period rhythmicity (Figure 4B); these behavioral cycles were also not that robust, in that the 7.2 transgenic flies did not exhibit very sharp distinctions between phases of activity versus rest (Hamblen *et al.*, 1986; Frisch *et al.*, 1994; cf. Figure 1A). Nevertheless, one imagined that *per* products encoded by this transgene would cycle, if not robustly, and both the "7.2-encoded" mRNA and protein did do (Frisch *et al.*, 1994). However, the regulation of these oscillating molecules appears to be *entirely* posttranscriptional (So and Rosbash, 1997). Thus, systematic fluctuation of mRNA degradation could be responsible for low-amplitude peak versus trough levels of the transgene's products over the course of a cycle. Once the production of PER is cyclical in flies of this genotype—given the mild oscillation in transcript abundance, albeit not its synthesis—the protein would go up and down simply by being turned over in a temporally constant manner.

However, the degree to which PER is phosphorylated fluctuates systematically (Edery *et al.*, 1994b), so even if the relevant proteolytic factors (Naidoo *et al.*, 1999) are functioning at the same level all the time (but not really; cf. Kloss *et al.*, 2001), the *turnover rate for a clock protein could vary* as the daily cycle progresses. There is some meaning to this possibility. Here, we revisit a fusion-transgenic type previously designed to overexpress PER and ask whether it acts in a basically negative manner (Zeng *et al.*, 1994). Further-modified forms of the relevant rhodopsin "promoter" (*Rh1*)–*per* fusion construct were applied to show that the transgene-encoded *per* mRNA cycles in photoreceptors maintained in LD (and, remarkably, that this is controlled by the *Rh1* regulatory sequences), but *per* transcript levels were temporally flat in DD (Cheng and Hardin, 1998). Nonetheless, the protein encoded by this *Rh1*-promoted *per* mRNA was able to go up and down over the course of two free-running cycles—a would-be circadian oscillator that operates molecularly at the *posttranslational* level alone. However, the PER-abundance fluctuations in this transgenic case were neither inherently robust nor phased like the wild-type protein cycling (Cheng and Hardin, 1998). Perhaps this caricature of normal molecular rhythmicity would be unable to mediate a biological cycle, but these investigators were not on that hook, because the *Drosophila* eye remains investigatorily barren in terms of any chrono*biology* that can be tied to expression of clock genes in this tissue.

To test further the behavioral regulatory capacities of constitutive *per* expression (Frisch *et al.*, 1994), Yang and Sehgal (2001) drove that gene's coding sequences with promoter regions of an actin gene and of one called *embryonic-lethal/abnormal-visual-system* (*elav*). Both drivers were presumed to be active in a non-temporally varying manner (this was confirmed by *in situ* hybridization for certain of the brain neurons to be mentioned later). The actin regulatory sequence is likely to be expressed in a spatially ubiquitous manner, that of *elav* in all neurons and only them (Yao and White, 1994). These regulatory sequences had been fused to those encoding yeast GAL4, the by-now-familiar transcriptional activator taken from yeast. Flies carrying *Act5C-gal4* or *elav-gal4* transgenes were crossed to ones newly made by Yang and Sehgal (2001), whose transgenes had GAL4-activatable UASs fused to either *per* or *tim* coding material. Behavioral tests of the various doubly transgenic types revealed appreciable free-running rhythmicity of flies whose genetic backgrounds were (as appropriate) per^{01} or tim^{01} (Yang and Sehgal, 2001). However, the percentages of locomotor-rhythm rescue were highly variable (depending, for example, on the UAS-clock-gene subline applied). The proportions of rhythmic individuals ranged down to zero, and the average periods were all over the place (between 20 and 28 hr among the groups of flies carrying a given *gal4/UAS* transgene combination). A histochemical corollary was obtained for certain of these arrhythmia-rescued flies: Rhythmic *elav-gal4/UAS-tim* individuals were re-entrained to LD cycles, then returned to DD; they exhibited cyclical variation of TIM immunoreactivity in certain brain neurons that normally express clock genes (described in Sections VI.A.1 and VI.B); in other such neurons the apparent level of this protein remained temporally constant (Yang and Sehgal, 2001). Before we delve into the details of these intrabrain expression patterns, we can extend the conclusion that anomalously constitutive primary expression of the *per* or *tim* clock genes can nevertheless be associated with cyclical variations of the final products (albeit not in all cellular locations) and that this is sufficient for one kind of biological rhythmicity (however unimpressive such behavioral cycles might be).

Remember that overexpression of *per*, mediated by application of eye-expressing transgenes, knocks out that gene's mRNA cycling in this tissue (Zeng *et al.*, 1994). Such results did not have biological significance, in the same way that rescue of PER eye cycling in the absence of transcript fluctuations so lacked (Cheng and Hardin, 1998). It was therefore useful to attempt overexpression of *per* and *tim* in the brain. Various *gal4* drivers (including those containing *per* or *tim* promoter regions) were combined with UAS-*per* or UAS-*tim* transgenes (as appropriate) in flies that carried the normal alleles of these genes. The various results included substantial attenuations of locomotor rhythmicity (in terms of proportions of periodically behaving individuals) and of eclosion cycles (Kaneko *et al.*, 2000b; Blanchardon *et al.*, 2001; Yang and Sehgal, 2001). These biological results were correlated with relatively high and temporally nonvarying expression of PER and

TIM proteins in certain lateral-brain neurons—those deemed significant for the control of circadian behavioral rhythms on other criteria that had been established previously (Section VI.A).

The final molecular subtopic within this section focuses on specific nucleotide sequences contained within the DNA that flanks the core of certain clock genes on their 5′ side, and it brings our attention back to transcriptional control. For this, let it first be reiterated that the positively acting CLK:CYC heterodimer appears to bind (Darlington *et al.*, 1998), or empirically does (Lee *et al.*, 1999), to a certain subset of the 5′-flanking regions of *per*. The sequence motif in question contains an E-box (Figure 9), a stretch of six consensus nucleotides that is a target of bHLH-containing transcriptional regulators in general (reviewed by Kyriacou and Rosato, 2000). Both *per* and *tim* possess E-boxes within their upstream regulatory regions (e.g., Hao *et al.*, 1997; Darlington *et al.*, 1998). For *per*'s part, and E-box is located about 500 bp upstream of the transcription-start. This 6-mer is embedded within larger "circadian transcriptional enhancer" (Figure 9), a.k.a. clock-regulatory sequence (CRS; Hao *et al.*, 1999). It was defined by performing mRNA timecourses on a series of engineered transgenes whose most salient experimental feature involved the presence or absence of various portions of 5′-flanking *per* DNA (Hao *et al.*, 1997). The CRS was thus found to be 69 bp long and is sufficient to drive normal levels and cycling of transcripts emanating from sequences fused to it (with the proviso that the CRS is juxtaposed with a heterologous "basal" promoter). The E-box located in the middle of the 69-bp sequence (Figure 9) was found to be necessary for quasi-normal levels of (reporter) mRNA, but not for transcript cycling (Hao *et al.*, 1997). Therefore, the *circadian*-enhancer sequences map elsewhere within the 69-bp DNA segment.

How well the *per* CRS and analogous material located to the 5′ side of the *tim* ORF (Figure 9) perform in their control of biological rhythmicity will be taken up later (including within Section VI, the next major one). Meanwhile, consider that the E-boxes of *per* and *tim* and their environs have been manipulated to ask further questions about the complexities of transcriptional regulation. *In vitro* mutageneses of sites 10 bp downstream of *per*'s E-box reduced transcript levels and knocked out their cycling in transgenic flies, whereas changes made over the same distance in the opposite direction had little effect (Lyons *et al.*, 2000). These results were largely consistent with earlier ones involving short deletions made on one side or the other of this E-box (Hao *et al.*, 1997). Another previous finding showed that multimerizing that 6-mer along with six bases to either side of it (which are present in the wild-type sequence) seemed sufficient for CLK and CYC to activate *per*-promoted reporter expression in transfected cells (Darlington *et al.*, 1998). Transforming such constructs into flies showed that four tandem arrays of the same 18-mer mediated high levels of cycling reporter-gene expression (Darlington *et al.*, 2000). Given that some of the apparently essential bases are located 3′ to the 18-bp stretch (Lyons *et al.*, 2000), these two sets of results are partly contradictory. Perhaps multimerization of the region in question compensates for sequences missing from Darlington and co-worker's artificial construct (as discussed further by Kyriacou and Rosato, 2000). In any case, these findings speak once again to the interaction of *trans*-acting factors, involved in turning on *per* and making its transcript cycle, with more than the E-box alone. Such a minimalist view is counterintuitive anyway, given that E-boxes are all over the *Drosophila* genome.

One or more of them is present within the *timeless* gene (Figure 9). These and other noncoding sequences at the locus have been manipulated; the results again revealed the actions of multiple *cis*-acting elements. However, one of the three studies of this sort made rather a crude start at delving into the potential regulatory complexity. For this, fusions of 5′-flanking *tim* sequences were made to *lacZ* (Okada *et al.*, 2001). Approximately 2,000 or 1,000 nucleotides upstream of transcription-start site (Figure 9) was found to be sufficient to mediate reporter mRNA cycling that had been brought under *tim* control, although the overall transcript levels were reduced in the *tim*(1000)-*lacZ* case. [The RNA was extracted from whole flies (Okada *et al.*, 2001), rarely done in these kind of experiments.] Both of these constructs contained an E-box located ca. 680 bases upstream of the transcription-start (Figure 9). In contrast to the expression of sequences driven by analogous *per*-reporter fusions in which the mRNA contains no PER-coding sequences, these two types of *tim* fusion transgenes produced cycling transcripts that *somewhat* faithfully tracked the timecourse of *tim*+ RNA dynamics (see later). However, in one of the four *tim*(2000)-*lacZ* and *tim*(1000)-*lacZ* lines the reporter mRNA exhibited an earlier-than-normal rise (Okada *et al.*, 2001), unlike what has been observed in analyses of upstream-only-*luc* fusions that lack the first intron of *per* (Brandes *et al.*, 1996; Plautz *et al.*, 1997b; Stanewsky *et al.*, 1997b, 2002). Deleting ca. 500 nucleotides from the 5′ end of the 1-kb *tim* sequence (Figure 9) abolished reporter mRNA cycling, which stayed constitutively at the trough level (Okada *et al.*, 2001), similar to a result obtained by McDonald *et al.* (2001). The upstream E-box (Figure 9) was removed in the *tim*(500)-*lacZ* construct of Okada *et al.* (2001).

However, that feature of the tactic and result is minimally interpretable, as was revealed in a contemporary set of experiments. Here, base-pair substitutions were engineered within the upstream E-box or these *tim* sequences were deleted (McDonald *et al.*, 2001). Such changes, effected in a ca. 760-bp promoter region fused to ca. 2 kb of UTR *downstream* of the transcription-start (Figure 9) and then to a complete *tim* coding sequences (as cDNA), left the *in vitro*-mutated transgenes functional in two ways: (i) The flies exhibited free-running rhythms 3–4 hr longer than control values; these were 26 hr when the 760 nucleotides worth of 5′-flanking material was left intact (obviously, the genetic background of the behaviorally tested transgenic types was tim^{01}). (ii) The strengths if *tim* (transgene) expression and reporter cyclings, observed in parallel types of *tim-luc* fusions in a tim^+ background, were correlated with those of the behavioral rhythmicities. Therefore, McDonald *et al.* (2001) had their attention drawn to regulatory motifs other than the upstream E-box Among the four or so candidates are two 11-bp *tim* E-box-like repeats (TERs) located a few base pairs upstream or a few dozen downstream of the canonical E-box (Figure 9). McDonald *et al.* (2001) engineered *luc*-reporter transgenics in which one or more of the TERs were deleted (in some cases accompanied by removal of other nearby sequences) and showed that they, along with the core E-box, function in *tim*'s transcriptional activation and cycling control (Figure 9).

 timeless-based molecular and behavioral results both similar to and different from those just summarized were obtained in further promoter-bashing experiments (Wang *et al.*, 2001). In these tests, a quadruple mutant engineered within the upstream E-box was *unable* to rescue tim^{01}-induced locomotor arrhythmicity (Figure 9). The base construct type her involved fusing 4.3 kb of DNA taken from upstream of the transcription-start *directly* to *tim* cDNA (Figure 9); when these regulatory sequences were left normal, 65% of the transgenic individuals were rhythmic (consistent with behavioral results reported by Ousley *et al.*, 1998), even though this control construct type exhibited no appreciable cycling of cDNA-encoded *tim* transcripts (Wang *et al.*, 2001). This result contradicts those of Okada *et al.* (2001), who reported that, in their *tim*(2000)-*lacZ* and *tim*(1000)-*lacZ* lines, 1 or 2 kb of 5′-flanking sequences (alone) was sufficient for RNA cycling [although with an anomalously early risetime and lower-than-normal amplitudes in some of these transgenic strains; perhaps there was body cycling of the reporter RNA (see above), and noncycling in the head (Wang *et al.*, 2001) brought the overall amplitude down].

 Returning to the behavioral side of Wang and co-workers' experiments, their rescue of tim^{01}'s arrhythmicity was much better when the 4.3 kb of 5′-flanking material was fused to intron-containing genomic *tim* DNA (cf. Stanewsky *et al.*, 2002), and cycling of mRNA emanating from this more complete construct was robust (Wang *et al.*, 2001). These investigators, along with McDonald *et al.* (2001), noticed a *tim* E-box within *tim*'s first intron (Figure 9), which was present in the upstream UTR used for the constructs applied in the latter study. The potential regulatory meaning of this intronic E-box was interrogated by transfecting a intron-*luc* construct, along with CLK-encoding sequences, into the aforementioned kind of CYC-constitutive cultured cells. However, no activation of reporter expression was observed (McDonald *et al.*, 2001; Wang *et al.*, 2001), compared with positive results obtained previously with *tim-luc* fusions containing several kilobases of 5′-flanking DNA or a multimer of 18 E-box-containing base pairs taken from this upstream (nontranscribed) portion of the gene. The negative results from applying the intron alone (no *tim*-like transcriptional activation) are consistent with those relatively 5′-sequences influencing *timeless* gene cycling at a posttranscriptional level (discussed by Stanewsky *et al.*, 2002). Finally, protein-DNA-binding assays, of the sort described earlier in analyses of 5′-flanking *per* DNA (Section V.C.1), were performed; both CLK and CYC could be detected as bound to a 59-bp *tim* sequence within which the upstream E-box is embedded (Wang *et al.*, 2001).

This result, involving as it does a very boiled down subset of the gene-regulatory material for *timeless*, harks back to the simple scheme presented in Figure 7—the part of it that shows only the CLK and CYC transcriptional activators associated at low resolution with *tim*'s 5′-flanking DNA. However, the ensemble of results from McDonald *et al.* (2001) and Wang *et al.* (2001) indicate that noncoding material contained within this gene needs to be conceptually broken down into several *cis*-acting regulatory components (Figure 9). Implicitly, *trans*-acting factors other than CLK and CYC are likely to be involved. In this regard, it may turn out to be significant that the 5′-flanking region of time contains a sequence similar to a cAMP-response element and a "half-site" for potential binding of a transcription-factor type known as PAR-bZIP (Okada *et al.*, 2001). The first of these putatively *cis*-acting sequences is located about 800 bp upstream of *tim*'s 5′-flanking E-box; the second (half PAR-bZIP) one is flanked by the 11-bp TER1 sequence (Figure 9) identified by McDonald *et al.* (2001) and the upstream E-box (Figure 9). In fact the latter investigators found that removal of 6 bp located between TER1 and this E-box (i.e., a deletion of the half-PAR-bZIP sequence) caused a further decrement of transcriptional activation beyond the ca. 20-fold decrease in *tim*'s overall expression level inferred to be caused by a TER1 deletion alone.

VI. PLACES AND TIMES OF CLOCK-GENE EXPRESSIONS

A. Temporal and spatial expression of *period*

1. *per* products in embryos through adults

Manipulation of regulatory sequences within genetic loci encoding clock factors could serve purposes beyond the dissection of transcriptional and cyclical control of these genes. *Cis*-acting factors involved in activating them at the appropriate stages of the life cycle and in tissues relevant to their chronobiological functions could also be established. Moreover, changes in or deletions of such regulatory sequences could lead to subsets of the normal expression patterns, which might be correlated with the presence or absence of rhythmic phenotypes.

First, the normal patterns of clock-gene expressions needed to be established. Studies of *per* are farthest along in this respect. The products of the normal form of this gene are expressed at essentially all stages of the life cycle (Table 4). *per* mRNA is detectable as early as the second half of the embryonic period

Table 4. Spatial and Temporal Expression Patterns of the *period* and *timeless* Gene Products[a]

Life-cycle stages	Tissues expressing	Remarks
per		
Embryo	Brain	Signals (here and in VNC) first appear ca. halfway through embryogenesis
	VNC	Ventromedia portion of each ganglion
	Salivary gland	Observed in one study, not others
Larva	Brain	Dorsal, lateral (relatively dorsal), and ventral regions
	VNC	Thoracic and abdominal ganglia (latter: examined by reporter only)
	Epidermis	Observed only in one reporter combination
	Mouthparts	Anti-PER: negative
	Ring gland	Mainly in corpus allatum; anti-PER and some reporter combinations: negative
	Salivary glands	Observed with some anti-PER's, not all; also positive in Northern blot of dissected tissue and in some reporter combinations
	Gut	Observed in three reporter combinations
	Malpighian tubule	Observed only in one reporter combination
	Trachea	Observed only in one reporter combination
	Spiracles	Observed only in one reporter combination
Pupa	Brain	Not all LN and DN cell groups exhibit (reporter) signals at all stages (e.g., limited to *small*-LN precursors early)
	Optic lobes	—
	VNC	—
	Ring gland	—
	Prothoracic gland:	
	Ovary	—
	Testis	—
Adult	Brain	Five or six discrete clusters of lateral and dorsal neurons; also many glia
	Optic lobes	Glial cells only, although centrifugal projections from certain LNs prominent
	VNC	Glia only by anti-PER; no such signals with reporter combinations: certain neuronal cell bodies and fibers instead
	Eyes	Compound eyes and ocelli
	H–B eyelet	—
	Proboscis	—
	Antenna	Signals associated with chemosensory structures
	Widely distributed cuticular structures	Near sensillae (bristles, hairs) on head capsule, thorax, abdomen
	Corpora cardiacum	—
	Salivary glands	—
	Gut	All regions of alimentary canal, from esophagus to rectal papillae and including thoracic crop and cardia [certain epithelial cells thereof (Kaneko *et al.*, 2000), but *not* "cardiac" (heart) cells as tabulated in that report]
	Fat body	—

continues

Table 4. *continued*

Life-cycle stages	Tissues expressing	Remarks
	Wing	Signals associated with chemosensory structures, especially at wing margin
	Legs	—
	Malpighian tubule	
	Female reproductive system	Ovarian follicle cells (not observed with some reporter combinations), uterus, seminal receptacle (SR), spermathecae; anti-PER: negative for SR and uterus; intracellular follicular signals cytoplasmic only, unlike nuclear PER in essentially all other cell/tissue types examined
	Male reproductive system	Testis (base of), seminal vesicle, accessory gland (AG), ejaculatory bulb (reporter signals); anti-PER: negative for AG; PER cycles in testis and SV (similar phase to head cycling)
tim		
Embryo	Brain	Assessed by *in situ* hybridization only
	VNC	—
Larva	Brain	Dorsal and lateral regions
	VNC	Very limited (posterior) expression and very likely nonspecific (same anti-TIM signals observed in *tim*^{01})
	Epidermis	Observed only in one reporter combination
	Mouthparts	—
	Ring gland	Mainly in corpus allatum; anti-PER and some reporter combinations: negative
	Gut	Observed in three reporter combinations
	Malpighian tubule	Observed only in one reporter combination
	Fat body	—
	Trachea	Observed only in one reporter combination
Pupa	Brain	Tendency for wider expression compared with *per*, e.g., late-larval DN clusters that are *per*-negative express *tim* (reported) in early pupae
	Nonneural tissues:?	Not examined
Adult	Brain	Five or six discrete clusters of lateral and dorsal neurons; also many glia
	Optic lobes	Glia only, although certain LN projections present
	VNC	Most expressing cells glia; possibly a few neurons
	Eyes	Compound eyes; ocelli and H-B eyelet not examined
	Proboscis	—
	Antenna	Signals associated with chemosensory structures
	Widely distributed cuticular structures	Near sensillae (bristles, hairs) on head capsule, thorax, abdomen
	Gut	Not examined in as much stem-to-stern detail as for *per*
	Fat body	—
	Wing	—
	Legs	—
	Malpighian tubule	—

Table 4. *continued*

Life-cycle stages	Tissues expressing	Remarks
	Female reproductive system	Ovarian follicle cells, uterus, seminal receptacle, spermathecae
	Male reproductive system	Testis, seminal vesicle, accessory gland, ejaculatory bulb; TIM cycles in testis and SV (similar phase to head cycling)

[a] The findings obtained before 1990 perforce involved *per* only. Those results (involving embryos, larvae, and adults) are cited and tabulated in Hall and Kyriacou (1990), and elements of them are retabulated here. Such data were obtained from *in situ* hybridizations, application of more than one kind of anti-PER, and by β-GAL activity encoded by a *per-lacZ* fusion transgene. With regard to stages of *per* expression, Northern blottings were also performed (augmented by such RNA-extract tests of salivary glands and ovaries, dissected from larvae and adults, respectively). The results from these separate kinds of *per*-expression assessments (and in some cases those from different investigators) were not always in agreement [as implied in certain of the table entries; see Hall and Kyriacou (1990) for further details]. More recent gene-expression studies have dealt with *per* more than *tim* (as implied by entries in the second half of this table and noted there under Remarks). The *per* determinations have relied on applications of antibodies against native PER protein (e.g., Zerr *et al.*, 1990; Ewer *et al.*, 1992; Liu *et al.*, 1992; Frisch *et al.*, 1994; Emery *et al.*, 1997; Giebultowicz *et al.*, 2001; Beaver *et al.*, 2002) and of reporter transgenics. For the latter, *per-lacZ* fusions were again applied (e.g., Ewer *et al.*, 1992; Kaneko *et al.*, 1997; Stanewsky *et al.*, 1997a), as were *per-luc* ones (e.g., Stanewsky *et al.*, 1997b, Krishnan *et al.*, 2001; Beaver *et al.*, 2002). Many additional *in situ* data came from "reporter combinations" in which *per-gal4* and UAS-marker transgenes were simultaneously present (most of which findings are in Kaneko and Hall, 2000; also see Plautz *et al.*, 1997a). Within mid-stage embryos, the presence of *tim*-encoded mRNA was determined by J. Blau and M. W. Young (unpublished observations). Various anti-TIM's have been applied mainly to adult-head specimens (e.g., Hunter-Ensor *et al.*, 1996; Myers *et al.*, 1996; Kaneko and Hall, 2000), but also to the larval (Kaneko *et al.*, 1997) and adult (Kaneko, 1999) CNS as well as to the Malpighian tubules and male reproductive system of adults (reviewed by Giebultowicz *et al.*, 2001; also see Beaver *et al.*, 2002). Many of the results of this gene's expression in neural and (especially) nonneural tissues were reported by UAS-marker transgenes combined with a *tim-gal4* driver (Kaneko and Hall, 2000); additional *tim* reporting was based on *luc* sequences fused to *tim* regulatory ones (Krishnan *et al.*, 2001; Beaver *et al.*, 2002). The larva subsections tabulate *per* and *tim* results from the 3rd-instar (L3) stage (Kaneko *et al.*, 1997; Kaneko and Hall, 2000). Only *per* expression has been examined in L1 and L2, and solely in the CNS; for this, not all lateral-neuronal (LN) and dorsal-neuronal (DN) cell groups express *per* (or *per*-driven β-GAL) in early larvae (e.g., precursors of the pupal and small/ventral LN's show signals only in L1). Certain cases of noncongruence of these genes's expressions in L3 (implied by the two listings for this stage, derived from Kaneko and Hall, 2000) reflect the absence of signals for one gene but positive results for the other (e.g., fat-body expression for *tim* but not *per*). In the pupa subsections, more limited expression patterns are implied by the listings, but most tissues have not been examined at this stage, especially for *tim*. With regard to adult patterns, fewer tissues are listed for *tim* because they have not been examined. Based on more extensive studies of *per* expression in mature flies, the only major categories of tissue found not to contain *per* mRNA, PER protein, or *per*-driven reporter signals are subcuticular muscles and trachea (as summarized in Hall and Kyriacou, 1990; Hall, 1995; Giebultowicz *et al.*, 2001). Other chronobiologically important genes–such as *Clk*, *cyc*, and *cry*—have had their expression patterns examined histologically in such limited locations (adult head only) that these results are not worth tabulating. Rhythm-related genes such as *dbt*, *lark*, and *vri* are also developmentally vital ones and could therefore be expected to exhibit broad expressions in terms of stages and tissues; thus, these patterns (insofar as they have been determined) are not tabulated either (but see the text for some paltry information about *dbt* and more extensive stage/tissue findings for *lark*).

(Bargiello and Young, 1984) and located within a portion of all ganglia of the developing nervous system (James *et al.*, 1986). These results are consistent with assessments of the temporal and spatial distribution of both native PER (Siwicki *et al.*, 1988) and of a PER-β-GAL fusion protein (Figure 10) encoded by a transgene (Liu *et al.*, 1988). No function for PER during this early stage is known. It is possible that that protein begins to influence developmental timing (Kyriacou *et al.*, 1990a) as early as embryogenesis. The would-be aging effects of certain *per* mutations (Section IV.C.6) may also be a consequence of abnormal PER products (Klarsfeld and Rouyer, 1998) or their absence (Ewer *et al.*, 1990) during development, even for its earliest stages. Deleterious features of adult morphology caused by neural mutations in *Drosophila* can have an embryonic etiology (e.g., Pflugfelder and Heisenberg, 1995). In this sense the shortened lifespans of *per* mutants (observed by some) would be interpreted as viability decrements associated with anatomical problems that may arise from impaired development. This supposition is made against a background of the many morphological mutants in *Drosophila* that show enhanced mortality (Gonzalez, 1923). In fact, two anatomical abnormalities are known for the per^{01} mutant: Certain neurosecretory cells in the supraesophageal ganglion of mutant adults tend to be scattered in abnormally dorsal locations (Konopka and Wells, 1980), and pericardial cells of the pupal heart are anomalously shrunken and wrinkled (Curtis *et al.*, 1999). The developmental stages at which an absence of PER causes initiation of these anatomical defects have not been determinded, nor is it known whether they might contribute to making the animals somewhat sick and short-lived.

In this regard, it has been argued from the results of a *per*-mutant phenocopy experiment that circadian rhythmicity is connected with *Drosophila*'s well-being. That there could be "some intrinsic value of circadian rhythms" (Sheeba *et al.*, 1999b) is based on the fact that rearing strains of *D. melanogaster* (almost) forever in LL left the cultures robustly rhythmic for LD or DD eclosion, once the descendants of the starting strains emerged from their constant-light hell. This rhythmic character remaining solid in such per^{01}-phenocopying conditions implied that functioning of per^{+} and other clock factors had not fallen by the wayside during the 600 generations in question (Sheeba *et al.*, 1999b). However, maintenance of wild-type clock-gene capacity may not speak to the adaptive significance of rhythmicity per se, but instead to the other biological problems suffered by *per*-null mutants (not only putative lifespan decrements, which did not connect temporally with the nature of a given clock malfunction, but also anatomical anomalies in the dorsal brain and in cells associated with the heart).

Continuing with the matter of this gene's pleiotropic expression, a nonneural tissue was shown to produce per^{+} products in maturing embryos. This is the salivary gland (Bargiello *et al.*, 1987). In larvae, that tissue was also found to contain *per* mRNA (Baylies *et al.*, 1993), possess PER immunoreactivity (Bargiello *et al.*, 1987), and express a *per*-controlled reporter enzyme (Kaneko and Hall,

A   B

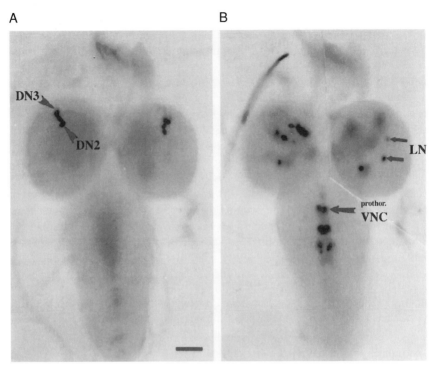

Figure 10. Larval CNS neurons of *Drosophila* marked by expression of a *per*-reporter transgene. A third-instar larva had its spherical brain lobes (near the top of each panel) dissected with the posterior ventral nerve cord (VNC) attached. The animal carried a fusion construct in which 5′-flanking regions of *period* along with approximately the 5′-half of the gene's coding region (cf. Figure 9) was fused to *lacZ* from *E. coli* (Liu *et al.*, 1988). The transgenic specimen was incubated in X-gal substrate, leading to deposition of a dark-blue reaction product that results from *lacZ*-encoded β-galactosidase (β-GAL) activity. (A) Dorsoanterior clusters of *per*-expressing brain neurons. Two types of cell bodies are pointed to, those within the DN2 and the DN3 groups (see Figure 12 for a diagrammatic view of all such larval brain clusters). It is the DNs that exhibit daily cycles of immunoreactivity in temporal monitoring of native PER fluctuations, compared with the staining cycles observed in DN1s (not shown here) or larval lateral neurons (see next panel). (B) Relatively ventral *per*-expressing neurons. The focal plane for the brain lobe on the left shows β-GAL-containing neurons in the ventral portion of the anterior CNS; the lobe on the right, which is in a more dorsal plane, shows a subset of the laterally located LN cell group (diagrammed more completely in Figure 12). Some of the *per*-expressing neurons near the midline of the VNC are in the (overall) focal plane, such as those pointed to within the anteriormost thoracic neuromere; cells in more posterior VNC regions (which stain as smaller clusters within abdominal neuromeres) are out of focus. Scale bar in A, 50 μm (same scale for B). This figure was adapted from one reported by Kaneko *et al.* (1997).

2000). These results were tacityl disputed by others (James *et al.*, 1986; Siwicki *et al.*, 1988; Liu *et al.*, 1988). Localization of native PER to the cell-boundary regions within salivary glands of third-instar larvae was once thought to connect with the effect of *per* mutations on intercellular communication among such cells (Bargiello *et al.*, 1987), before this issue crumbled like a piece halvah (Section IV.C.4).

Albeit unaccompanied by guesses as to their functional meanings, reported *per* expression has also been observed in seven additional different kinds of nonneural tissue (Table 4) during this last (L3) larval stage (Kaneko and Hall, 2000). In such animals, most attention has been paid to localization within the CNS of the gene's final product or of reporter activity carried by PER-β-GAL fusion proteins (Figure 10). Depending on the reporter construct applied, *per* was inferred to be expressed within a few small clusters of brain cells in L1 and L2 larvae, or within such anterior neurons *and* in cells distributed along the ventral nerve cord (VNC) of L2 specimens (Kaneko *et al.*, 1997). The former pattern was revealed by the aforementioned N-terminal-2/3-PER-βGAL fusion and best reflects the CNS expression of native PER; the latter, more extensive pattern was reported by β-GAL fused downstream of PER's N-terminal-1/2 (Figure 10). These differences between the transgenic types were maintained into L3, the only stage during which detailed assessments of PER itself were performed in terms of "where and when" expressed. Thus, both the native protein and PER-β-GALs were readily detected in four brain clusters of third instar larvae. In one such group of cells, which is located dorsally, and in ventral brain regions with weak signals, PER immunoreactivity was constant at different times of day; the same was shown for cells in the VNC (Figure 10). However, in the more intensely staining larval brain clusters—two of them dorsal, the other lateral—the staining exhibited daily cycling in LD and in DD. The peak was during the late nighttime (or subjective night) for one dorsal cluster and the lateral one, but occurred at the end of the day in the other dorsal cluster (Kaneko *et al.*, 1997).

The biological significance of certain *per*-expressing neurons cycling out of phase with other L3 brain cells is unknown. Nevertheless, it may be that circadian clocks are running in the larval CNS. Why? Might a central pacemaker influence the fluctuation of salivary-gland membrane potential (Weitzel and Rensing, 1981)? This is the only biological rhythm purported to be exhibited by larvae, which includes the fact that rhythmicity of larval locomotion could not be detected (Sawin *et al.*, 1994). Another possible function for a larval clock is that it is *keeping time* internally within the brain, but without elaboration into an overt rhythm. Such a presumption stems from demonstrations that pulsatile light exposures delivered to *Drosophila* as early as during L1 can cause cultures to eclose rhythmically, several days later (Brett, 1955). This light input to the developing animal's circadian system is likely to involve, in part, a structure called Bolwig's organ; certain axons projecting from it (in Bolwig's nerve, BN) terminate close to the lateral neurons (LNs). The initial such observations were made using brain preparations from third-instar larvae (Kaneko *et al.*, 1997), and it was subsequently shown that BN → LN contacts initially form during late embryogenesis (Malpel *et al.*, 2002). Examining the effects of visual mutations on light-to-clock inputs suggested that an "adult-blinding" *norpA* mutation (introduced in the foregoing text in various contexts) can eliminate synchrony of clock-protein cycles among LNs within the brain of third-instar larvae. The "can" hedge just stated means that this *norpA* dependence was not revealed unless a *norpA* mutation was combined with one at a *D. melanogaster* locus encoding a blue-light-absorbing substance called cryptochrome (introduced briefly in Sections II.A and IV.C.6; dealt with properly later in Section IX.B.3).

So, light *can* "get into the circadian system" of developing *Drosophila* and that this has chronobiological meaning (Brett, 1955) implicitly would involve "time memory," a process required to *keep a group of developing animals in synchrony* during the post-pulse time they are in DD. Otherwise, when they reach the final stages of metamorphosis, eclosion of a light-pulsed → DD culture could not be periodic. Time memory, moreover, can be maintained *through* metamorphosis: Again, subjecting L1 larvae to a light pulse, or delivering such stimuli during later larval stages, led to adults whose behavioral rhythms are largely in phase with one another, whereas "all-DD" culturing resulted in flies that behave in a periodic, but asynchronous manner (Sehgal *et al.* 1992). Certain features of these results are disputed by others (Dowse and Ringo, 1989; Power *et al.*, 1995a,b), but one component of Sehgal and co-workers' findings—the adult-synchronizing effects of L3 light pulses—was reproduced in a study that went on to ask whether functions of the *period* gene are involved in larval time memory (Kaneko *et al.*, 2000a). If that were the case, then a period-altering mutation would change the cycle duration of a clock that is keeping time, even though its functions are not output into rhythmicity as such during this formative stage. Indeed, it was strongly inferred that *per*^{S} larvae operate a time-memory clock which races ahead to a phase at which light pulses result in the opposite shift of the eventual adults' locomotor peak, compared with the behavioral phase-shifting effects of an equivalent stimulus delivered to wild-type L3 animals (Kaneko *et al.*, 2000a).

Clock genes should produce their products during the 4-day pupal period, given that several of the mutations that identified them were isolated by way of eclosion-rhythm abnormalities. The interplay between *Drosophila's* developmental state and clock function, as altered by application of *per*^{S}, have been analyzed (Qui and Hardin, 1996b). Indeed, *per* is expressed during metamorphosis (Bargiello and Young, 1984; Reddy *et al.*, 1984; Saez and Young, 1988). Production of this gene's products *in situ* (Table 4) has been assessed mainly within the CNS (Konopka *et al.*, 1994; Kaneko *et al.*, 1997; Kaneko and Hall, 2000). Thus, some of the prominent larval-brain clusters persist in their expression of this gene into and throughout the pupal period—the lateral group (LN) and two of the dorsal ones (DN). Two additional LN clusters arise about half-way through metamorphosis, along with three DN groups that appear either early or late in the pupal period (Kaneko *et al.*, 1997). Expression of this gene within DNs at these metamorphosing stages represents either a resurgence of product production (which is essentially not observed after L3 ends) or involves "new *per* neurons" (this quoted hedge referring to the pupal births of these cells or expression of the gene within preexisting neurons).

Transcripts complementary to *per* probes have been observed in the pupal ring gland (RG) by Saez and Young (1988), which presaged the gene's (marker-reported) expression in the larval form of this tissue (Kaneko *et al.*, 1997), but no reporter signals were observed in the third-instar larval RG by Kaneko and Hall (2000). An antibody against the PER protein permitted pupal expression in this glandular tissue to be assessed in more detail; daily oscillations of immunoreactivity within the RG's prothoracic gland (PG) were observed starting on day 2 of metamorphosis (Emery *et al.*, 1997). This molecular cycling may participate in the control of ecdysteroid synthesis in *Drosophila*'s PG (Baehrecke, 1996); this is a rhythmic process in other insects (e.g., Vafopoulou and Steel, 1991).

Various nonneural tissues have been monitored during metamorphosis for expression of a *per-lacZ* transgene. The reported signals appeared at different pupal stages: an early one in the rectal pads of the posterior gut; mid-pupae for the male's seminal vesicles; late pupae for the hindgut (the remainder of the rectum); and pharate adults for the Malpighian tubules (Giebultowicz *et al.*, 2001). During the late stages of the latter tissue's development, an "autonomous timer" was suggested to be operating, by virtue of an upswing in urate oxidase activity in the Malpighian tubule being synchronized with the time of adult emergence and that transplantation of this pupal structure into already emerged flies left the enzyme's appearance within the implanted tubule on its own schedule (Friedman and Johnson, 1977, 1978). Conceivably, the pupal presence of PER along with certain of its companion proteins influences the Malpighian tubule's late-developmental timing of this cell-differentiation event, although the clock-regulation logic proposed by Friedman and Johnson was disputed by Jackson (1978), in advance of any knowledge about whether a gene like *per* is expressed in this tissue during this metamorphosing stage.

But it is the posteclosion expression of clock genes in *Drosophila* (Table 4) has perhaps the highest potential for being experimentally connected to biological rhythmicity. In this respect, the various developmental stages during which the *period* gene is expressed could be reflected in brain damage to (eventual) *per*-mutant adults that are trying to behave periodically. In fact, there is some damage of this sort in the per^{01} mutant (Konopka and Wells, 1980). However, the persistent functioning of this clock factor in adults seems to be necessary for their locomotor rhythmicity, and such *per* expression is sufficient insofar as the various life-cycle stages are concerned. These conclusions stemmed from application of a construct in which a heat-inducible promoter (hsp) was fused to per coding sequences; activation of the gene during development did not allow per^{01}-bearing flies to behave rhythmically, whereas turning it on after development was over (and even after the per^{01}; hsp-per^{+} flies had proceeded from LD into DD) caused a rapid appearance of periodic behavior (Ewer et al., 1988, 1990). These free-running rhythms were weak and anomalously long, but this may have been because the only postinduction cycling of the clock molecules possible was that which is controlled posttranscriptionally. This is because per sequences carried by this fusion transgene were identical to those in the aforementioned 7.2-kb, 5'-flanking-less DNA fragment (Figures 4B and 9). The hsp-7.2 construct did exhibit weak PER cycling in situ, but only in cells that usually express the normal gene (Frisch et al., 1994), that is, against a staining background of constitutive expression elsewhere (when activated, hsp drives transcription in a spatially promiscuous manner). A second-generation hsp-per^{+} construct (designed to allow for higher inducibility of product expression) was applied in different kinds of experiments, whose results again suggested that the actions of this gene involve pacemaking function as such: Rapidly induced appearance of PER led to permanent phase shifting of the adult's locomotor rhythm, and the shifts obtained could not be accounted for solely by a heat-pulse PRC that was obtained from equivalent treatments of wild-type flies (Edery et al., 1994a).

Therefore, one would like to know from where the ongoing presence of per products regulates the flies' rest–activity cycles. For tissues within such adults (Table 4), among the early studies of this gene's expression were those based on in situ hybridization (Liu et al., 1988; Saez and Young, 1988), but little could be resolved in terms of where the signals were present, especially within the CNS. Better appreciations of the gene product's spatial distribution were obtained by applying per-lacZ transgenes and anti-PER reagents (e.g., Liu et al., 1998, 1991, 1992; Saez and Young, 1988; Siwicki et al., 1988; Zerr et al., 1990). Ganglia within the head express this gene in certain clusters of neurons; analogous to the larval pattern, the rather small number of cell groups is divided into lateral and dorsal groups, called LNs and DNs, respectively (Figure 11; see color insert). PER immunoreactivity and β-GAL-reported signals are also found in many small cells distributed throughout the neural cortex and neuropil of anterior CNS ganglia (e.g., Siwicki et al., 1988). And per-expressing cells of this sort are observed throughout the VNC of adults (e.g., Ewer et al., 1992). Further refinements of these patterns (Figure 11) showed that the LNs and DNs, which consist of two lateral groups and three dorsal ones, are indeed neurons (by coexpression of a neuronal-specific marker); the more widely distributed, smaller cells are glia (thus, no "per neurons" were detectable within the VNC); and PER immunoreactivity is most prominent within the nuclei of both cell types (Siwicki et al., 1988; Liu et al., 1992; Ewer et al., 1992; Frisch et al., 1994; Kaneko and Hall, 2000).

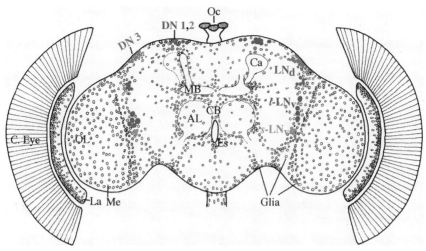

Figure 11. Expression of clock genes in the brain of adult *Drosophila*. This diagram is adapted from Helfrich-Förster (1996). A frontal view of the *D. melanogaster* brain is shown, such that dorsal is at the top. The distance between the distal regions of the optic-lobe (OL) complex is ca. 450 μm. The other basic structures labeled are the compound eye (C. Eye), simple-eye ocelli (Oc), lamina (La) and medulla (Me) optic lobes, dorsoposterior mushroom bodies (MB) and calyces (Ca) thereof, central body (CB, a.k.a. central complex; cf. Table 2), the ventroanterior antennal lobes (AL), and esophagus (Es, which runs through a relatively ventral brain region in an anterior → posterior direction). Cells expressing the *period* gene: dorsal neurons (DNs), which are located in relatively posterior brain regions (cf. Figure 13) and defined by two rather small and medial clusters plus the larger, more lateral DN3 cluster; and lateral neurons (LNs), which are located in relatively anterior regions (Figure 13) and composed of two or three clusters, depending on whether the relatively ventral (v) group is best regarded as consisting of the large (*l*) LNv's and small (*s*) LNv's. Axonal projections from the *l*-LNv's (running across the brain's midline, also out into the medulla) and from the *s*-LNv's (into the dorsal brain, terminating near the Ca) are marked by the product of the *Pdf* neuropeptide-encoding gene (Helfrich-Förster, 1996; cf. Table 6). These neurites are not shown here (but see Figure 12 for a diagram that includes them), along with projections from the relatively dorsal LNd's and the DN cell bodies. The large number of open circles shown in the present figure represent the ca. 1,800 *per*-expressing glial cells that are located in several cortical and neuropile regions of the central-brain proper and those of the OLs. Many of the *per* neurons depicted, and probably such glia as well, express the *timeless* gene (Kaneko and Hall, 2000), making it likely that most LNv cells coexpress *per*, *tim*, and *Pdf* (see Figure 13 and Helfrich-Förster, 1995; Renn *et al.*, 1999; Park *et al.*, 2000a). It is also possible that at least some of the *per/tim* LNs coexpress the *Clock* and *cycle* genes (So *et al.*, 2000) along with *double-time* (Kloss *et al.*, 1998, 2001) and *cryptochrome* (Emery *et al.*, 2000b).

In conjunction with these appraisals of CNS expression, one could not help notice that a host of other tissues was stained (Table 4). Thus, *per* products and reporter signals are observed in photoreceptors of the compound and simple eyes, in other sensory structures (notably appendages), the imaginal gut, and other internal organs, especially those within the abdomen, including reproductive tissues (Liu *et al.*, 1988; Siwicki *et al.*, 1988; Saez and Young, 1988; Frisch *et al.*, 1994; Beaver *et al.*, 2002). In fact, this gene is expressed in essentially all types of adult tissues except exoskeletal muscles, the epidermis, and the tracheal epithelium. These findings have been catalogued by Hall and Kyriacou (1990), Hall (1995), and Giebultowicz *et al.* (2001); the latter study also noted that anti-PER signals cycled, with late-night peaks, in various regions of the alimentary canal, fat body, and spermathecae of the female reproductive system. Moreover, both *per* mRNA and PER protein cycle (in temporally typical fashion) within the male reproductive system (Beaver *et al.*, 2002).

per products tend not to be uniformly distributed throughout a given tissue or organ, as exemplified by the spatial distribution of a *per*-promoted reporter displayed in various regions of the Malpighian tubule (Hege *et al.*, 1997), and inasmuch as *per* mRNA and PER are localized within the proximal regions of the male testis (Beaver *et al.*, 2002). In almost all cell types for which anti-PER leads to staining, the signals are nuclear. [This includes cells within the pupal RG (Saez and Young, 1988; Emery *et al.*, 1997).] A notable exception is that somatically derived cells of the adult female's reproductive system exhibit cytoplasmic staining (Liu *et al.*, 1988, 1992; Saez and Young, 1988). Recall that a biological function associated with this tissue is affected by *per*-null variants, but not in terms of the relevant clock parameter (Saunders *et al.*, 1989; Saunders, 1990), and, perhaps in this regard, that *per* mRNA extracted from the ovary does not exhibit daily oscillations in its abundance (Hardin, 1994).

This brings us within a hair's breadth of considering experiments in which the functional meanings of *period* gene products within more anteriorly located tissues have been analyzed or least inferred. First, contemplate that daily cycling of PER and of *per*-reporting factors occurs within the aforementioned brain neurons (e.g., Zerr *et al.*, 1990; Frisch *et al.*, 1994; Stanewsky *et al.*, 1997a), even though temporal variations of the relevant immunoreactivity are perhaps most salient in the compound eye (e.g., Siwicki *et al.*, 1988) and implicitly most prominent within that peripheral tissue in timecourses established for extracts of head homogenates.

The relentlessly annoying visual system comes into play in a second consideration, which involves *anatomical* outputs from the brain neurons in question (as we will see, some of these nerve fibers project into the optic ganglion). Portions of this projection pattern, referring to axons elaborated from PER-containing neuronal cell bodies, can be appreciated by staining for a certain clock-output substance known as the PDF neuropeptide (the *Drosophila* form of which is discussed in Section X.B.1). However, *per* itself has also been manipulated such that neurite-filling markers would in principle reveal the axonal projections from all of the brain neurons that express the gene. For this, a fairly long stretch of *per*-regulatory material (the same 5'-flanking region used in the best-rescuing *per*$^+$ DNA fragments and in fusion constructs for which portions of the gene were fused directly to *lacZ*) was ligated to sequences encoding GAL4 (Plautz *et al.*, 1997a; Emery *et al.*, 1998). Flies carrying the resulting *per-gal4* fusions were crossed to those in which one or another transgene included UASs (GAL4 regulatees); the latter had been fused to marker encoding factors, in this case (mammalian) tau protein or green-fluorescent protein (GFP).

The complex neurite patterns, revealed by applying anti-tau or taking advantage of GFP's light-inducible fluorescence (Kaneko and Hall, 2000), included the following (Figure 12, color insert):

(i) Relatively large, ventrolateral *per* neurons project axons that ramify over most of the optic lobes, except the distalmost lamina.

(ii) These l-LN$_v$ cells, or perhaps certain of their companions within this neuronal cluster, also send fibers *across the brain midline that terminate near the contralateral LNs* (it is not known whether certain individual l-LN$_v$'s project contralaterally as well as centrifugally).

(iii) Smaller s-LN$_v$'s, whose direct precursors are the LNs observed as early as the L1 larval stage (Kaneko *et al.*, 1997; Helfrich-Förster, 1997b), send axons into a *dorsomedial brain region*.

(iv) A distinctly separate LN cluster of cells alluded to earlier, which is in a more dorsal brain region (hence LN$_d$), sends nerve fibers to a similar dorsomedial target.

(v) Three clusters of DNs—two of which consist of cells that are in the dorsal brain by definition and also rather near the midline—also project at least some of *their* neurites to dorsomedial locations; such DN nerve endings are in the vicinity of those emanating from the s-LN$_v$'s and LN$_d$'s; one of the dorsal cell-body clusters was newly identified in this study (Kaneko and Hall, 2000); this third group consists of a few dozen small neurons in each brain hemisphere; the longer-reaching fibers projecting from these "DN3" cells (which are located more laterally than the DN1's and DN2's) display dorsomedial axonal termini.

(vi) Other DN1 cells (the majority of such neurons) send their neurites toward or into more-ventral brain regions.

(vii) The numbers of neurons expressing *per* within these five clusters (or six, if the l- and s-LN$_V$'s are considered separately) sums to about 70–75 pairs of bilaterally symmetrical cells (not all of which are diagrammed in Figure 12; cf. Figure 11).

Recall that the expression of this gene within several of these neurons appears during pupation; thus, the majority of these imaginal cells may establish

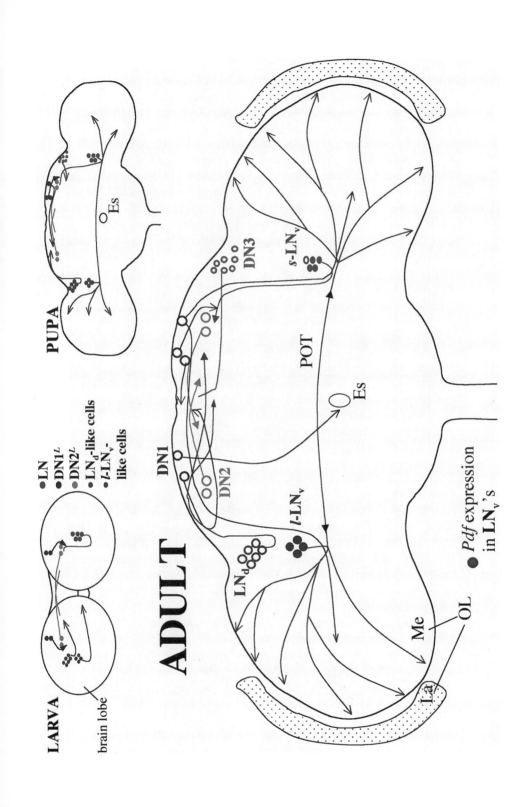

LARVA

brain lobe

- LN
- DN1^L
- DN2^L
- LN$_d$-like cells
- l-LN$_v$-like cells

PUPA

Es

ADULT

DN1
DN2

DN3

s-LN$_v$

POT

Es

l-LN$_v$

LN$_d$

Me

OL

La

- *Pdf* expression in **LN$_v$**'s

their full-blown differentiations relatively late. However, stainings of neurites in *per-gal4*;UAS-marker pupae (and even larvae) display projection patterns similar to those just listed for the adult brain (Kaneko and Hall, 2000). With reference to the third (and last) larval stage, the neuronal clusters putatively corresponding to those expressing clock genes in adults are called LNs (the aforementioned precursors of s-LN$_V$'s), l-LN$_V$-like, LN$_d$-like, DN1^L and DN2^L (no forerunners of the DN3's can be surmised to be present during L3). Thus, many features of the adult's interneuronal connectivity pattern may begin to be established about half-way through development, even though *per* expression within certain cell bodies takes some time off at the end of larval life.

Figure 12. Neuronal projection patterns of CNS cells expressing clock genes in brains of *Drosophila* at three life-cycle stages. The diagram at the top left summarizes neuroanatomical results obtained from third-instar larvae (L3) whose CNS has been analyzed in a fairly extensive manner for cells in which the *period* gene (cf. Figure 10) and the *timeless* one make their products. This L3 brain pattern stemmed in the main from stainings of CNS carrying *per-gal4* or *tim-gal4* transgenes designed to drive neurite-filling markers, as in Kaneko and Hall (2000), from which these diagrams were adapted. Additional *in situ* findings are implied by elements of the larval diagram: LNs strongly expressing native PER and TIM are shown in the brain lobe on the right; LN- and l-LNv-like cells with weak PER and TIM immunoreactivities are shown in the left lobe. The diagram at the top right shows *per*- and *tim*-expressing neuronal cell bodies—and axons projecting from them—as the pattern changes during metamorphosis. The emerging imaginal cells are color-coded in the same manner as in the ADULT diagram at the bottom. (See additional version of this figure in the group of colored inserts.) For this imaginal pattern, the three clusters of lateral neurons (LNs) and the three groups of dorsal ones (DNs) have their cell bodies shown as in Figures 11 and 13 (however, because of a space limitation, many fewer DN3s are diagrammed here than the actual number of ca. 40 pairs of such cells). The neurite projections in the adult brain are also diagrammed, based on stainings for the PDH and PDF immunoreactivity (see Figure 13 and Table 6) that is present in all of the l-LN$_v$'s and in of most s-LN$_v$'s, as well as by application of neurite markers (transgene-driven and -produced, as noted above). Focusing on the projection patterns: l-LN$_v$ perikarya send axons across the brain midline (via the POT, posterior optic track) to contralateral LN regions and also project into the optic lobes (OL), but only as far as the medulla (Me), which is underneath the distalmost lamina; it is unknown whether a given l-LN$_v$ cell body might elaborate both POT neurites as well as centrifugal fibers. s-LN$_v$ cells project fibers into a dorsomedial brain region, with these nerve terminals being near the mushroom-body calyces (cf. Figures 11 and 13). These patterns are congruent with those determined either from application of anti-(crab)PDH (summarized by Helfrich-Förster, 1996) or anti-(fly)PDF (although note that one s-LN$_v$ cell within each bilaterally symmetrical cluster is PDF-less, and that the heterologous antibody spuriously stains certain dorsal-brain neurons); or from marker expression driven by a *Pdf-gal4* transgene (Renn *et al.*, 1999; Park *et al.*, 2000a). The (PDF-nonexpressing) LN$_d$ perikarya send their (transgenically marked) axons into the dorsal brain. Most of the DNs, already within that brain region by definition, also (and in the main) project locally to regions relatively near the termini of s-LN$_v$ and LN$_d$ axons. However, certain DN1 cells also (or instead) project fibers to the vicinity of s-LN$_v$ perikarya or to that of the relatively ventral esophagus (Es).

2. *per* variants and others used to dissect the neural substrates of behavioral rhythmicity

The functional meaning of *per*-expressing cells in the nervous system has been assessed, mostly with regard to adult locomotion, in the following disparate ways:

(i) Transplanation of a per^S brain into the abdomen of a per^{01} host caused about 7% of the latter to exhibit free-running rhythms characteristic of the donor genotype (Handler and Konopka, 1979).

(ii) Genetic mosaics that were each partly per^S, with their remaining tissues being $per^S/+$, showed a correlation between the *per* genotype of head tissues and behavioral cycle durations (Konopka *et al.*, 1983).

(iii) A subsequent series of mosaics—whose alternative intrafly genotypes were per^{01} or $per^{01}/+$ and whose internal tissues were marked as to whether they harbored per^+—indicated that expression of the normal allele in the brain proper was necessary for rhythmicity; moreover, the patches of per^+ cells included those in a lateral brain region, on one side of the brain or the other, in the vast majority of rhythmic individuals (Ewer *et al.*, 1992).

(iv) The largely arrhythmic *disco*[1] mutant (Dushay *et al.*, 1989) was found to be devoid of PER immunoreactivity in most LNs (Zerr *et al.*, 1990), but appears normal for expression of the clock gene in brain glia and DNs (Zerr *et al.*, 1990; Kaneko and Hall, 2000); typically, the rare *disco* individuals exhibiting robust rhythmicity possessed at least one LN_V cell, which, via marking by the PDF neurochemical factor mentioned earlier, projected fibers into the dorsomedial brain (Helfrich-Förster, 1998).

(v-a) Behavioral and brain-expression analyses of *per* transgenics carrying fragments of this gene missing 5'-flanking DNA showed that its expression in subsets of the brain is sufficient for rhythmicity; when the 5' end of the transgene is approximately in the middle of the first intron (Figure 9), brain expression is relatively broad (Figure 13, see color insert) includes the LN region, and allows flies from all transgenic lines to exhibit weak, slightly long period behavioral cycles (Liu *et al.*, 1991).

(v-b) When the transformed per^+ sequences are missing a further ca. 800 bp of this intron, in the aforementioned "7.2" transgenic (Figures 4B and 9), only certain lines are similarly rhythmic (with cycle durations ca. 1 hr longer compared with the "8.0" transgenic type); one rhythmic 7.2 line exhibited PER immunoreactivity in only subsets of DNs and LNs (as well as in the eye and gut); the other one was PER-positive only in LNs, within both the LN_v and LN_d groups (Frisch *et al.*, 1994) [These faithful mimics of subsets of the wild-type spatial pattern prompted the aforementioned notion that insertion of the 7.2-kb *per* transgene in these two strains "trapped a promoter," as opposed to invoking "enhancer trapping;" the

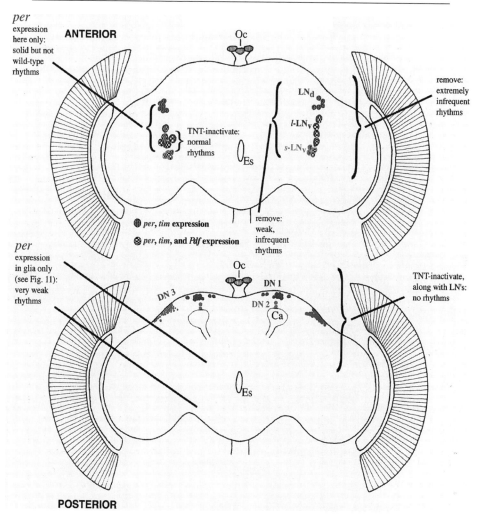

per expression here only: solid but not wild-type rhythms

ANTERIOR

Oc

LN$_d$

l-LN$_v$

s-LN$_v$

TNT-inactivate: normal rhythms

Es

remove: extremely infrequent rhythms

⊕ *per, tim* expression

⊗ *per, tim,* and *Pdf* expression

remove: weak, infrequent rhythms

per expression in glia only (see Fig. 11): very weak rhythms

Oc

DN 3

DN 1

DN 2

Ca

TNT-inactivate, along with LN's: no rhythms

Es

POSTERIOR

Figure 13. Neural substrates of *Drosophila*'s circadian locomotor rhythm. The basic diagram is from Hall (1998b). Added to it are indicators of clock-gene-expressing brain cells that contribute to regulation of rest–activity cycles exhibited by *D. melanogaster* adults, as deduced from the application of various mutants and molecularly manipulated genes. Essentially all LN cells are "removed" in a given *disco* mutant individual (Table 2). *per* expression occurring only in a subset of the LNs is mediated by a "promoterless" transgene (Frisch *et al.*, 1994) functioning in flies whose genetic background included the *per^{01}* null mutation (Figure 4B), or (less definitively demonstrated) in a (*per^{01}*-background) transgenic type carrying *glass*-gene regulatory sequences fused to PER-coding ones (Vosshall and Young, 1995). Specific elimination of *l*-LN$_v$ and *s*-LN$_v$ cells was mediated by *Pdf*-promoter driving of cell-killer factors (Renn *et al.*, 1999). The same *Pdf-gal4* driver

continues

latter is often taken to mean that the insertion site reveals activity of a factor involved in temporal and spatial regulatory features of "some gene," but when a chromosomal landing site for the 7.2 transgene allows for "any" PER production, a random enhancer seems uninvolved—because it would seem to be a miracle if such a factor mediated spatial expression that is a well-defined subset of *per*'s domain and only that. Stated another way, this 7.2-kb *per* fragment seems as if it must possess a spatial enhancer of its own. An alternative is that the two 7.2's did trap a standard enhancer that drives PER production in some of its usual cells and in several others; however, this protein would not accumulate unless the latter "random" cells also normally produce TIM (Zerr *et al.*, 1990; Price *et al.*, 1995).]

(vi) Although the *per* 5′-flanking region (Figure 9) is not necessary for components of the gene product's spatial distribution, subsets of this upstream regulatory material are sufficient: The 69-bp CRS element (Hao *et al.*, 1999) or a multimerized E-box-containing one (Darlington *et al.*, 2000) can drive *lacZ* expression in all the normal tissue locations (Figure 9 shows where *per*'s CRS and the E-box within it are situated). Testing the aforementioned mutations that were engineered within the CRS revealed correlations between grossly determined expression of the driven sequences (see above) and behavior as it was in turn correlated with tissue expression. The most informative *in vitro*-mutated transgenics for our current purposes were those in which bases 11–20 or 48–57 of the CRS (upstream and downstream of the E-box) were altered: Ostensibly normal *lacZ* reporting within the head was observed only in LNs and fat-body tissues and, in the "57-mut" type only, the antenna as well (β-GAL signals were said to be "limited" in the two other head locations examined: glia and photoreceptors); 4/9 of the corresponding 57-mut lines and 1/3 of the "20-mut" lines, in which the mutated CRS was fused to PER-encoding sequences, gave fairly solid rescue of *per*[01]'s arrhythmicity. In the

---

(*continued*) was used by Kaneko *et al.* (2000b) to cause local expression of tetanus toxin (TNT) transgenes in which sequences encoding the toxin were fused to *cis*-acting targets (UAS) of the GAL4 transcription factor. These same investigators applied *per-gal4* or *tim-gal4* transgenes, in combinations with UAS-TNT ones, putatively to ruin synaptic release of neurotransmitters (but perhaps not of the PDF neuropeptide) from all LNs and DNs that express either clock gene (bottom panel of the present figure). *per* glia (Figure 11) are inferred to be involved in behavioral-rhythm regulation (bottom panel) owing to rare *per*[+]//*per*[01] genetic-mosaic individuals (Ewer *et al.*, 1992) in which no LN expression of the normal allele was detectable; these flies nonetheless exhibited weak, long-period rhythms, correlated with the presence of *per*[+] in certain glia only (depending on the brain region where that allele was retained in the mosaic individual in question). However, *disco* flies exhibit robust and normally cyclical expression of PER in brain glia (Zerr *et al.*, 1990), which therefore is insufficient for routine behavioral rhythmicity in the almost-certain *anatomical* absence of the LNs in this mutant (see above).

companion reporter lines, no distinctions were attempted between marking of cells in the various LN clusters (Lyons *et al.*, 2000).

(vii) An occult type of fusion transgene led to brain–behavioral correlations similar to those inferred from the *per* constructs (in which 5′-flanking material had been removed or altered); thus, the promoter region of our old friend the *glass* (*gl*) gene (Section II.C) was fused to the *per* ORF (Vosshall and Young, 1995) and found partially to rescue per^{01}-induced arrhythmia (for about one-fourth to two-thirds of the individuals tested within a given transgenic line, which exhibited ca. 28- to 34-hr free-running periods); a *gl-lacZ* reporter construct drove marker expression in external photoreceptors, naturally, and also in a few brain neurons—possibly small subsets of the LN and DN clusters, as well as in some subesophageal ganglionic cell. However, Helfrich-Förster (1996) disputes the inference that *gl-per* drives expression in the laterally located brain neurons that classically contain clock-gene products: "All three groups of these PER-expressing neurons [in the *gl-per* transgenic brains] appear to be clearly different from the LN's. . .; they are located in the lateral posterior brain instead of the lateral anterior brain [cf. Figure 13]. This suggests that other still unknown neurons [although see Kaneko and Hall (2000)] might contribute to rhythmicity of *Drosophila*" (Helfrich-Förster, 1996). Another matter arising in the study of Vosshall and Young (1995) is that *gl* mRNA transcribed from the native gene was temporally flat (in whole-head extracts), but PER protein encoded by the *gl-per* transgene cycled in compound-eye photoreceptors. The latter issue was shown by these investigators to be irrelevant for *gl-per*'s behavioral rescue by genetic elimination of the external eyes (Section II.A). The results of Vosshall and Young (1995) foreshadowed those of Yang and Sehgal (2001): no clock-gene transcript cycling (as could be inferred from the formers' *gl* abundance measurements, even though a timecourse for *glass*-promoted *per* mRNA was not performed), → weak PER cycling, → variable occurrence of flabby, long-period behavioral rhythmicity.

The neural-substrate/behavioral-function results summarized in Figure 13 indicates that functioning of the *period* gene in the brain is responsible for *Drosophila*'s rest–activity rhythm. This is as would be expected. Detailed aspects of the results just listed, as well as certain additional findings and considerations, point to further features of correlations between clock-gene expression and biological rhythmicity.

(i) The presence of *per*-expressing LNs seems necessary for behavioral periodicity; DN clock functioning may be insufficient or irrelevant.

(ii) However, it is unknown whether the LNs are truly missing from *disco* brains or instead fail to activate their *per* gene: *disco* is in general an anatomical mutant (Steller *et al.*, 1987), but the gene encodes a putative transcription factor (Heilig *et al.*, 1991; Mahaffey *et al.*, 2001) that may influence cell differentiation of neural structures as opposed to their formation only.

(iii-a) *per* neurons other than the s-LN$_v$ cells seem to contribute to rhythm regulation, because a small proportion of *disco* flies that behaved rhythmically (equal to the percentage of those that were unambiguously rhythmic and turned out to possess s-LN$_v$'s and their projections into the dorsal brain) did not display the *per*-neuronal marker in any LN$_v$ cells and thus may have lacked these particular lateral neurons [marking of LN$_d$'s or DNs was not possible in this study, and the rhythmic LN$_v$-less individuals exhibited anomalously "complex" behavioral patterns (Helfrich-Förster, 1998)].

(iii-b) Transgenically mediated cell killing of all LN$_v$'s leads to severe decrements in behavioral rhythmicity (Renn *et al.*, 1999) (discussed further in Section X.B.1, where the LN$_v$-specific driver derived from the PDF-encoding gene is described); compared with most assessments of *disco* behavior (Dushay *et al.*, 1989; Dowse *et al.*, 1989; Hardin *et al.*, 1992a; Helfrich-Förster, 1998), LN$_v$ ablations caused a less oppressive effect than those manifested by *disco*, mutation of which seems to eliminate all three categories of lateral neurons (Zerr *et al.*, 1990) [In a separate study, which applied a putatively LN$_v$-specific, "enhancer-trapping," *gal4*-containing transgene, the behavioral effects of killing these neurons were found to be similar to those of a *disco* mutation; although in this one study (Blanchardon *et al.*, 2001) unusually high proportions of mutant individuals (ca. 20–35%) were determined to be rhythmic.]

(iii-c) *per*-promoted expression of UAS transgenes that encode tetanus toxin (TNT), such that synaptic communication at nerve terminals involving all LNs and DNs could be wrecked, led to sharp drops in the proportions of rhythmic individuals (Kaneko *et al.*, 2000b)—overall, a worse array of behavioral phenotypes than those observed in flies specifically ablated of their LN$_V$ cells (Renn *et al.*, 1999); driving TNT in the LN$_v$'s *alone* (Kaneko *et al.*, 2000b; Blanchardon *et al.*, 2001) led to almost *no* decrement in periodic behavior.

To let the cat out of the bag—in advance of a fuller discussion of the neurochemical content of these ventrolateral brain neurons (Section X.B.1)—here is device for assimilating the complex results described in the foregoing paragraphs: Release of a neuropeptide called PDF from LN$_v$ nerve terminals is believed to participate in regulation of *Drosophila*'s behavioral rhythmicity (Renn *et al.*, 1999; Park *et al.*, 2000a). The deleterious effects of TNT, when the toxin is produced in *all* PER neurons (Kaneko *et al.*, 2000b), suggest that "standard" neurotransmitter release from some of these axons is important for sending clock outputs along their way (toward the control of daily locomotor cycles by the thoracic ganglia). However, TNT's presence in the LN_v's would not seem to impinge on PDF release, assuming that that is what happens to this substance.

With regard to putative *per* functions in glial cells:

(iv) PER protein in brain glia is neither necessary (Frisch *et al.*, 1994) nor sufficient (Zerr *et al.*, 1990, taken together with Dushay *et al.*, 1989) for robustly rhythmic behavior.

(v) However, a small number of $per^{01}//per^+$ mosaics, which were weakly rhythmic, expressed the normal allele only in a subset of the normal brain-glial pattern, not in any detectable neurons (Ewer *et al.*, 1992).

(vi) Several such glia have been observed to decorate fiber tracts projecting from *per* neurons (Helfrich-Förster, 1995), as if the former cell type may modulate or even mediate crude features of rhythmic outputs traveling along these axons.

(vii) $per^S//per^+$ mosaics studied for both circadian behavior and courtship-song rhythmicity suggested, once again, brain control of the former phenotype (albeit with less intrabrain resolution compared with Ewer *et al.*, 1992); yet the effects of a given *per* allele on the ultradian behavioral rhythm (Figure 6B) mapped to the VNC (Konopka *et al.*, 1996).

(viii) A conceivable corollary, although purely a molecular one, is that *per* mRNA content in the VNC-containing thorax is different from that in the brain [the alternative splice form A is less abundant than that of B' in former body region, but both forms are apparently equal in their brain concentrations (Cheng *et al.*, 1998)].

(ix) Glial chronobiology may be connected with a nonbehavioral phenotype: daily fluctuations in the sizes and shapes of certain interneuronal axons within the lamina (Pyza and Meinertzhagen, 1999a), the distalmost optic lobe to which the centrifugal fibers of l-LN_v neurons fall short of projecting (Figure 12). However, there are *per* glia in every optic lobe of wild-type flies (Figure 11), as discussed by Meinertzhagen and Pyza (1996); they go on to deliberate about the manner by which certain such cells may influence laminar rhythms, mostly from the perspective of studying analogous (time-based) studies of houseflies. In *Drosophila*, all glia are PER-negative within all head ganglia of the behaviorally rhythmic, LN-PER-positive 7.2 transgenics (Frisch *et al.*, 1994); no laminar anatomical rhythms were detectable either in these per^+-transformed flies or in fully mutant per^{01} ones (Pyza and Meinertzhagen, 1995, 1999a).

Whereas this visual-system rhythmicity is unappreciable for its biological sgnificance:

(x) There is a sensory rhythm in another per-expressing peripheral tissue: the antenna, which exhibits a nighttime peak in odor sensitivity (discussed further in Section VI.D); this mild oscillation free-runs and is eliminated by per^{01} (Krishnan et al., 1999); it may be consequential that marker-reported expression of the gene was undetectable within antennal sensory neurons of per-gal4 or tim-gal4 transgenics (Kaneko and Hall, 2000).

(xi) Additional cellular rhythms might be regulated by outputs from per-expressing brain cells, in this case neurons: There are bimodal daily rhythms of nuclear size in corpus allatum (CA) and pars-intercerebralis (PI) cells (Figure 17, see color insert) and the latter (dorsal-brain) structure exhibits similar rhythmicity in its content of neurosecretory material (Rensing, 1964). These phenotypes may bear a relationship to innervation of the anterior-thoracic CA region (which also contains the corpus cardiacum, CC) by certain neurites projecting from DN1 cells (also note that PER is expressed within the CC; Frisch et al., 1994), and to the fact that several of the axons stemming from all DN groups appear to terminate near the PI (Kaneko and Hall, 2000).

B. *timeless* products in developing and mature *Drosophila*

Expression of tim at different life-cycle stages and in various tissues has been examined and manipulated in a less comprehensive manner compared with findings obtained for per and genetic analyses of the latter's chronobiological roles. Early assessments of TIM's spatial distribution showed the protein to be present in compound-eye photoreceptors and lateral-brain neurons (e.g., Hunter-Ensor et al., 1996; Myers et al., 1996; Yang et al., 1998), with little in the way of distinguishing among the separate LN clusters; these staining studies were performed mainly in conjunction with effects of environmental stimuli on this protein's immunoreactivity (see Section IX.B.2).

In embryos, in situ hybridizations to detect tim RNA led to a pattern similar to that of per expression during this stage (Table 4): segmentally arranged signals present in most or all of the developing CNS ganglia (J. Blau and M. W. Young, unpublished). In larvae, tim products or marker reporting are observed in several neural and nonneural tissues. For the latter, it is intriguing that the spatial pattern does not fully overlap with that of per expression [e.g., marking of tim, but not per, was observed in ring gland and fat body; and conversely for salivary gland and spiracles (Kaneko and Hall, 2000)]. In the L3 CNS, TIM immunoreactivity is essentially the same as that of PER and colocalized within the various brain neurons (Kaneko et al., 1997). However, VNC expression of tim was inferred (from reporter signals) to be in fewer cells compared with those exhibiting *per*-driven marking (Kaneko and Hall, 2000). Yet, within the larval brain, the daily cycles of apparent TIM levels are similar to PER ones, including late-night peaks for LN and DN1 staining (Figure 10), but a daytime or late-day maximum for the DN2 cells (Kaneko *et al.*, 1997). As these L3 animals proceed into metamorphosis, *tim* and *per* expression within the CNS remain spatially similar; however, *tim*-driven marker signals in early-pupal DNs are observed in advance of the *per* expression that reappears later in these dorsal-brain regions (Kaneko and Hall, 2000). It seems, therefore, that the larval *per* DNs may not degenerate (to be replaced by similarly situated pupal and imaginal cells); instead they could turn down expression of their *period* gene, but not the putatively coexpressed *timeless* one.

In adults, the profusion of body parts and organs in which *per* products have been directly observed or inferred by reporter factors (Table 4) is essentially the same as in the case of *tim* (congruence was noted for approximately a score of tissues, distributed throughout the head, thorax, and abdomen; Kaneko and Hall, 2000). In two nonneural cases, intratissue coexpression of the two genes was nailed down, that is, within cells of the imaginal Malpighian tubule (Giebultowicz and Hege, 1997) and within the male's internal reproductive organs (Beaver *et al.*, 2002).

Turning our attention to neural sites of *tim*'s adult expression, the patterns are grossly like those of *per*. However, double-labeling tests revealed a less-than-perfect spatial congruence: Among the *tim*-expressing neuronal clusters examined (Kaneko and Hall, 2000), only the *l*-LN$_v$'s were invariably labeled by a *per*-reporting marker. In other neuronal groups (*s*-LN$_v$, LN$_d$, DN1, DN2) some of the cells labeled by anti-TIM did not express *per*. Within most of the neuronal groups exhibiting such noncongruence, cells containing the *per* marker only were in close proximity to those exhibiting TIM immunoreactivity alone (Kaneko and Hall, 2000).

Scrutiny of the doubly labeled brain hemispheres also uncovered a novel TIM-immunoreactive cluster of three cells within the posterior lateral cortex of the protocerebrum. In some of these specimens, one or two of these cells exhibited weak expression of a marker driven by *per-gal*4 (Kaneko and Hall, 2000). With regard to cells within the so-called main neuronal groups—the three sets of LNs and three of DNs (Figures 11–13, color insert)—it cannot be said that there are specialized subsets of neurons in terms of those that do not coexpress the two clock genes. For example, it could have been that the connectivity patterns (Figure 12) are different, but they were at least grossly the same for neurite-marker expressions driven by either *per-gal*4 or *tim-gal*4 (Kaneko and Hall, 2000). Moreover, TIM immunoreactivity within the head cycles largely in the same manner as does that of PER (Hunter-Ensor *et al.*, 1996; Myers *et al.*, 1996), possibly with a higher amplitude for the former protein (Stanewsky *et al.*, 1998). This difference was observed for photoreceptors as well as brain cells. Another anterior sensory structure is notable for its lack of detectable *tim* expression within a subset of the appendage: Antennal sensory *neurons* were devoid of marker signals drivable by *tim-gal4*, as was found for the companion *per-gal4* transgenics (Kaneko and Hall, 2000). The potential significance of these particular findings is taken up in a subsection after the next one with regard to a circadian rhythm of the fly's odor responsiveness.

Considering tissue expression of *timeless* more broadly, functional correlations between the action of this gene within a given neural structure and its chronobiological meaning can barely be surmised. There is of course no TIM in the brain neurons of behaviorally arrhythmic tim^{01} flies, but all the rest of the gene's expression pattern—which incidentally includes many CNS glia (Stanewsky *et al.*, 1998; Kaneko, 1999; Kaneko and Hall, 2000)—is absent as well. Similarly, global brain–behavioral correlations were educed from *tim-gal4*-driven expression of TNT, which led to nearly arrhythmic behavior (Kaneko *et al.*, 2000b). This effect was more onerous than those observed in locomotor tests of the companion *per-gal4*;UAS–TNT flies (perhaps because *tim* expression is in general higher than that of *per*; e.g., Zeng *et al.*, 1996) and clearly more severe than the consequences of LN_v-specific cell killing (Renn *et al.*, 1999; Blanchardon *et al.*, 2001). However, no additional "dissection" information could be inferred from the toxin experiment (Kaneko *et al.*, 2000b), given that the *tim-gal4* fusion is presumably transcribed in all neuronal and glial cells that normally express this clock gene (Kaneko and Hall, 2000). A final fillip to these *tim*-variant/phenotypic relationships is that the anatomically determined rhythm in the lamina optic lobe (Pyza and Meinertzhagen, 1999b) is eliminated in tim^{01} as well as in per^{01} flies (Pyza and Meinertzhagen, 1999a).

C. Preliminary assessment of four additional clock genes' expressions

A tacit assumption that has accompanied molecular analyses of interactions among the products of all five of the fly's clock genes is that they would be co-expressed at least within some of the rhythm-relevant cells. This is not observed universally for *per* and *tim* expression (as noted previously), and those of the remaining three factors have only begun to be determined. For *Clock*, Northern blottings that used mRNA taken from separate fractions of the adult detected signals in the head extracts (naturally), the rest of the fly's body, and in appendages (Darlington *et al.*, 1998). *In situ* hybridization of *Clock* and *cycle* probes to head sections (from flies sacrificed near the end of the night) showed signals broadly distributed throughout the cortical regions of brain and optic ganglia; nuclear regions of compound-eye photoreceptors stained as well (So *et al.*, 2000). *Clk* expression was also found the male reproductive system, for which temporally controlled *in situ* appraisals revealed this gene's mRNA to be present within the base of the testis and in the seminal vesicle; the signals cycled out of phase with those of *per* mRNA (Beaver *et al.*, 2002; cf. Sections V.B and V.C).

Recall that Clk^{Jrk} and cyc^0 mutations cause dramatic lowerings of *per*- and *tim*-product levels. However, mRNAs and proteins produced by the latter two genes are not eliminated by homozygosity for null alleles of *Clk* or *cyc*. Correlative findings came from *in situ* studies (Kaneko and Hall, 2000): *per*- or *tim-gal4* transgenes were each put in Clk^{Jrk} or cyc^{01} genetic backgrounds; some cell types exhibited poor or no marker expression, if not in cell bodies, then in fiber tracts. Yet, most LN cell bodies (in all three clusters) and DN1 perikarya still showed signals (the small DN2 group was not examined). Clk^{Jrk} versus cyc^{01} differences were observed for certain neuronal types (e.g., almost no LN_d or DN3 signals were driven by *tim-gal4* in Clk^{Jrk}, but solid signals were seen within these two groups in cyc^{01}). There was nothing like global or uniform decrements of reported *per* or *tim* expression in these two mutants (although perhaps amplifications associated with GAL4 driving maximized detectability of mutationally lowered activation of these genes). Thus, factors other than the CLK and CYC activators can be inferred to participate in *per* and *tim* regulation within some brain cells; in certain of them, the normal alleles of *Clk* and *cyc* may not even be coexpressed with per^+ or tim^+.

The *double-time* gene product is expected to be widely present during life-cycle stages and within tissues: Several *dbt* mutations cause early developmental lethality, accompanied by an imaginal "discless" phenotype for null alleles (Zilian *et al.*, 1999). Whereas disclessness alone would not be expected to kill late embryos (the lethal period for certain *dbt* mutants), spatial breadth of the gene's expression is inferred from mutationally induced abnormalities of imaginal discs and the resulting adult structures in a variety of body regions (Zilian *et al.*, 1999). Within the head of wild-type flies, a *dbt* nucleic-acid probe hybridized to mRNAs in several regions of CNS cortex and near the nuclei of photoreceptors (Kloss *et al.*, 1998). This spatial pattern was similar to the labeling determined in parallel for *per* by *in situ* hybridization. Transcripts complementary to *per* and *dbt* probes gave denser and more broadly distributed mRNA signals in the head (notably throughout the CNS cortex) compared with those elicited by *tim* sequences (Kloss *et al.*, 1998). Now, *per* expression is broad within the adult head of *Drosophila*, but the gene products are not found everywhere (e.g., the 1,800 glia immunoreactive for PER compose a very small subset of such cell types in the central brain and optic ganglia). Yet, application of an antibody against DBT was said to elicit "ubiquitous" staining of head tissues (Kloss *et al.*, 2001), which therefore may include several PER-less locations. This may fit with the fact that the *double-time* gene (a.k.a. *discs-overgrown*) has broad biological functions, which could include nonchronobiological ones in the mature fly in addition to those mediated during development (Zilian *et al.*, 1999) by the kinase it encodes.

In none of these assessments of the spatial distributions for *Clk*, *cyc*, or *dbt* transcripts in the CNS was cellular colocalization with *per* or *tim* expression explicitly examined. However, the manner by which retinal cells are displayed in tissue sections through the compound eye makes is very likely that all five genes generate their products within a given photoreceptor. Another issue is that genetic controls for the specificity of the antibody-mediated stainings have been few and far between for the more recently discovered clock-gene products [cf. Siwicki *et al.* (1988) for the requisite performance of such PER-localization controls, and see Section X.B.1 for further examples and discussion].

This problem may pertain also to the case of *Timekeeper*'s spatial expression. Yet, elements of this pattern was dealt with in a slightly more intensive manner in terms of where the relevant casein kinase 2α is found within the brain (and other kinds of staining-specificity determinations were made). Thus, application of an antibody against this catalytic subunit to adult-brain specimens revealed signals that colocalized with a neuropeptide (Section X.B.1) which marks some of fly's lateral neurons (again, those that coexpress PER, TIM, and the marker). The signals were cytoplasmic within these neuronal cell bodies, and the neurites shown previously project from these LN perikarya (Figure 12) were also stained by administering anti-CK2α. Additionally, this antibody also led to weak labeling of two cells in the dorsal brain of adults (Lin *et al.*, 2002a), and expression of this kinase subunit in the compound eye was inferred from a ca. 20% reduction of CK2α activity in a flies expressing an *eyes-absent* mutation. In larval brains, the *Tik*-defined gene again produced its product in LNs containing the PDF neuropeptide (Kaneko *et al.*, 1997) as well as in a putative neuron that did not contain this "clock-cell" marker (Lin *et al.*, 2002a).

D. Autonomous functions of clock genes in peripheral tissues

period, timeless, and *Clock* are the three such factors known to generate their products in many adult tissues located on the fringes of the head or posterior to it. The aforementioned cyclical expression of *per* in the prothoracic gland was shown to occur in explants of this pupal gland, although in reality the cultured material consisted of "CNS–RG complexes" (Emery *et al.*, 1997). The PG oscillations of PER immunoreactivity remained robust in the presence of tetrodotoxin, indicating no involvement of sodium-dependent action potentials in the output from CNS neurons that could be imagined to regulate *per* expression in more posterior structures such as the prothoracic gland. PER cycles in cultures of this neuroendocrine tissue were synchronized to LD cycles and those in which the L versus D phase was reversed. An additional experiment took advantage of *per*-driven luciferase expression to infer autonomous cycling of the fusion gene's expression in real time, although it was not possible to know the portions of these CNS-RG specimens from which the reporter signals emanated (Emery *et al.*, 1997).

Other *per-luc* experiments homed in more closely on the matter of tissue autonomy: Dissociated body segments as well as a variety of head and thoracic appendages were shown to exhibit glow cycling (Plautz *et al.*, 1997a). The clarity of these molecularly determined rhythms diminished as most specimens proceeded from LD to DD, but a given *luc*-reported oscillation could be nicely reestablished upon return to light–dark cycling conditions. The implied resynchronization of intratissue oscillators suggested that the presumed cellular clocks had not been winding down in DD, and, just as important, that these separated body parts possess autonomous light-responsiveness.

Internal tissues monitored for *per-luc* expression in this study (Plautz *et al.*, 1997a) included the Malpighian tubules (MTs), although the reporter's autonomous cycling was not documented. Quasi-autonomous rhythmicity for this abdominal organ was demonstrated by decapitating adults and showing that the apparent levels of both PER and TIM in MT nuclei went up and down in a systematic manner (Giebultowicz and Hege, 1997). These cycles continued in DD, and their phases responded appropriately to shifts of an LD cycle. Reporter-detected expression of these two genes in either headless flies (*per-lacZ*) or cultured MTs (*per-luc* and *tim-luc*) gave similar results (Hege *et al.*, 1997; Giebultowicz *et al.*, 2000), which were extended to the rectum (another part of the fly's excretory system) in the cultured-specimen tests (Giebultowicz *et al.*, 2000). Designing an MT transplant experiment required non-real-time monitoring of *tim* expression. Here, it was shown that MTs pre-entrained to an LD cycle in donor flies maintained synchrony to this cycle when removed and placed within the abdomens of hosts that had been "reverse-phase-entrained," then sent into DD; the latter's MT (within a given fly) exhibited TIM immunoreactivity fluctuations appropriate to the host's entrainment regime (Giebutowicz *et al.*, 2000). Thus the implanted and endogenous organs cycled 180° out of phase with one another.

Are there any *biological* rhythms running in these far-flung tissues? The only information that has been obtained in the direct context of clock-gene cyclings within peripheral tissues stemmed from monitoring *Drosophila* antenna for sensitivity to applied odorants: Systematic fluctuations of electrophysiologically recorded responsiveness were observed, with a peak near the middle of the night-time; this rhythmicity has a low amplitude in wild-type flies, but it was entirely flattened in *per*[01] or *tim*[01] flies (Krishnan *et al.*, 1999). That the sensory rhythm is mediated locally was inferred from antennal recordings of the *per* "7.2" transgenic type, which exhibits expression of the clock gene only within some of the lateral-brain neurons (Frisch *et al.*, 1994); no sensitivity rhythm was observed in recordings of the PER-less antennae (Krishnan *et al.*, 1999). It seems noncoincidental that this appendage runs molecular rhythmicities of *per* and *tim* expression when antennae are dissected away from the fly (Plautz *et al.*, 1997a; Krishnan *et al.*, 2001; also see Section IX.B.3). However, *per*- or *tim*-driven marker signals were not detectable in sensory neurons (Kaneko and Hall, 2000). Therefore the rhythm of odor responsiveness affected by mutations in these genes, as measured by electrodes inserted into the antenna (Krishnan *et al.*, 1999, 2001), do not seem to be the direct consequence of fluctuating electrical output from the sensory neurons projecting centripetally from this appendage. Instead, clock-controlled oscillations of the physiological readout may depend on sensitivity to the odor *inputs*, which would be influenced by PER and TIM in antennal chemoreceptive factors that feed into its sensory neurons.

VII. NATURAL VARIANTS OF CLOCK GENES, INCLUDING INTERSPECIFIC STUDIES

Studying tissues that compose the rhythm systems of insects other than *D. melanogaster* has a longer history compared with chronobiological investigations of the fruit fly. Therefore, one imagines that cloning clock genes in various insect species might, for example, lead to informative descriptions of tissues where these factors produce their products in long-studied forms such as moths or cockroaches. Interspecific molecular genetics necessarily results in identifying variant forms of rhythm-related genes, especially when one moves from a dipteran insect to a lepidopteran or an orthopteran. Most studies of clock factors originally known in *Drosophila melanogaster* have been devoted to *period*. This gene was poised to be analyzed for interspecific divergences by virtue of *intra*specific variations that were demonstrated previously.

A. Naturally occurring *per* variants

The threonine–glycine (T–G) repeat located roughly in the middle of PER (Figures 3 and 14; see color insert) was found to be 17, 20, or 23 T–G pairs among various strains of *D. melanogaster* (Yu *et al.*, 1987a). Another common *per* variant makes 14 such pairs (Costa *et al.*, 1991). These and the other three types make up ca. 99% of naturally varying forms with respect to this region within the gene (Costa *et al.*, 1992); the very rare variants include T–G × 15, T–G × 18, and T–G × 21 cases.

One perspective from which this intraspecific variation has been scrutinized is phenomenologically (Table 5), with regard to mutational mechanisms operating in tandem nucleotide repeats of this sort (Rosato *et al.*, 1996, 1997a). Just as interesting is the fact that the T–G length varies in a geographically systematic manner: Above the equator, more northerly located natural populations of this species tend to possess *per* alleles encoding 20 T–G pairs; the variation grades into a tendency for 17 such pairs to exist in the PERs of more southern strains (Costa *et al.*, 1992). This cline is one of the strongest cases of such geographical variation known for genes in *D. melanogaster*.

per sequences have been determined for all or part of the gene from several close relatives of this species, all members of the *melanogaster* subgroup (Thackeray and Kyriacou, 1990; Wheeler *et al.*, 1991; Peixoto *et al.*, 1992). Some studies of the different species within this group, and of sister ones, focused on implied speciation events, without necessarily worrying about the fly's chronobiology (e.g., Kliman and Hey, 1993; Lachaise *et al.*, 2000). The same is so for certain of the wider ranging interspecific comparisons (e.g., Gleason and Powell, 1997; and see later discussion).

Other such analyses scraped up against the edge of *per*'s functional significance. First, consider that PER subsequences can be readily divided into conserved regions and those that are distinctly variable; this was initially discovered by sequencing the *per* relatives cloned from *D. pseudoobscura* and *D. virilis* (Colot *et al.*, 1988). Right around the time the potential meaning of the PAS domain (Figure 14) began to emerge, this motif was shown to be among the conserved stretches of PER amino acids. Intriguingly, however, there is geographically based PAS variation within semispecies of the *D. athabasca* complex (Ford *et al.*, 1994): 7 aa's worth of coding sequence, corresponding to a PAS region between its A and B repeats, is missing in "eastern" flies in North America, but present in "western–northern" forms (and all other known *per*-gene sequences).

Other comparisons involving thoroughly different species showed that nucleotide sequences of *per* corresponding to the T–G-repeat region are more wildly variable with regard to their coding capacities, length, or both (Colot *et al.*, 1988; Costa *et al.*, 1991; Peixoto *et al.*, 1993). In fact, there is no T–G repeat as such in the PER protein of several species, although *D. pseudoobscura* produces both T–G pairs as well as a degenerate pentapeptide repeat in this region of the polypeptide (Figure 14), and in some such subgroups this centrally located part of *per* is minimally variable among the taxa analyzed (Zheng *et al.*, 1999). Whatever is the amino-acid composition and sequence for this middle portion of the protein, its putative predicted secondary structure—a series of β-turns— appears to be conserved (Castiglione-Morelli *et al.*, 1995; Guantieri *et al.*, 1999).

Further analyses of this region of the gene suggested that repeat-length divergences have coevolved with sequences immediately 5′ (Figure 14) and 3′ of nucleotides encoding the repeated residues (Peixoto *et al.*, 1993; Nielsen *et al.*, 1994). In the latter region, which encodes residues C-terminal to *D. melanogaster*'s T–G repeat, there is a small number of amino-acid differences compared with PER in *D. simulans* (Wheeler *et al.*, 1991; Peixoto *et al.*, 1992). Intragenic mapping of the *per* sequences responsible for *melanogaster* versus *simulans* song-rhythm periodicity suggested that one or more of these diverged residues are responsible for the interspecific difference in this ultradian rhythm. In this regard, the T–G repeats within these two species's worth of PER are barely different, and intra-*melanogaster* repeat-length polymorphisms were uncorrelated with song-rhythm variations among males of different strains (Wheeler *et al.*, 1991). However, a complete deletion of the repeat engineered within a *melanogaster per* gene shortened the song-cycle duration from its usual ca. 1 min to about 40 sec (Yu *et al.*, 1987a).

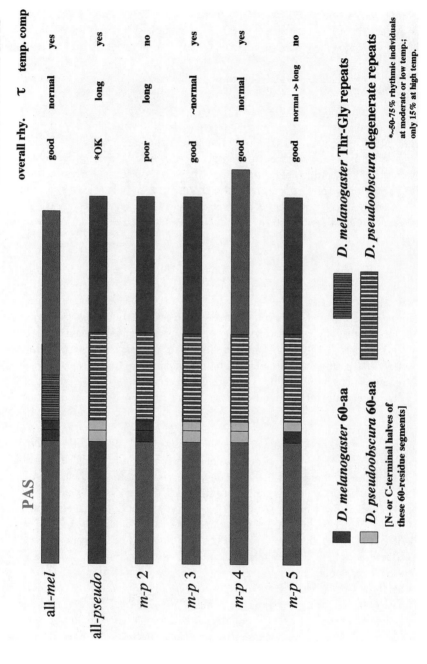

LOCOMOTOR RESULTS

	overall rhy.	τ	temp. comp
all-*mel*	good	normal	yes
all-*pseudo*	*OK	long	yes
m-p 2	poor	long	no
m-p 3	good	~normal	yes
m-p 4	good	normal	yes
m-p 5	good	normal -> long	no

PAS

D. melanogaster 60-aa
D. pseudoobscura 60-aa
[N- or C-terminal halves of these 60-residue segments]

D. melanogaster Thr-Gly repeats
D. pseudoobscura degenerate repeats

*~50-75% rhythmic individuals at moderate or low temp.; only 15% at high temp.

Figure 14. Behavioral effects of *period* gene sequences from *D. pseudoobscura* transformed into *D. melanogaster*. The PER proteins from these two species are diagrammed in the top two rectangles: *D. melanogaster* (*mel*) mostly red, *D. pseudoobscura* (*pseudo*) mostly blue (PER in the latter species is ca. 20 aa longer than *mel* PER). (See color version of this diagram in group of inserted such figures.) The PAS (interprotein-binding) motifs are located as shown roughly (cf. Figure 3); these regions of the polypeptides make up approximately the middle one-third of the N-terminal red or blue boxes on the left. Downstream of PAS is a ca. 60-aa stretch of amino acids that is inferred to be important for behaviorally meaningful function of PER, as revealed by the application of "fusion transgenes" designed by Peixoto *et al.* (1998) to produce the four types of chimeric proteins shown. These contain aa sequences encoded by portions of *per* cloned from *mel*, with other such sequences stemming from part of *pseudo*'s gene. The repeat region within *mel* PER consists of ca. 20 pairs of threonine–glycine (Thr–Gly = T–G) dimers (see Figure 3); whereas the *pseudo* intra-PER repeat contains a few such T–G pairs and a degenerate repeat of about 35 5-mers. When *per* coding sequences are entirely from *D. melanogaster* and transformed into a *per*[01] host of that species, "overall rhy." (locomotor rhythmicity), reflecting the proportion of periodically behaving individuals, is "good" (left column under Locomotor Results); and the cycle durations (τ) for this free-running behavior (in DD) are in the normal ca. 24-hr range (middle column), provided that *mel-per*'s 5'-flanking region (cf. Figure 9) is fused to the gene's coding sequences. Indeed, all the transgenes implied in this figure (including the interspecific chimera's) contained such *mel–per* regulatory sequences. When they were fused upstream of an all-*pseudo* ORF, rescue of *per*[01] (in a *D. melanogaster* host) was "OK"; and τ's were ca. 29 hrs, as inferred from relatively sensitive spectral analyses of the free-running behavioral records (Peixoto *et al.*, 1998). Cruder periodogram analyses of analogous data (cf. Figures 1, 2, and 4) led to the inference that a *mel–per–regulatory/pseudo–per–coding* construct mediated rather poorly such rescue (Petersen *et al.*, 1988). The latter study included an intriguing finding that the *pseudo–per*-encoded protein, functioning in *D. melanogaster* (and especially discernible when the genetic background of such a transgenic host was *per*[+]), mediated locomotor cycles that tended to be qualitatively like those of wild-type *D. pseudoobscura* (i.e., lower activity-to-rest ratios compared with those typically observed in the daily behavioral cycles of *D. melanogaster*). For the chimeric PERs (Peixoto *et al.*, 1998), juxtaposing the 60-aa protein segment from *mel* to the *pseudo* repeat in the *m-p* 2 type led to the worst overall rhythmicity observed among the transgenic types shown. It follows that chimeric constructs encoding relatively few *mel* amino acids in the N-terminal half of PER (*m-p* 3 or 4) paradoxically worked better than did the *m-p* 2 type. In these "good rhy./normal τ" *m-p* 3 and *m-p* 4 cases (although the periods were a bit long in *m-p* 4 at 25–29°C, hence the ~), juxtapositions of the *pseudo* 60-aa sequence with the *pseudo* repeat were inferred by Peixoto *et al.* (1998) to carry the day; the presence of *mel* PAS in this chimeric protein may also be important and the reason that it functions better than does all-*pseudo* PER. In contrast, the diverged 60-aa sequence from *mel* makes the protein function less well when it is next to the *pseudo* repeat in *m-p* 2. Functioning "well" included good temperature compensation of locomotor τ's (rightmost Results column) in the *m-p* 3 or 4 cases, whereas the *m-p* 2 transgenic flies gave free-running periods that varied from ca. 24 to 30 hrs over an 18–29°C range. The final chimeric *per* type, *m-p* 5, was temperature-compensated even worse: Sequences encoding N-terminal residues from *mel*—including PAS and extending through the N-terminal *half* of the 60-aa segment)—were fused to the remainder of the *pseudo* protein (i.e., its yellow C-terminal half of the 60-aa region (green)—then the *pseudo* repeat, and so forth; this transgene gave good rescue of *per*[01], but as long as ca. 35–40 hrs in relatively warm conditions (Peixoto *et al.*, 1998). Temperature effects on the all-*pseudo* transgenic type occurred only with respect to proportions of rhythmic individuals (as indicated in the asterisked remark), in that the τ values were steadily ca. 29 hrs (hence "long") over the 11-degree range. This figure was adapted from Peixoto (2002).

Table 5. Homologs of *Drosophila melanogaster*'s *period* and *timeless* Genes in Other Insect Species[a]

Species	Insect type	Complete or Partial clone	Remarks
per			
D. simulans	mel. subgroup within Sophophora subgenus	C	Analyzed for intraspecific variation (mainly for Thr-Gly-(T–G)-repeat-encoding region), as has mel. per; sim. form effects solid rescue of melanogaster's per^{01}-induced arrhythmic B and effects sim.-like song rhythm period; "C" means entire gene cloned (for transformations), but complete sequence never reported
D. yakuba	mel. subgroup	C	Solid rescue of mel. per^{01}-induced B arrh.
D. santomea	mel. subgroup; close to yakuba	P	Phylo. only
D. mauritiana	mel. subgroup	P	Phylo. only, but including neutrality tests of PAS-region variations
D. sechellia	mel. subgroup	P	Ditto
D. erecta	mel. subgroup	P	Ditto
D. orena	mel. subgroup	P	Ditto
D. tessieri	mel. subgroup	P	Ditto
D. ananassae	mel. group	P	Phylo. only
D. serrata	mel. group	P	Phylo. only
D. athabasca	Group within Sophophora	P	Three semispecies analyzed for sequence variation and phylogeny by restriction enzymes (thus "P" barely applicable); this group is polymorphic for 21-bp deletion within PAS-encoding region
D. willistoni	Group within Sophophora	Almost C	T–G-repeat-encoding region unusually conserved, within this species and within and among other closely related ones of this group (see footnote)
D. saltans	Group within Sophophora	P	Phylo. only
D. pseudoobscura	obscura group within Sophophora	C	Weak rescue of melanogaster's per^{01} induced arrhythmic B; also used for studies of intragenic repeat region coevolving with neighboring sequences (Figure 14)
D. miranda	pseudo. sibling species	P	Phylo. only
D. persimilis	pseudo. sibling species	P	Phylo. only
D. p. bogatana	pseudo. sibling species	P	Phylo. only
D. virilis	Within Drosophila subgenus (as opposed to Sophophora)	C	First demonstration (along with analysis of D. pseudoobscura) of blocks of conserved vs. nonconserved PER-coding sequences
D. lummei	virilis phylad	P	Phylo. only
D. a. americana	virilis phylad	P	Phylo. only
D. a. texana	virilis phylad	P	Phylo. only
D. novamexicana	virilis phylad	P	Phylo. only
D. albomicans	nasuta subgroup within Drosophila	P	Phylo. only; also see footnote

Table 5. *continued*

Species	Insect type	Complete or Partial clone	Remarks
D. immigrans	nasuta subgroup	P	Phylo. only; but also see Remarks attached to D. pseudoobscura (about intra-per repeat)
D. mediostriata	Within Drosophila	P	Ditto
D. mojavensis	Within Drosophila	P	Ditto
D. robusta	Within Drosophila	P	Ditto
D. pictiventris	Large Hawaiian picture-winged	P	Ditto
D. picticornis	Large Hawaiian picture-winged	P	Ditto
Zaprionus tubercultasus	Small (mel.-sized) Hawaiian fruitfly	P	Ditto
Chymomyza costata	Drosophilid	C	Encodes a transcript that exhibits cycling in LD and dampening thereof in DD; a handful of intragenic mutations identified in a strain of this species that exhibits anomalous nonphotoperiodic diapause and eclosion arrhythmicity (although these per mutations are not the main etiology of such defects)
Musca domestica	Housefly	C	Solid rescue of melanogaster's per^{01}-induced arrhythmic B; complementary to ca. 9-kb mRNA
Rhagoletis completa	Walnut huskfly	P	Phylo. only
Loxecera albisata	Muscid-type fly	P	Phylo. only
Beris vallata	Soldierfly	P	Phylo. only
Syritta pipiens	Hoverfly	P	Phylo. only
Luclila cuprina	Blowfly	C	encodes ca. 1000-aa protein (200 residues smaller than melanagaster PER), which, along with the encoding mRNA, exhibits daily abundance cycle
Lucilia sericata	Blowfly	P	no geographical variation of the short T-G repeat (same as negative result for L. cuprina)
Apis mellifera	Honeybee	C	complementary to ca. 8.5-kb, cycling mRNA which exhibits increased abundance in foragers
Manduca sexta	Hawkmoth	P	nucleic-acid and immuno probes used to assess per expression within brain and compound eyes; per PAS-domain immunogen used to make antibody inferred to cross-react with C. pomonella PER
Anthereae pernyi	Silkmoth	C	3' one-third of sequence (with reference to melanogaster's) absent; complementary to ca. 8-kb mRNA; weak rescue of mel. per^0
Hyalophora cecropia	Saturnid moth (as is A. pernyi)	P	complementary to ca. 8-kb mRNA

continues

Table 5. *continued*

Species	Insect type	Complete or Partial clone	Remarks
Bombyx mori	Silkmoth	P	among 26 Lepidopteran forms whose sequences are available, including this species and the 3 preceding, for which P data only reported except in the case of *A. pernyi*
Cydia pomonella	Codling moth	P	gene's presence and expression of mRNA and protein (which cycle in LD) also inferred from application of heterologous probes
Periplaneta americana	Cockroach	P	complementary to ca. 9.5-kb mRNA
tim			
D. simulans	*mel.* subgroup	P	initiating Met codon at 5′ end of ORF, downstream of where one of two polymorphically usable such codons located in *melanogaster*
D. yakuba	*mel.* subgroup	P	see *D. simulans*
D. virilis	*Drosophila* subgenus	C	initiating Met codon as in *D. simulans* and *yakuba*
D. hydei	*Drosophila* subgenus	P	—
A. pernyi	Silkmoth	C	sequence available at GENBANK accession number AF132032

[a]The interspecific homologs of these two clock genes are listed (within each section of the table) in an order that roughly corresponds to decreasing evolutionary relatedness to *D. melanogaster*. The majority of the *per* relatives have been sequence-analyzed solely from phylogenetic perspectives (indicated by "Phylo. only" entries under Remarks), as opposed to chronobiological ones as well. *per's* from several members of the *willistoni* group in addition to *D. willistoni* itself have been partially sequenced by Gleason and Powell (1997), but the nucletotide data (from the middle of the ORF into its 3′ half) were reported in a verbal-only manner for *D. equinoxialis, D. paulistorum, D. pavlovskiana, D. tropicalis, D. insularis, D. neulosa, D. sucinea,* and *D. capricorna*. Similarly, *per's* from several members of the *nasuta* subgroup other than *D. albomicans* and *D. immigrans* have been partly sequenced, but conceptual amino-acid (aa) sequence data were reported only for these two species, by Zheng *et al.* (1999) and Peixoto *et al.* (1993), respectively; the former investigators (verbally) noted high conservation within a central region of *per* for *D. nasuta* (itself), *D.s. neonasuta, D. kepulauana, D. kohkoa, D. niveifrons, D. pallidrifrons, D. pulaua, D.s. sulfurigaster, D.s. albostrigata, D.s. bilimbata,* and *D.s. albostrigata*). *per's* from 22 lepidopteran species other than those tabulated were partially sequenced by Regier *et al.* (1998); the only four listed involve moths from which additional results were obtained (see Remarks and text) or which have been studied in one chronobiological manner or the other (see text and Table 6). The other moths and butterflies whose *per's* were partly sequenced ("Phylo. only;" relatively close relatives bracketed) are *Plodia interpunctella;* {*Phaedropsis alitemeralis, Mesocondyla dardusalis*}; *Rupela albina; Asterocampa clyton; Phyciodes tharos; Colias eurytheme;* {*Campaea perlata, Plagodis fervidaria, Eutrapela clemetaria, Biston betularia*}; *Anagrapha falcifera; Lacosoma chiridota; Apatelodes torrefacta;* {*Lapara coniferarum,* close to M. *sexta* (see body of table)}; *Paonias myops;* {*Antherina suraka,* close to *A. pernyi* (see body of table)}; {*Callosamia promethea,* close to H. *cecropia* (see body of table)}; *Citheronia sepucralis, Automeris io, Janiodes* sp. The *per* sequence for *A. mellifera* was not reported in Toma *et al.* (2000), but is available at GenBank accession number AF 159569 (prefix AmPer). The *tim* sequence for *A. pernyi* was determined by S. M. Reppert and co-workers and is "published" only via the GenBank number tabulated. Abbreviations: C for complete *per* or *tim* open-reading-frame (ORF) sequence; P for partial such nucleotide sequence; bp, base pairs; others (e.g., B, f-r) are as in previous tables.

Subsequences of *per* that spread out beyond the repeat-encoding nucleotides in both directions to encompass most of the ORF have been analyzed to ask whether departures from neutrality can be inferred in terms of evolutionary divergence. For some of the intragenic regions involving certain closely related taxa, reductions in nucleotide variations forced rejections of the "neutral hypothesis" as applied to explain the base-pair substitutions observed (Rosato *et al.*, 1994, 1997a; Ford *et al.*, 1994; Hilton and Hey, 1996; Wang and Hey, 1996). Several of these molecular comparisons usefully (if unwittingly) concentrated on certain proportions of *per*'s sequence that happen to connected with known features of the protein's expression (i.e., PER's PAS region, extending in a C-terminal direction to the T–G repeat). These results imply that natural selection has been acting on *per* (or sequences near that locus), probably with regard to the adaptive significance of the gene's function and of subdomains of the encoded protein (an issue discussed by Costa and Kyriacou, 1998).

One of the studies referred to in the foregoing passages exemplifies how part of the *per* story in insects has been moving entirely away from *Drosophila* (Table 5). For example, in more widely ranging diptera, sequences corresponding to the T–G region have been shown to be short and not as variable (both in length and amino-acid composition) compared with *per*'s in the *Drosophilidae* (Nielsen *et al.*, 1994). This relative lack of variation for this part of the gene in large flies includes none observed geographically in one study that took this factor into account (Warman *et al.*, 2000).

B. Chronobiological implications of ways that *per*'s evolutionary variants are expressed

Another large fly is the house one. Its *per* sequence as a whole conforms to the comparative pattern just stated (Piccin *et al.*, 2000). However, generation of a "PAS tree" that included comparisons among this region of the protein (Figure 14) among *Musca domestica, D. melanogaster, D. pseudoobscura,* and *D. virilis* placed the housefly's PAS domain closer to that of *D. melanogaster* than to those of latter two *Drosophila* species. These investigators inferred a potential functional significance for these sequence comparisons from behavioral bioassays of *Musca per*-coding sequences, fused to a *D. melanogaster per* "promoter" and transformed into *per*[01] hosts of that *Drosophila* species: Rescue of arrhythmicity was much better than the biological activity of *per* sequences cloned from *D. pseudoobscura,* which had been previously applied in the same kind of interspecific gene-transfer experiments (Petersen *et al.*, 1988; Peixoto *et al.*, 1998).

Expression of *period* gene products has been examined in additional insect species, the discussion of which will soon move us away from flies altogether. First, however, it is notable that *per* mRNA cycles in extracts of the blowfly, *Lucilia cuprina;* the peak time was at least a couple of hours earlier compared with the daily maximum for *D. melanogaster.* Consistent with these temporal patterns, a PER peak for *L. cuprina* (inferred by application of an antibody made against the *D. melanogaster* protein) was observed relatively early in the nighttime (ZT15), 3 hr after the transcript maximum (Warman *et al.,* 2000). A slightly less extensive temporal study of *per* expression in a drosophilid fly species, *Chymomyza costata,* was performed subsequent to molecular identification of this gene by Shimada (1999). Thus, Kostál and Shimada (2001) inferred abundance fluctuations of *C. costata*'s *per* transcript (by applying a reverse-transcriptase/PCR method) and detected one strong mRNA peak at about ZT15 in an LD cycle. A similarly timed, but lower amplitude peak was observed during the first DD cycle (during which RNA continued to be extracted from adult heads), but the apparent cycling disappeared by day 2 of constant darkness (Kostál and Shimada, 2001). Nevertheless, these data provided a useful molecular control for assessment of how a rhythm mutant in *C. costata* might cause defective *per* expression (see Section VII.E).

Earlier, the *period* gene had been cloned from the silkmoth, *Antheraea pernyi* (Reppert *et al.,* 1994), by investigators who had its chronobiology in mind (at least in terms of gene-expression studies in addition to sequence comparisons). The informational content of silkmoth *per* observed led to certain conclusions similar to those stemming from sequencing several fly species's worth of the gene: The usual conserved versus diverged regions were determined, and this conclusion was bolstered by analysis of partial *per* fragments obtained from another lepidopteran species, from housefly (previewing Piccin *et al.,* 2000) as well as from a cockroach (Reppert *et al.,* 1994). Later this kind of analysis was extended to 26 lepidopteran species; the main impulse here was phylogenetic analysis of family and superfamily relationships within this insect order (Table 5), as deduced from sequencing a 0.9-kb "insect-conserved" subset of *per* from these species (Regier *et al.,* 1998).

A. pernyi's *per* was sequenced in its entirety (Reppert *et al.,* 1994). Notwithstanding the conservation noted above, silkmoth *per* was revealed to make a protein appreciably shorter than those of flies (which tend to be composed of 1,000–1,200 aa); this lepidopteran PER is "missing" ca. 400 residues present in the C-terminal one-third of the *D. melanogaster* protein. Yet all was well, for the moment, when expression patterns of *A. pernyi* PER were preliminarily examined: Application of anti-*D. melanogaster* PER, generated against a conserved stretch of 14 aa near the site of the fly's *per*[S] mutation (Siwicki *et al.,* 1988), showed strong signals in the moth's photoreceptor nuclei soon after lights-on in a 17L:7D cycle; there was no such staining 8 hr after that transition point (Reppert *et al.,* 1994). Concomitantly, extracts of *A. pernyi* heads revealed a *per* mRNA-abundance peak in the late nighttime (1 or 4 hr before lights-on, depending on the timecourse plotted).

However, further analyses of these temporal and spatial patterns caused certain conundrums to rear their ugly heads insofar as molecular regulation of insect rhythmicity is concerned. Thus:

(i) Moth-PER immunoreactivity in the adult brain was determined by application of an antibody made against the *pernyi* protein; the signals were solely cytoplasmic in—and in fact present in neurites projecting from—the four pairs of brain cells that stained (Sauman and Reppert, 1996a); this is completely different from what is observed in D. *melanogaster*, including that *per* neurons in the fly are much more numerous (Figures 11–13).

(ii) The cytoplasmic PER signals in moth-brain cell bodies and axons cycled in their apparent abundance, being relatively low or bottoming out during the first half of a 17-hr L period, then rising to a maximum between ZT12 and 17 and staying high throughout the short (7-hr) nighttime.

(iii) *per* mRNA cycling in brain alone (not whole-head extracts as in Reppert *et al.*, 1994) was defined by minimum levels in midday and the same kind of ostensible plateau as that reached in the immunoreactivity timecourse during the late daylight hours; these temporal patterns are different (at least in terms of PER) from the photoreceptor ones in this moth and from either the eye or brain cyclings observed in *Drosophila*—including a lack of apparent mRNA/protein delay in A. *pernyi* brains (Sauman and Reppert, 1996a).

(iv) TIM expression could be inferred in these lepidopteran tissues by application of two antibodies made against the *Drosophila* protein; the signals were colocalized with PER immunoreactivity, including subcellularly, and a staining cycle within these eight brain cycles was found to be temporally congruent with that determined immunohistochemically for PER.

(v) A naturally occurring *per* antisense transcript was inferred by application of the appropriate A. *pernyi* probe (encompassing PAS-encoding sequences and those 3′ of that intragenic region); both this material and the encoding sense mRNA colocalized in brain cells with PER, but the antisense RNA cycled essentially 180° out of phase with the sense substance (Sauman and Reppert, 1996a).

(vi) Extension of the moth studies to other tissues and life-cycle stages showed PER to be characterized not only by cycling signals, but also nuclear ones in the embryonic and the larval gut (Sauman *et al.*, 1996; Sauman and Reppert, 1998); embryonic brains coexpressed PER protein with TIM-like immunoreactivity in four pairs of cells, within which the signals were neither nuclear nor cyclically varying (the same was so for fat-body cells of embryos).

(vii) Injection of antisense *per* RNA—involving *experimental* such material designed to flank the translation-start site—into pharate larvae disrupted hatching behavior; this phenotype was determined to be circadianly gated in normal (late) embryos, but became aperiodic after the antisense-RNA injections (Sauman et al., 1996). Therefore, it would seem as if there are (in the phrase of these authors) "novel mechanisms" by which clock proteins regulate rhythmicity in silkmoth (Sauman and Reppert, 1996a).

Studies of another moth species, which overlapped those performed in A. pernyi in terms of the neural tissues examined, were in a way designed to break the tie. Thus, Wise et al. (2002) inspected period expression in the hawkmoth, Manduca sexta. mRNA and protein were shown to be produced by per in eyes and in several CNS regions, including the optic lobes. Many of these cells (aside from compound-eye photoreceptors) were judged to be glia, but among the per-expressing neurons were scores of cells near the "accessory medulla," in a region similar to that which contains per-LNs in Drosophila. Almost all the PER immunoreactivity in the (anterior) CNS and PNS of M. sexta was nuclear (Wise et al., 2002). However, per mRNA was found to exhibit daily cycles of in situ-hybridization signals only in glia. So, if apparently not exerting a feedback effect on the encoding transcripts, what is PER doing in the nuclei of neurons and photoreceptors? Moreover, that kind of subcellular localization was not universal within the hawkmoth brain: Neurosecretory cells were stained by anti-(M. sexta) PER, which happen also to express a peptide called corazonin (mentioned in another context later; see Table 6). These four pairs of neurons exhibited both nuclear and cytoplasmic signals and were judged by Wise et al. (2002) to be "homologous" to the small number of similarly located "cytoplasmic PER cells" within the silkmoth brain (see earlier). Thus, the tie was not broken, because the spatial pattern for this (presumed) clock factor in M. sexta "combines some features of per expression in both D. melanogaster and A. pernyi" (the bottom line of Wise et al., 2002).

Further studies in this arena spoke to additional ways that the lepidoptera may part company with the fly's chronobiology: Manipulations of neural tissues in the context of per expression in nonneural ones showed that "PER movement into the nuclei" of larval gut cells in A. pernyi—which can be inferred per se to be rhythmic, perhaps accompanied by protein-abundance oscillations (Sauman et al., 1996a)—is dependent on signals emanating from the brain (Sauman and Reppert, 1998). This nonautonomy would be different from the self-contained molecular oscillations that operate in a variety of Drosophila tissues off-shooting from a fly head or well posterior to it (Section VI.D).

All in all, the functions of clock genes, and how they may mediate output rhythms, seem no longer appreciable by considering the abundance of findings obtained in fruit flies alone—including the gene-product abundance oscillations that are such a prominent and nearly ubiquitous feature of the per and tim patterns of temporal expression in Drosophila. However, bear in mind that something like cyclical PER depends on the cellular context: This protein remains cytoplasmic and at temporally constant levels within ovarian cells of *Drosophila* females (e.g., Liu *et al.*, 1992; Hardin, 1994). There is context dependence for A. *pernyi* PER as well, even within a given insect CNS. This conclusion stems in part from transformation experiments in which the silkmoth gene put into *per^{01} D. melanogaster* mediated weak and anomalous rhythmicity (Levine *et al.*, 1995). The biological efficacy of A. *pernyi per* in *Drosophila* is superficially about as good as that of the *D. pseudoobscura* gene in the same circumstance (Petersen *et al.*, 1988), although Peixoto *et al.* (1998) tacitly dispute this (see legend to Figure 14). In any case, an ostensible corollary to the biological activity of the entire silkmoth sequence functioning in *D. melanogaster* is that the protein produced by the incoming gene was observed to be *nuclearly located* within at least some of fly's lateral-brain neurons (Levine *et al.*, 1995). PER in A. *pernyi* itself, whose posttranslational qualities may be different in a lepidopteran versus a dipteran milieu, can display different subcellular localizations (Sauman and Reppert, 1996a; Sauman *et al.*, 1996; Wise *et al.*, 2002), depending on the cell type (compound-eye photoreceptors vs. certain brain neurons vs. other such cells).

Additional findings involving the silkmoth form of this gene have cast doubt on certain inferences as to how its expression *may* regulate circadian rhythmicities: The natural *per* antisense RNA is not produced by the main gene, which is located on A. *pernyi*'s Z chromosome, in (ZZ) males and (ZW) females; there is an intraspecific relative of *per* located on the W chromosome, which produces a truncated form of PER and from which the female-specific antisense RNA may be transcribed (Gotter *et al.*, 1999). Given that males are biologically rhythmic in general in this species, the antisense *per* material cannot be globally required for clock functioning. With regard to the products of this moth gene and its one piece of experimentally determined chronobiological significance, the *application of exogenous antisense RNA proved to be irreproducible*, in terms of its effects on both the hatching rhythm and endogenous levels of PER protein (Sauman *et al.*, 2000).

Even as elements of the maybe-meaningful components of this heretical story unravel, we need to keep wondering about how a protein like PER might function as a cytoplasmic entity, either in neurons (Sauman *et al.*, 1996) or elsewhere (Liu *et al.*, 1992; Sauman *et al.*, 1996). Further considerations in this regard have stemmed from application of the aforementioned PER antibody that was made against a small fragment of the protein, encoded by an evolutionarily conserved subset of the gene. However, Regier *et al.* (1998) found this region to be more widely variable than might have been expected, by sequencing part of *per* across the several lepidopteran species referred to above. That aside, it seemed at least temporarily intriguing that this antibody stains putative rhythm related neurons in a variety of species:

(i) In the beetle *Pachymorpha sexguttata*, a network is formed by the labeled structures, including those that could be argued to include the neural substrates of the animal's behavioral rhythmicity; most of these immunohistochemical signals were in the cytoplasm of the relevant neuronal cell bodies as well as their neurites, and no cyclicity was observed (Frisch et al., 1996); these results from P. sexguttata are similar to those from moth, although a boiled down and (even) less interpretable version of them. [There was no evidence that the immunoreactivity observed in beetle specimens reflected bona fide PER; this was not known as such in silkmoth either, but seemed likely because signals from in situ mRNA hybridization colocalized with the immunohistochemical staining (Sauman and Reppert, 1996)].

(ii) Presaging the results from anti-PER application to neural tissues of P. sexguttata, the same reagent led to staining of molluscan pacemaker structures, such that the relevant eye neurons in Aplysia and Bulla showed cytoplasmic perikaryal and axonal signals (Siwicki et al., 1989).

(iii) Crossing the boundary between invertebrates and bone-ridden animals, this anti-PER material detected proteins in eye or brain extracts of Xenopus (García-Fernández et al., 1994); none of the Western-blot bands exhibited day versus night differences in apparent polypeptide concentrations; however, immunohistochemical signals suggested that elements of the anti-PER detected proteins may function within the frog's rhythm system (e.g., labeling of the hypothalamus and of immunoreactive fibers that enter the pineal gland).

(iv) This antibody even produced immunochemical signals in the rat's suprachiasmatic nucleus (SCN) and in extracts of hypothalamic brain tissue (Siwicki et al., 1992b); one band in the Western relevant blottings appeared to cycle in its abundance (Rosewell et al., 1994), although the peak time was later than those of bona fide PER proteins (two gene products's worth) in mouse (Field et al., 2000); the in situ side of the earlier study revealed little in the way of neuronal-process signals, but cell-body staining within the SCN appeared cytoplasmic (Siwicki et al., 1992b).

Inferences one might wish to draw from these out-of-insect findings are minimal or dubious: Only for Aplysia has the antigen reactive with this particular anti-PER material been identified, and the protein seems nothing like PER as we know it (Strack and Jacklet, 1993). As for mammalian forms of bona fide PER relatives, application of antibodies against these proteins to SCN specimens or extracts (e.g., Field et al., 2000) led to nuclear stainings and apparent molecular weights different from than that inferred for putative PER-like material detected by the Drosophila antibody (Rosewell et al., 1994).

Returning to the insect comfort zone, two further per-expression studies were performed in nondipteran forms. These investigations were based on tissue and temporal issues related to the biology or behavior of the large insects in question. In one of them, the codling moth, *Cydia pomonella*, a rhythm of sperm release operates autonomously in reproductive systems removed from males and placed in culture (reviewed by Giebultowicz, 1999, 2000, 2001). This phenomenon probably prompted similar experiments performed with various *Drosophila* tissues, albeit mostly from the limited perspective of autonomous *per*-product cyclings (Section VI.D). Inferences about expression of the moth gene have been obtained, connected with a hope that the molecular timecourses would correlate with the biological one. Thus, a probe designed to detect *C. pomonella*'s *per* mRNA (Regier *et al.*, 1998) showed cycling of a transcript (whose size suggested it is PER-encoding) in extracts of head and testis (Gvakharia *et al.*, 2000). Maximal levels were observed at the end of the night (heads) or throughout most of that timespan in a 16-hr L:8-hr D cycle, but the mRNA rhythm was very weak or nonexistent in DD. *In situ* hybridization led to a picture similar to those observed in assessments of the spatial distribution for *per* products in nonneural tissues of *Drosophila*: In these moth males, LD-cycling signals were observed only in portion of the testis, that is, within epithelial cells that form the wall of the upper vas deferens (UVD) and in others located in the terminal epithelium; application of a PER antibody (designed to detect PAS amino-acid sequences in the protein of the distantly related moth *Manduca sexta*; cf. Wise *et al.*, 2002) showed an oscillation of LD immunoreactivity within nuclei of the corresponding intra-testis locations; the upswing of the protein's apparent abundance lagged behind mRNA accumulation in these epithelial cells (Gvakharia *et al.*, 2000). Fortunately, *per* is expressed in the posterior tissue that manifests biological oscillations, and the increase in testis *per* mRNA happens to coincide with accumulation of sperm in the UVD, preparatory to their gated release. However, it would seem as if the transcript's abundance oscillation is not necessary for the free-running rhythmicity of this piece of male biology. Yet it is possible that PER in the moth testis cycles in DD, given that such a protein rhythm can occur (in *Drosophila*) in the absence of abundance fluctuations for the encoding mRNA (Section V.C.2).

One chronobiological study of the male reproductive system in moths may have brought the biological rhythmicity exhibited by these tissues one step closer to the clock. Thus, Bebas *et al.* (2002) found circadian cycling of pH in the UVD's lumen. The moth species in which these observations were made is *Spodoptera littoralis*. As in *C. pomonella*, sperm are released in a daily rhythm (Bebas *et al.*, 2001); it is abolished by constant light, a treatment that sterilizes *S. littoralis* males by causing reductions in sperm quantity and quality (Bebas and Cymborowski, 1999). The trough for the pH rhythm correlated with an increase in concentration of a vacuolar H^+-ATPase's B subunit in the apical portion of the UVD epithelium (Bebas *et al.*, 2002). Therefore, this rhythm of the proton pump's expression could be underpinned by a clock-controlled gene (Section X.C), such that the physiological rhythm is a short step downstream of the circadian pacemaker than runs autonomously in this tissue (Giebultowicz *et al.*, 2001). Cyclical production of at least one portion of the multisubunit H^+-ATPase would result in acidification of the UVD lumen at the "right time"; this may be required for sperm maturation, as Hinton and Palladino (1995) hypothesized is operating with regard to the low pH observed in the sperm ducts of mammals. Bebas *et al.* (2002) went on to show by drug treatments that apical translocation and enzymatic activity of the proton pump are necessary for the nighttime pH drop that usually occurs in the distal region of *S. littoralis*'s UVD. Therefore, this system seems nicely poised to determine how one or more clock genes, presumably expressed in the male reproductive system of this moth (Gvakharia *et al.*, 2000; Beaver *et al.*, 2002), specifically regulate the fluctuating transcription of at least the proton-pump's B-subunit gene.

   A very different kind of temporal parameter was considered in the context of clock-gene activity in the honeybee. For this, the entire ORF of the *period* gene was cloned from *Apis mellifera* and sequenced (Toma *et al.*, 2000), although its similarities to other insect forms of the gene were noted only in a verbal manner (see Table 5). Regulation of *per* transcript levels within a given daily cycle was found to occur as usual (if you will), exhibiting nighttime peaks. The additional temporal twist involved determining of average (cycling) levels of *per* mRNA extracted from bee brains of different ages (Toma *et al.*, 2000): Young adult females exhibited lower abundance of the (cycling) transcript compared with progressively older ones; this was at least roughly correlated with the maturation of free-running locomotor rhythmicity (<5% of 1- to 4-day individuals were rhythmic compared with ca. 75% for 13- to 16-day bees). As these insects age, many of them become foragers, and the level of *per* mRNA was about threefold higher in those bees compared with 4- to 9-day-old ones. This was not merely an age effect, because precocious foragers (induced to take on that quality by the absence of older bees) had transcript levels similar to the relatively high abundance determined from normally maturing foragers (Toma *et al.*, 2000). Further manipulations of this system attempted to uncouple behavioral rhythmicity, age, and socially related tasks in conjunction with monitoring *per* mRNA levels (Bloch *et al.*, 2001). No link between such abundance measurements and either locomotor rhythmicity or age were detectable; nor were any molecular corollaries found with nursing versus foraging activities in experimental situations that socially promoted precocious or reversed behavioral development. Nursing bees and other relatively young hive-bound ones were shown to exhibit high or low levels of brain-extracted *per* mRNA depending on the social environment, whereas both foraging bees or others that had progressed to that age of behavioral development exhibited similarly high abundances of this gene product (Bloch *et al.*, 2001). Moreover, foragers maintained in LD exhibited a higher amplitude of *per* RNA cycling compared with nurses, regardless of age (Toma *et al.*, 2000).

C. Variations of *timeless* gene sequences within and among species

Descriptions of different forms of *tim* can be finessed rather quickly because of the paucity of data collected so far (Table 5). However, one feature of the variability for this gene is special: In the four non-*melanogaster* species of *Drosophila* for which *tim* sequences are available, usually involving certain fragments of the genes (Myers *et al.*, 1997; Ousley *et al.*, 1998; Rosato *et al.*, 1997b), the translation-initiating methionine (Met) codon is approximately 20 aa's worth of nucleotides downstream of the start codon originally determined for *D. melanogaster* (Myers *et al.*, 1995). These data are interesting to ponder in the context of intraspecific comparisons: Sequencing *tim* from various stocks of *D. melanogaster* showed that flies of a given strain can use either the upstream initiating codon and the downstream one—to make two TIM forms of slightly different lengths—or only the one that is 23 aa shorter; a nucleotide substitution and a deletion between the genomic source of these two codons in the "short-TIM" strains would cause translation initiation at the upstream Met codon soon to encounter an in-frame stop one (Rosato *et al.*, 1997b). The latter property (albeit not always the same in terms of which specific nucleotides vary or are deleted) was observed in all *tim* sequences from the 5′ regions of this gene in *D. simulans*, *D. yakuba*, *D. hydei*, and *D. virilis*. The potential meaning of "one-TIM" in these four species, compared with the capacity to produce two forms of the protein in some individuals of *D. melanogaster*, has been speculated to be biologically significant. Preliminary findings were noted (but not documented) that showed the protein-length polymorphism to be "ubiquitous [in] natural European populations"; moreover, they revealed a "latitudinal cline in its geographical distribution" (Costa and Kyriacou, 1998). If a population at a given latitude can thus be inferred to contain some individuals with the inter-Met stop codon, this would imply intra*strain* variation of a sort. In contrast, the six laboratory stocks of *D. melanogaster* initially surveyed for this molecular property were such that all flies within two of them could make short-TIM only, whereas each individual within the other four strains had the potential to make both forms of this protein (Rosato *et al.*, 1997b).

The complete sequence of *tim* from one non-*melanogaster* species suggested greater conservation of this gene compared with *per*. TIM in *D. virilis* exhibits about three-fourths overall "identity" with the *melanogaster* form of the protein, whereas the corresponding value for PER is only about one-half (Ousley *et al.*, 1998). A further feature of these results is that the chronobiologically functional version of *tim* known in *Drosophila*, a mutation in which is sufficient to make *D. melanogaster* arrhythmic, is quite off by itself in terms of informational content and the manner by which its product performs. This is because the aforementioned *tim2* (a.k.a. *timeout*) gene of this species is not very similar to the *tim* that was originally mutated in *D. melanogaster* (Benna *et al.*, 2000; Gotter *et al.*, 2000); and in this respect *tim2* is closer to would-be *timeless* relatives in mammals (which are reviewed by Reppert and Weaver, 2001). Bringing elements of PER into the picture shows that intrapolypeptide regions involved in protein interactions with "TIM-1" (as it were) involve relatively conserved stretches of amino acids, when both genes's worth of products are taken into account in comparisons between *D. melanogaster* and *D. virilis* (Ousley *et al.*, 1998). In other words, the TIMs of these two species are bit more similar in the interaction domains (Section V.C.1) compared with the overall identity value; however, the two PERs are some 30% more similar in these domains compared with the ca. 50% figure for the *D. melanogaster* versus *D. virilis* forms as a whole.

One highly conserved region of the *tim* gene among *D. melanogaster*, *D. virilis*, and *D. hydei* was dealt with experimentally. The attention of Ousley and colleagues was drawn to sequences near the 5' end of the ORF that encode 32 amino acids in a TIM segment of the nuclear-localization sequence (Figure 3)— but downstream of the N-terminal amino acids present in the polymorphically shorter form of TIM (i.e., the 23 extra aa made by the alternative ORF that is longer at its 5' end is not at issue here). Transforming sequences that would make full-length short-TIM into tim^{01} gave reasonably good behavioral rhythmicity (Rutila *et al.*, 1998a; McDonald *et al.*, 2001; Wang *et al.*, 2001), in that about three-fourths of the flies were periodic with normal τ's; however, a minus-33-aa version of the coding sequences led to essentially no rhythms or (depending on the transformed line) relatively infrequent ones with ca. 37- to 38-hr periods (Ousley *et al.*, 1998). One could infer that this evolutionarily conserved portion of TIM is important for some unknown feature of the protein's actions. As usual, another interpretation would be that elimination of any intrapolypeptide segment of about this residue length causes a nonspecific abnormality of overall protein structure, as opposed to involving removal of a motif that functions in some particular manner.

D. Potential genetic applications of rhythm factors identified outside *Drosophila*

The *tim* transformation experiment just described, and the interspecific gene-transfer tests mentioned in other of the passages in Section VII (e.g., Petersen *et al.*, 1988; Levine *et al.*, 1995; Peixoto *et al.*, 1998), prompt questions that suggest themselves in the context of these many species's worth of clock genes (Table 5). Is any but a small fraction of such factors going to be tractable chronobiologically— other than by drawing on interspecific sequence comparisons for *in vitro* mutagenesis or by taking a rhythm-related gene from somewhere (even mouse! Shigeyoshi *et al.*, 2002) and putting it into *D. melanogaster*? If the answer to this negatively formulated question is "yes," might most interspecific studies stay stuck at the level of describing nucleotide sequences, augmented a bit by assessing where and when given forms of genes like *tim* and *per* are expressed within some insect form or the other? Now, cloning clock factors from species well beyond *D. melanogaster* or other fly types (Table 5) facilitates determinations of spatial and temporal patterns of gene-product productions. The nucleic acids isolated by virtue of sequence similarity to fruitfly material can instantly lead to probes for *in situ* hybridization, and these sequences (as such) or knowledge of their informational content allow for antigen generation, such that the native clock proteins in whatever insect is of interest can be detected immunochemically. Another operation permitted by cloning rhythm-related genes and recombinantly effecting synthesis of their products is to bioassay such proteins by injecting them into whatever insect form is at hand (more about this in Section X.A.1).

However, the applications of material cloned from interspecific chronobiological perspectives are farther ranging in their potential, so answers to the questions posed nonoptimistically in the previous paragraph are "no." This is because transgene technology and usage is spreading out to many insect species, way beyond *D. melanogaster* (Atkinson and James, 2002). Therefore, if *genomic* clock-gene sequences were to be isolated from a non-fruit fly to augment gatherings of the coding material cloned so far (as is implicit throughout most of Table 5), then a variety of manipulable genes could be applied to *experimental* molecular-genetic studies of species for which much is already known in terms of their normal chronobiology. Thus, one can anticipate the creation of factors in which clock-gene regulatory sequences are designed to drive a heterologous transcription factor (viz., GAL4), such that cell-ablating or cell-inactivating transgenes have their damaging effects placed specifically in locations where these genes carry out their normal rhythm functions. The genetic backgrounds of such animals need not contain anything but wild-type alleles, so that the lack of rhythm mutants in hardly any species other than *D. melanogaster* (although see the next subsection) will not hinder these kinds of genetic manipulations in far-flung insect species.

Moreover, mutants themselves can be engineered. For this, *in vitro* mutagenesis of a gene such as muscid *per* or lepidopteran *tim* can be effected, drawing on the qualities of intragenic mutations known in *D. melanogaster* (Figure 3). When an amino-acid substitution is molecularly manufactured into a clock gene and gets transgenically introduced into the moth or housefly in question (cf. Hediger *et al.*, 2001), it will almost certainly be semidominant in terms of the primary mutational effect (Table 1 and Figure 3). This will permit meaningful assessments of whatever chronobiological phenotype one desires to analyze. Finally, if decrement- or loss-of-function mutations become compelling to generate in "any" insect species, then the aforementioned RNAi technology is available (Martinek and Young, 2000); the pertinent transgene-encoded RNAi would be introduced into insects with normal genetic backgrounds to impede the function of transcripts produced by whatever wild-type allele is being targeted. *Chromosomally* hypomorphic and null rhythm mutants can in principle be generated as well, without having to rely on chemical mutagenesis and screening the progeny of (for instance) vast numbers of huge insects that may take a long time to develop. Instead, the *gene replacement* technology that is emerging by way of transposon manipulations carried out in *D. melanogaster* (Rong and Golic, 2000, 2001; Rong et al., 2002) is likely to spread experimentally throughout this class of arthropods. All that one needs to do to initiate such an operation is to take an extant chronogenetic clone (Table 5) and alter elements of its sequence such that the incoming (replacing) transgene will generate a poorly functional protein (cf. dbt^{ar} in Table 1 and Figure 3) or none at all (cf. per^{01}, tim^{01}, or cyc^{01}).

E. Genetic variations causing rhythm anomalies in insects other than *D. melanogaster*

Compared with *Drosophila melanogaster*, *D. pseudoobscura* was for years much better studied with respect to several features of its chronobiology, mainly involving eclosion rhythmicity (Saunders, 1982). An occasional investigation involving the latter species dealt with genetic changes that alter its rhythms. Directional selection for eclosion time led to "early" and "late" strains (Clayton and Paietta, 1972). An analogous strain difference, in which eclosion of *D. pseudoobscura* cultures was earlier than usual, included a demonstration of an anomalously long period in constant conditions (Pittendrigh, 1981). This is counterintuitive, but happens to correlate with the entrained and free-running eclosion properties of the *psi-2* and *psi-3* mutants in *D. melanogaster* (Jackson, 1983). Some similarly underanalyzed rhythm mutants were induced in *D. pseudoobscura* (Table 1): five such variants, which were arrhythmic for eclosion and said to define two complementation groups on the *X* chromosome (Pittendrigh, 1974). This was documented subsequently (for the four mutants then remaining) by way of locomotor testing, which showed that one pair of recessive single-gene mutations failed to complement for arrhythmicity in DD, and the same was so for another pair (Hall, 1998a). The biarmed *X* chromosome in *D. pseudoobscura* corresponds to the *X* plus half of the third chromosome in *D. melanogaster* (the latter's "left" arm, a.k.a. 3L). It is therefore conceivable that the original mutagenesis of the former flies hit the *per* gene of *D. pseudoobscura*, as well as either the *Clk* or *cyc* ones; as shown in Figure 18 (see color insert), these two loci are within 3L of *D. melanogaster*. Recall that in per^{01} adults of *D. melanogaster*, certain neurosecretory cells tend to be located in ectopically dorsal brain regions (Section VI.A.1). In two of the *D. pseudoobscura* arrhythmics, a similar anatomical anomaly was observed (Konopka and Wells, 1980). However, such mutants happened to be from one of the two complementation groups in question (Table 1), prompting the speculation that these are loss-of-function *period* mutants in *D. pseudoobscura*.

Two eclosion variants cropped up in another *Drosophila* species. The *hpa* and *hra* strains of *D. jumbulina* neither entrained to light:dark cycles nor free-ran in constant darkness with regard to periodic adult emergence (Joshi *et al.*, 2002). Temperature cycles entrained rhythmic eclosion of both wild-type and *hpa* cultures developing and emerging in either DD or LL. Release from this thermoperiodic condition into constant temperature accompanied by DD allowed for eclosion rhythmicity of wild-type *D. jumbulina*, but not of *hpa* cultures; uniquely and intriguingly, the latter exhibited free-running emergence rhythms in a temperature-cycling → LL test. The other variant strain, *hra*, was thermoperiodically entrainable for eclosion in DD, but not LL, and temperature cycles were unable to synchronize these mutant cultures to exhibit free-running rhythmicity in any photic condition (Joshi *et al.*, 2002). The blowfly species mentioned earlier gave also rise to an eclosion mutant: A recessive, autosomal mutant of *L. cuprina* was identified owing to aperiodic adult emergence (Smith, 1987). This *ary* mutation was also shown to cause arrhythmicity of spontaneous flight activity in DD, although such behavior was periodic in LD.

In the melon fly, *Batrocera cucurbitae*, artificial-selection protocols led to strains in which total developmental time was reduced by ca. 15% (two lines) or increased by ca. 25% (two lines). Flies from the former pair of lines exhibited free-running locomotor periods ca. 2 hr shorter compared with a "massreared" strain ($\tau \simeq 25$ hr), and those from the slow-developing lines were ca. 2 or 6 hrs longer than normal for their behavioral τ's (Shimizu *et al.*, 1997). These correlations between developmental time and circadian rhythmicity are reminiscent of certain multiple effects of *period* mutations on temporal phenomena in *D. melanogaster*—those operating on a scale of either 1 day (Table 1) or more than 1 week (Table 3). In *B. cucurbitae*, the results of crosses between the "26-day/31-hr" line and the "17-day/23-hr" one suggested (owing to ready re-extraction of short and long individuals in the F2) that a "major" autosomal gene difference might be the etiology of these phenotypic variations (Shimizu *et al.*, 1997). Furthermore, the F2 segregants with rather short or long locomotor cycles tended to have developed relatively rapidly or slowly, respectively. However, these genetic suppositions are somewhat soft, given the particulars of the F2 τ-value spreads and the fact that the re-extracted developmental plus circadian values were not as sharp as in the starting strains.

Among the many phenotypes of *X*-linked *period* mutants in *D. melanogaster* (Table 3) are those associated with ovarian diapause. Harbingers of these results (Section IV.C.6) came from studies of other *Drosophila* species. In *D. littoralis*, two strains originating from the Caucusus, USSR, and Oulu, Finland, were identified for which the critical day length (CDL) for diapause induction was 12 versus 19 hr, respectively (Lumme and Oikarinen, 1977). The genetic etiology of this difference was found to be monogenic (Lumme and Oikarinen, 1977; Lumme and Pohjola, 1980), and the locus responsible for it (*Cdl*) was mapped to an autosomal linkage group (Lumme, 1981). A genetic difference influencing diapause-enabled *D. lummei* females vis à vis the lack of this phenotype in *D. virilis* was shown to be linked to the *X* chromosome by interspecific crosses (Lumme and

Keränen, 1978). Multigenic differences were found to responsible for strains of
D. triauraria in which ovarian diapause was either inducible by ca. half-day photoperiods or could not be elicited at all (Kimura and Yoshida, 1995). Crosses
between these strains (which led to diapausing F1 females) and analysis of recombinant inbred (RI) lines generated subsequently suggested that the distinction
between the original strains is caused by factors at three or four genetic loci,
one of them on the *X* chromosome. Performance of Nanda–Hammer experiments
(Section IV.C.6) led to "circadian peaks" for three diapausing *D. triauraria* types—
the original such strain and two of the RI lines (Kimura and Yoshida, 1995). The
three female types differed in their CDLs, indicating that that diapause parameter
can be modified independently of the relevant circadian oscillation. In a way, this
fits with the results obtained from applying *period* mutations to diapause analysis
in *D. melanogaster*, wherein none of the *per* alleles applied affected the photoperiodic clock's cycle duration, although *per*-null variants altered CDL in females of
this species (Saunders *et al.*, 1989; Saunders, 1990). Nothing is known about the
genes defined by allelic differences affecting photoperiodism in these particular
non-*D. melanogaster* drosophilids. However, certain of the diapause-affecting mutations in the other species, at least those that influence CDL, may have occurred
at their *period* loci.

 An involvement of *per* in a diapause defect identified in another
drosophild species, *Chymomyza costata*, has been entertained. This substory started
with the identification of a mutant strain in larvae entered developmental diapause
at 11°C irrespective of the photoperiod (Riihimaa and Kimura, 1988). Cultures of
this strain are also arrhythmic for eclosion, and a single-gene autosomal mutation
called *non-photoperiodic-diapause* (*npd*) was shown to cause this circadian phenotype and nonphotoperiodism as well (Riihimaa and Kimura, 1989; Lankinen and
Riihimaa, 1992, 1997). Curiously, the *X*-chromosomal *period* gene in flies from the
npd strain of *C. costata* varies from the normal allele, exhibiting a 6-bp deletion
that removes two amino acids as well as other point mutations (Shimada, 1999).
However, Kostál and Shimada (2001) showed this *per*^{npd} allele "not to be primarily responsible for the loss of eclosion rhythm" in LD conditions (even though
application of an index related to the tightness of emergence within a one-third-
day "gate" showed *per*^{npd}/Y; *npd*$^{+}$/*npd* and *per*^{npd}/*per*$^{+}$; *npd*$^{+}$/*npd* flies to be weakly
rhythmic). As introduced earlier, *per* mRNA cycles in extracts of *C. costata*, at
least in LD, but the abundance of this transcript in *npd* heads was chronically low
(Kostál and Shimada, 2001). Therefore, this gene could regulate *per* expression
"*in trans*," given that the two genes are unlinked and that we are now familiar
with autosomal mutations in *D. melanogaster* that lead to strong abnormalities of
per-product levels and cyclings thereof. In turn, the putative *npd* effect on *per*$^{+}$
may go on to disrupt not only eclosion rhythmicity, but also clock-controlled
measurement of photoperiodic time in *C. costata* (Yoshida and Kimura, 1995;
Kostál *et al.*, 2000). The "point of contact" between *period* gene functions in this

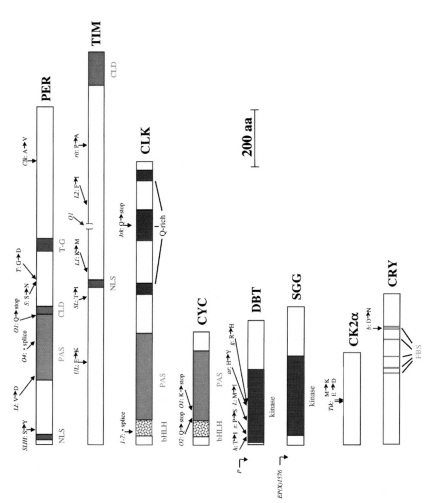

Figure 3. Proteins encoded by major rhythm-related genes in *D. melanogaster*.

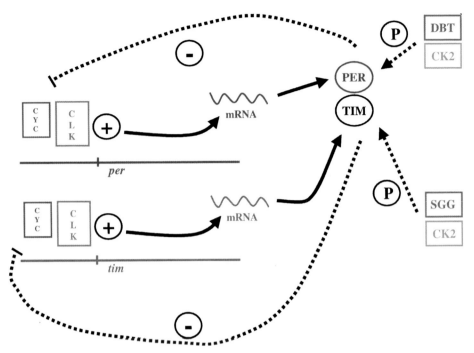

Figure 7. Rudiments of clock-gene actions in *Drosophila* and interactions among their products.

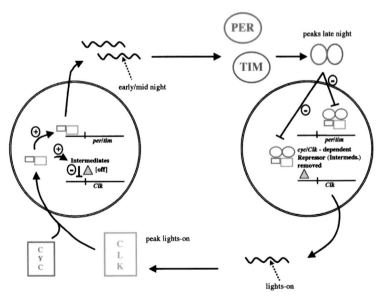

Figure 8. Expansions of the basic pacemaker mechanisim: the interlocked feedback loop hypothesis.

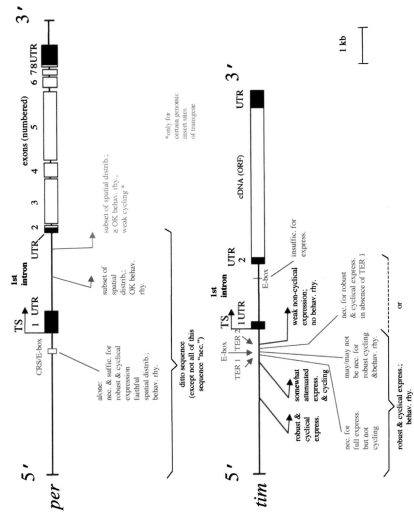

Figure 9. Dissection of gene-regulatory roles played by noncoding sequences within the *period* and *timeless* loci of *D. melanogaster*.

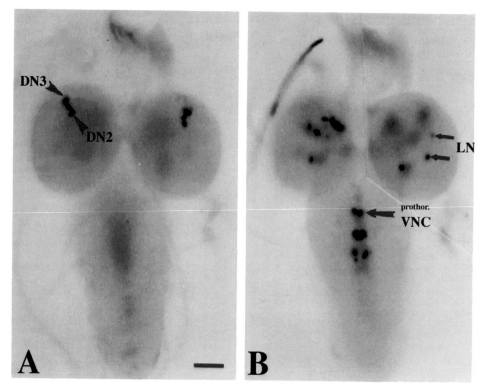

Figure 11. Larval CNS neurons of *Drosophila* marked by expression of a *per*-reporter transgene.

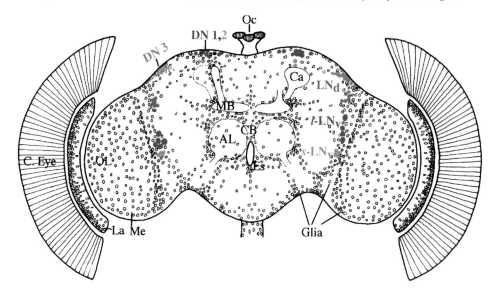

Figure 11. Expression of clockgenes in the brain of adult *Drosophila*.

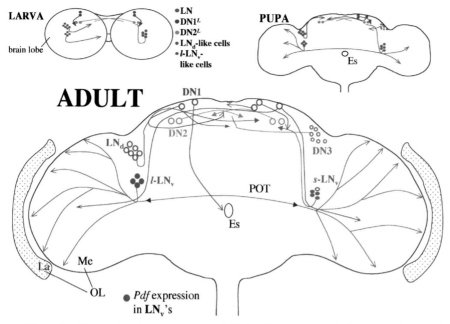

Figure 12. Neuronal projection patterns of CNS cells expressing clock genes in brains of *Drosophila* at three life-cycle stages.

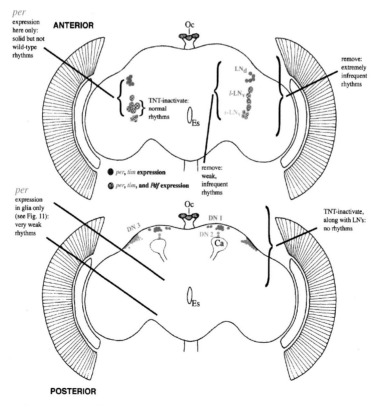

Figure 13. Neural substrates of *Drosophila's* circadian locomotor rhythm.

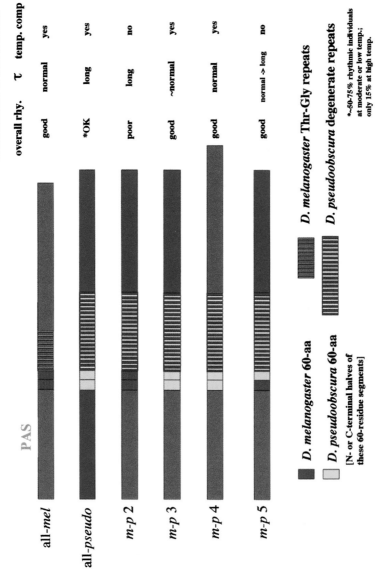

Figure 14. Behavioral effects of *period* gene sequences from *D. pseudoobscura* transformed into *D. melanogaster*.

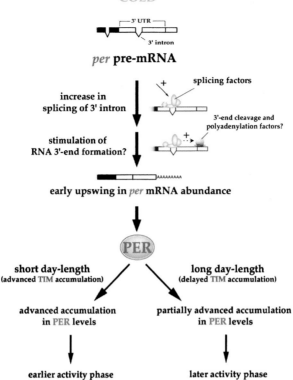

Figure 15. Thermal and photoperiodic changes combining to modulate the expression of *Drosophila's period* gene.

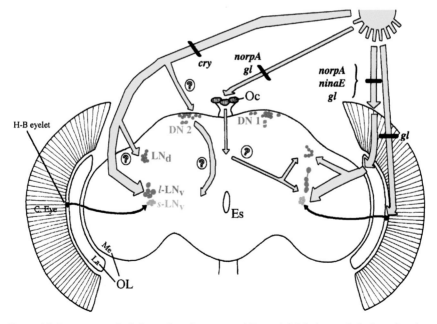

Figure 16. Input routes for light-mediated resetting of *Drosophila's* behavioral clock, and molecules functioning in these anatomical pathways.

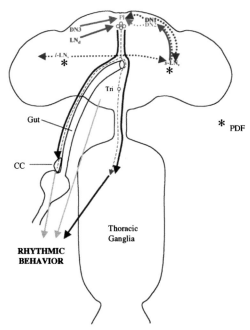

Figure 17. Anatomical and physiological outputs from neurons expressing clock genes in the brain of *Drosphila* adults.

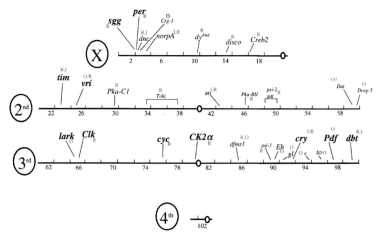

Figure 18. Mutations and genes involved in the biological rhythms of *D. melanogaster*.

drosophilid (quoting a conclusional phrase in Kostál and Shimada, 2001) may be stronger than that which is surmised (almost dismissed, really) from genetic analysis of photoperiodism in *D. melanogaster* (Saunders *et al.*, 1989; Saunders, 1990; discussed in Section IV.C.6).

A final set of non-*D. melanogaster* rhythm variants harks back to the geographical variations discussed near the beginning of this section and also sets up elements of the next one. Thus, Lankinen (1993a) found that Scandinavian strains of *D. subobscura* exhibited earlier-phased eclosion rhythms and shorter free-running periods compared with (five) strains from the Canary Islands— located far to the south of Finland, Norway, and Sweden—the source of the seven early/short strains studied. [Certain wild-caught *D. annanassae* lines also show a lower-latitude correlation with relatively early locomotor phases in LD and shorter τ's in DD (Joshi, 1999).] One of the Scandinavian *subobscura* strains exhibited rather low-amplitude eclosion rhythmicity in LD or DD, which had de- teriorated to arrhythmicity when it was reassessed 7 years later (Lankinen, 1993a). This line was dubbed *linne* (Table 1), the further analyses of which (Lankinen, 1993b) showed: (i) non-entrainability in LD cycles; but (ii) synchronization to a 4°-amplitude temperature cycle, running either in DD or LL; (iii) such entrain- ment to cause periodic eclosion in subsequent constant conditions, in which the temperature was held steady, and the cultures were either in DD or LL; and (iv) other strains (apprehended as normal ones) to be arrhythmic in conditions of constant temperature and light. Genetic crosses between a normal (LD-rhythmic) strain and *linne* showed that the etiology of the anomalies just listed is recessive and autosomal and may involve monogenic inheritance (Lankinen, 1993b). The latter inference could only be teased out with some difficulty from the eclosion profiles of F2 cultures. It is nonetheless intriguing that "coupling between light signals and the pacemaker" is apparently disrupted in the *linne* mutant (Lankinen, 1993b), a point that dovetails with subjects presented in the next two sections about changing environmental stimuli. A variety of rhythm variants will come into play, including certain of the naturally occurring ones introduced here in Section VII.

VIII. TEMPERATURE CHANGES AND HOW CLOCK-GENE PRODUCTS ARE INVOLVED IN COMPENSATING FOR THEM

A. Clock mutants with aberrant temperature compensation

Temperature cycles can entrain locomotor rhythmicity as well as periodic eclosion (Figure 5B). In another kind of manipulation of this environmental variable, stepped temperature changes leave both kinds of free-running rhythms running at their characteristic periods. Most such studies of eclosion have been performed in *D. pseudoobscura* (Saunders, 1982); those involving adult behavior have been devoted mainly to *D. melanogaster* (e.g., Konopka *et al.*, 1989; Majercak *et al.*, 1999). For this, the effects of rhythm mutations brought with them a bonus of sorts: Several of the mutants turned out to be temperature-sensitive in terms of their cycle durations. Thus, *per*[L] (Konopka *et al.*, 1989; Ewer *et al.*, 1990; Tomioka *et al.*, 1998; Matsumoto *et al.*, 1999; Rothenfluh *et al.*, 2000a) and mutations of this gene engineered to possess amino-acid substitutions at the site defined by that classic long-period mutant (Curtin *et al.*, 1995; Huang *et al.*, 1995) cause behavioral periodicities to lengthen as the temperature is raised. Inasmuch as the clock slows down in these circumstances, the abnormality exhibited by such mutants is formally termed "overcompensation" to temperature changes. In this regard, recall that the lower the level of *per* products (or their inferred efficacy), the longer is the period (Sections II.A.3 and V.B.2). Therefore, one may view *per*[L] and its analogs as not being defective in temperature compensation per se— instead, exhibiting heat-induced protein enfeeblement. Yet other *per* alleles cause the clock to run faster as the temperature is raised: *per*[S] (Konopka *et al.*, 1989; Ewer *et al.*, 1990; Tomioka *et al.*, 1998; Matsumoto *et al.*, 1999; Rothenfluh *et al.*, 2000a), *per*[T] (Konopka *et al.*, 1994), and *per*[SLIH] (Hamblen *et al.*, 1998; Rothenfluh *et al.*, 2000a) exhibit shorter periods in heated conditions. These temperature dependences are not dramatic (especially in the case of *per*[S]), but they do smack of temperature-compensation defects: Thermometer-like, the process is ratcheted up as heat is applied.

One of the *timeless* mutants brings us quickly back to overcompensated phenotypes: The *tim*[SL] mutation not only suppresses the basic longness of *per*[L], but also blocks the period-lengthening effects of higher temperatures associated with the latter mutant (Rutila *et al.*, 1996). Another *tim* allele causes an opposite kind of formalistically viewed interaction with *per* function. First, note that *tim*[rit] flies exhibit almost normal periods at 15°C but 35-hr ones (or arrhythmicity) at 30°C; second, there is a synergistic defect of this slow-clock mutation (referring to its effects in midrange thermal conditions) when it is combined with *per*[L]: Some such flies gave 45- to 50-hr periods as the temperature approached 30°C (Matsumoto *et al.*, 1999). The mutational effects in this doubly mutant type could imply a near collapse of PER–TIM associations as the temperature is raised, if the concentration of such dimers (hence effective function of the cooperating proteins) is inversely correlated with free-running period (see above). However, *tim*[rit] is not mutated in a region of TIM that interacts with its partner (Matsumoto *et al.*, 1999; cf. Section V.B.2). It is notable that other long- or short-period *tim* mutants—two of the former harboring amino-acid substitutions (Figure 3) in or near TIM's PER-interaction domains—are not temperature-sensitive for locomotor period (Rothenfluh *et al.*, 2000a). The same was found for the slightly long period *Toki* mutant (Matsumoto *et al.*, 1994).

B. Naturally occurring and engineered *per* variants with modulated or anomalous responses to thermal changes

The latitudinal cline observed for naturally occurring *per* alleles (Section VII.A) was inferred to be connected with the temperature responsiveness of *D. melanogaster*'s clock in northern versus more southerly climes (Costa *et al.*, 1992). Leaving aside a possible relationship of this geographical variable to the system's photoperiodic responses, these investigators appraised the *per*-allelic variation in question by monitoring locomotor periodicity at different (laboratory) temperatures (Sawyer *et al.*, 1997). They remembered that a deletion of the *per* region that encodes its threonine–glycine repeat (Yu *et al.*, 1987a) leads to temperature sensitivity: *per^L*-like, although not as marked an effect, the circadian clock slowed down as the "ΔT–G" transgenic flies were heated (Ewer *et al.*, 1990). The natural variants exhibit more subtle variations in T–G length, with strains from northern-caught *D. melanogaster* tending to encode 20 such dimers compared with the 17 that are more characteristic of the PER protein made by their southern relatives (Costa *et al.*, 1992). Flies carrying the "T–G/20" isoallele were found to exhibit more solid insensitivity to temperature changes than did shorter types, whose repeat lengths varied from 14 to 17 pairs; other rarely occurring natural variants whose PERs include 21, 23, or 24 T–G pairs were also not that great at holding their locomotor periods steady in conditions of 18°C versus 29°C (Sawyer *et al.*, 1997). These results were bolstered by engineering intragenic *per* deletions, starting with a T–G/20, proceeding downward to T–G/17 and T–G/1, and including a retest of the ΔT–G type. All the resulting transgenic strains with relatively short or no T–G pairs were temperature-sensitive for circadian periods; only half of the T–G/20 lines exhibited slower clock function at 29°C compared with 18°C (Sawyer *et al.*, 1997), and their temperature dependences were smaller than those associated with the other three genotypes. [Wise drosophilists generate and characterize more than one transgenic line of a given type because varying chromosomal-insertion sites can differentially influence the efficacy of a transgene's expression (exemplified in histological detail by Kaneko and Hall, 2000).]

There are potentially two pieces of significance to these findings, one of which is argued to be connected with certain molecular investigations: (i) A northern population of *D. melanogaster* faces wider swings of temperatures throughout the year than does a southern one; it could follow that the former type would evolve clock-gene functioning that is relatively impervious to this environmental variable, if it were the case that the mechanism of temperature compensation resides with the clockworks as they are defined by products of at least some of the relevant genes (discussed with regard to a fungal clock as well as that of *Drosophila* in the review of Hall, 1997). (ii) Structural studies of T–G dimer sets in solution showed that the dominant conformation is a series of β-turns, but the structures exhibited lability, including that which occurred as a function of temperature (Castiglione-Morelli *et al.*, 1995).

These authors and others speculated that the T–G repeat and analogous structures in the middle of PER from other *Drosophila* species (Guantieri *et al.*, 1999) act as "spacers" which maintain various domains of PER in an orientation suitable for the protein's functions, including and especially those involving interactions with other polypeptides. What if the temperature goes up? Energy would be added to the system, and the orientation of PER domains could be altered, such that they would no longer interact optimally with heterologous molecules or possibly with themselves intramolecularly (see later). However, what, in turn (or literally in *β*-turns), if the *flexibility* provided by a T–G repeat and/or structurally similar pentapeptide repeats (Guantieri *et al.*, 1999) permitted the overall structure to adapt in different temperature environments? This could somehow maintain the functional constraints of the molecule. Under such a scenario, perhaps a PER molecule with 20 T–G pairs possesses more optimal spacing and flexibility in the face of varying temperatures compared with polypeptides containing shorter or longer such repeats in their middle regions (Sawyer *et al.*, 1997).

These suppositions remain somewhat soft with regard to exactly what happens to the molecules as they strive to maintain normal functioning as the temperature changes. In this respect, chronobiologists have long speculated that hypothetical intraclockwork reactions are in fact temperature-*sensitive*. Yet, somehow *two* molecular effects (let us say) of temperature changes are *opposing* and would balance one another out when it is either hot or cold. Some experimental force supporting such a conjecture seemed to arise when it was found that the PAS domain of PER could interact with a more-C-terminal region of the polypeptide (here called "C-domain"). The latter is an evolutionarily conserved stretch of amino acids that happens to include the site of *per*[S]'s amino-acid substitution (Figure 3) and others causing similar clock abnormalities (Section V.B.1). The possibility of intramolecular interactions was inferred from tests of PER fragments manipulated in the yeast two-hybrid system (Huang *et al.*, 1995). The experiments included tests of interactions between two PAS fragments, necessarily involving *inter*molecular associations in actual fly cells. Strikingly, either type of interaction was accentuated at higher temperature (Huang *et al.*, 1995). PER fragments substituted in a *per*[L]-like manner mediated heat-labile intermolecular interactions, but the inferred intramolecular associations between PER[L]–PAS and the C-domain were stronger than normal at higher temperature. These data allowed the inference that a PAS/C-domain association, which would compete with a PAS–PAS one, leaves the overall process insensitive to temperature changes. The relative blocking effects of PAS/C-domain associations on the intermolecular ones would be compensated for by heat-induced enhancement of PAS–PAS binding. The latter dimerizations would somehow be associated with robust functioning of the gene product, which seems consistent with the biological and molecular effects of per^L: weak, long-period rhythmicity, a further clock slowdown at higher temperatures, too much of an intramolecular interaction, and too little of an intermolecular one.

However, these kinds of PER-o-centric experiments did not lead to further tests of the temperature-compensating mechanism implied. The initial ones were necessarily performed in advance of appreciating PER–TIM interactions. Not long after *tim* was cloned and its product began to be analyzed, TIM-with-PER heterodimers were found to be far more prominent in extracts of fly tissues compared with PER–PER dimerizations (Zeng *et al.*, 1996), and in yeast two-hybrid testing of PER–TIM interactions, it was shown that adding relatively C-terminal material to a PAS fragment derived from the former polypeptide did not affect its ability to associate with TIM (Gekakis *et al.*, 1995). Thus, the presence of the latter protein seems to overcome the interactability of PER's PAS with a C-domain to which it is attached. These two regions of the *per* product are of course fastened to each other *in vivo*. However, the meaning of their ability to associate in yeast cells, in the context of temperature compensation as manifested by live flies, remains obscure.

There is one more region of the PER protein to be considered in conjunction with temperature manipulations of the organism. For this, recall that a 60-aa region (Figure 14) just upstream of the T–G repeat (but downstream of the C-domain just dealt with) was determined to coevolve with evolutionary divergences within the T–G region or that of its structural relatives (Section VII.A). The biological significance of these relatively N-terminal "60-aa variations" was tested by generating chimeric *per* transgenes, using material cloned from *D. melanogaster* and *D. pseudoobscura* (Figure 14). If a *D. melanogaster* host carries the latter's *per* as its only functional such gene, the flies exhibit infrequent and non-robust rhythmicity (Petersen *et al.*, 1988). One might imagine that *per* chimeras with increasing amounts of *D. melanogaster* sequence would work better. This is not what happened (Peixoto *et al.*, 1998): Fusing a 5′ fragment of *mel–per* to the 3′ 60% or so of *pseudo-per* led to mediocre rescue of per^{01}-induced arrhythmia, in particular when sequences encoding the 60-aa region from *D. melanogaster* were juxtaposed to those that specify the T–G region (a short stretch of such dimers) and pentapeptide repeats of *D. pseudoobscura*'s PER protein (Figure 14). The next type of chimeric transgene gave the most telling result. This *per* construct contained *fewer* sequences from *D. melanogaster* (the host species for the incoming transgene) than did the *mel*-60-aa/*pseudo*-repeat chimera just described. Thus, a construct encoding *pseudo*-PER's 60-aa region was made in which these sequences were juxtaposed upstream of *mel*'s repeat-encoding nucleotides (Figure 14). This transgene gave proportions of rhythmic individuals comparable to controls in which the entire incoming *per*⁺ allele was derived from D. *melanogaster* (Peixoto *et al.*, 1998). As expected, the latter transgenic type exhibited solid temperature compensation of locomotor periods (Figure 14); the same was so for the low proportion of rhythmic flies whose *per*⁺ coding sequences were entirely those of D. *pseudoobscura* and in others designed to make *mel*-PER only through the PAS region (such that the 60-aa one, the repeats, and the rest of the protein were all derived from *pseudo* sequences). However, the two transgenic types carrying constructs designed to generate *pseudo*-PER repeats juxtaposed with a 60-aa region that was all or half *mel*-like showed substantial temperature sensitivity (60-aa all *mel*: 26- to 30-hr periods over an 18–29°C range; 60-aa N-terminal half *mel*, all the rest *pseudo*: 25- to 35-hr periods over the same range).

To summarize these complex results (Figure 14): Any chimeric protein in which the 60-aa region, upstream of a D. *pseudoobscura* repeat domain, was all or in part derived from D. *melanogaster* sequences gave poor temperature compensation, whether or not such a chimera led to good overall rhythmicity (Peixoto *et al.*, 1998). The authors interpreted these findings to mean that the 60 residues N-terminal to repeat-containing regions within various forms of PER have coevolved to participate in fine-tuning structural adaptations of the protein as its conformation changes in accordance with temperature variations.

A broader summary of these clock-gene variants, considered from the perspective of temperature compensation, leaves us without a deep appreciation of the mechanism that underlies this process. One does sense that the canonical clock factors are of paramount importance, compared with the notion that other, undiscovered genes would specify all features of the compensating actions. In the latter conception, two types of temperature compensators might modulate core pacemaking by being grafted onto the periphery of the clock; the former pair of hypothetical functions would get to the pacemaker molecules by the consequences of the aforementioned "opposing reactions" they mediate as the flies are heated. Instead, we can now infer that molecules such as PER and TIM may modulate their functions (interactions between them, effects on CLK and CYC activities) by responding directly to temperature changes. However, how PER so responds could involve a myriad of intramolecular entities: The temperature-sensitive *per* mutants involve various amino-acid substitutions extending from a region N-terminal to PAS to one that is not that far upstream of the T–G repeat (Hamblen *et al.*, 1998); the naturally occurring plus molecularly engineered variants with putative temperature-response significance run from PAS down through the repeat region. Thus, fully half of the PER molecule needs to be dealt with further if this is the way to proceed toward understanding temperature compensation more fully. It would also not be surprising if functions specified by *tim* (Rutila *et al.*, 1996; Matsumoto *et al.*, 1999), by other the genes known to be "in the loop" (Figures 7 and 8), and by factors that have neither been mutated nor cloned turn out to be involved in this mechanism.

IX. MOLECULAR GENETICS OF CLOCK RESETTING BY ENVIRONMENTAL STIMULI

A. Temperature changes

That fluctuating temperatures throughout a daily cycle can entrain *Drosophila*'s rest–activity cycles (Figure 5B) essentially demands that pulses of altered temperature would have phase-shifting effects. This is the case [also see Saunders (1982) for analogous effects on *D. pseudoobscura* eclosion]. The resulting phase response curve for *D. melanogaster* behavioral resets is similar to those obtained by delivering a series of light pulses (Edery *et al.*, 1994a). However, temperature entrainment comes out different from that of the flies to LD cycles (the late-day peak of locomotion is distinctly earlier in the former case), and these two types of stimuli cause dissimilar affects on the clockworks at certain times of a daily cycle.

To set up a description of the relevant results, it is necessary to preview the subject of the following subsections (IX.B) in which *light*-induced "disappearance" of the *timeless* gene product will be prominently featured. A similar sharp drop in TIM abundance occurred after flies were *heat*-pulsed during any stage of the cycle (Sidote *et al.*, 1998). [In contrast, exposing the flies to a heat step-up, of the type described in the next paragraph, causes daily *tim* mRNA levels to be higher than those measured over the course of a cycle in chronically colder conditions (Majercak *et al.*, 1999).] Rapid decreases in the level of PER were also induced by such heat pulses [cf. the effects of a cold step-down, which accentuates *per* mRNA levels (Majercak *et al.*, 1999)], whereas similarly transient light stimuli do not markedly affect the abundance of *period* products. [However, see Lee *et al.* (1996) and Section IX.B.2.] To recapitulate briefly a salient feature of these results: Elevated temperatures are associated with a lowering of *per*-product levels, but not of *tim* ones.

The heat-pulse component of these experiments (Sidote *et al.*, 1998) showed that only very hot temperatures (37°C, outside the organism's physiological range) were effective for shifting operation of the clockwork's feedback loop appropriately. Thus, 37°C pulses delivered to the flies delayed the abundance peaks for *per* and *tim* products compared with the molecular phases of undisturbed flies (Sidote *et al.*, 1998). This thermal effect is similar to that of light pulses delivered at similar cycle times (Section IX.B). Something quite different occurred during the "advance phase" of the daily cycle: Whereas stable advances of the PER and TIM oscillations are induced by late-night light, only a transient increase in the "speed" of this molecular program accompanies the heat-induced decrements of PER and TIM that occur at this (and all other) cycle times (Sidote *et al.*, 1998). There is a behavioral corollary to these results, in that heat pulses delivered late in the nighttime were found not to cause steady-state shifts in the locomotor rhythm, so the light versus temperature PRCs are not really that similar (Sidote et al., 1998; cf. Edery et al., 1994a).

Responsive of the rhythm system to changing temperatures was examined in two further ways. One such experiment was performed perhaps to knock down a wooden soldier: Would the effects of a "heat-shock transcription factor" (HSF) impinge on the clockworks? The answer was no: Heat-induced decreases in PER and TIM occurred in the usual manner in flies expressing an HSF mutant (Sidote and Edery, 1999), suggesting that the "heat-shock pathway" does not contribute to the temperature responsiveness of Drosophila's clock (although it is involved in fly sleep; Section IX.C.2). The second set of experiments brings us once more to the interface between the effects of temperature and of photic stimuli on the system.

Thus, flies were manipulated in a manner that mimics wintertime. It was shown that daytime locomotion is accentuated in cold-temperature, short-day conditions (Majercak et al., 1999). Molecular corollaries were obtained: In the cold, the peaks of per and tim mRNA abundances occurred at earlier-than-normal times in an LD cycle, although the phase advances for PER and TIM were less marked (no doubt because of a light-delaying effect on TIM accumulation, as described in Section IX.B.2). Cold conditions enhanced splicing of the afore-mentioned alternative intron in the 3' UTR (Figure 15, see color insert) of per's primary transcript (see Section V.C.2), such that production of "type B" mRNA (missing this short stretch of noncoding sequence) was boosted relative to that of the slightly longer type A material. Behavioral assays of transgenics whose only PER-encoding mRNAs are unspliceable in the 3' UTR or from which the per-tinent intron had been pre-removed (cf. Cheng et al., 1998) showed both types to lack the (usual) anticipation of light-to-dark transitions and exhibited delayed evening-activity peaks in 6-hr L:18-hr D cycles running in the cold (Majercak et al., 1999). In such short-day conditions the amplitude of per's mRNA rhyth-micity was dampened, and PER cycling was delayed in these variant transgenics (relative to a "genomic" type that produces both splice forms). Putting these results together (Figure 15) suggested that splicing (per se) of per's 3' intron plays a significant role in gene-product cycling parameters, which are involved in cold adaptation of the fly's rest–activity cycles, especially in days with short photoperiods.

Focusing on what fluctuates environmentally with regard to the latter variable, it is important to register that light influences the cold-induced advances in per-product cyclings. This is because TIM accumulation is so heavily dependent on photic conditions (as the current section keeps alluding to), and the level of this protein in turn affects the abundance of its partner, PER. Bearing in mind that the cyclicities of per- and tim-product cycles involve coordinating the effects of temperature and photic signals (Majercak et al., 1999), we now turn to a consideration of the manner by which light alone influences the clockworks.

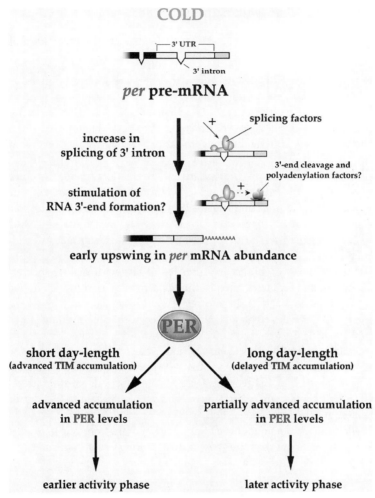

COLD

3' UTR

3' intron

per **pre-mRNA**

increase in
splicing of 3' intron

+ **splicing factors**

3'-end cleavage and
polyadenylation factors?

stimulation of
RNA 3'-end formation?

early upswing in *per* mRNA abundance

PER

short day-length
(advanced TIM accumulation)

long day-length
(delayed TIM accumulation)

advanced accumulation
in PER levels

partially advanced accumulation
in PER levels

earlier activity phase

later activity phase

Figure 15. Thermal and photoperiodic changes combining to modulate the expression of *Drosophila*'s
period gene. This scheme, adapted from Majercak *et al.* (1999), shows what happens to
the quality and quantity of *per* products at a relatively cold temperature and in photic
conditions defined by either relatively short durations of daytime or by photoperiods
comprising higher proportions of a given 24-hour cycle. Subjecting flies to 18°C causes
enhanced RNA splicing within a 3'-terminal intron of the primary *per* transcript; that
sequences in *per*'s 3' UTR are alternatively spliced (irrespective of a temperature effect)
was first reported by Cheng *et al.* (1998). The "splicing factors" and others operating on
this end of the RNA molecule, as depicted here by ovoid protein symbols, are discussed
in terms of the presumptively relevant catalytic processes by Majercak *et al.* (1999).
As shown in the top half of this diagram, these features of *period* RNA processing result in higher levels of (mature) *per* mRNA at low temperatures—in particular, a cold-
induced advance in the accumulation of such material and an earlier-than-normal

B. Light inputs to *Drosophila*'s rhythm system

1. Basic light-responsiveness and involvement of the *period* gene

It has long been known that fairly strong constant light induces arrhythmia in insects (Saunders, 1982). With regard to the workhorse rhythm studied in *D. melanogaster*, it was repeatedly demonstrated that adult flies are behaviorally arrhythmic in LL (e.g., Zerr *et al.*, 1990; Matsumoto *et al.*, 1994; Power *et al.*, 1995a; Tomioka *et al.*, 1997, 1998; Emery *et al.*, 2000a; Helfrich-Förster *et al.*, 2001), with the proviso that the light intensity is above 10 lux or so (Konopka *et al.*, 1989).

Less well known in *Drosophila*, and not rationalizable in molecular-genetic terms, are the aftereffects on locomotor activity of *prior* exposures of *Drosophila* to LL [see Barrett and Page (1989) and Page *et al.* (2001) for examples of analogous experiments in another insect]. In one study, maintaining wild-type cultures of *D. melanogaster* in this condition for several generations, then placing the ultimately tested flies in DD led to periods that were the same as those in LD-reared controls; however, LL *per*[S] cultures led to flies with slightly shorter than control periods, and *per*[L] responded in the opposite direction to such anomalous rearing (Power *et al.*, 1995b). In another study (Tomioka *et al.*, 1997), rearing cultures in LL for one developmental cycle led to shorter free-running (DD) periods of the resulting wild-type adults compared with the aftereffects of DD rearing or that which occurred in short-photoperiod cycles. *per*[S] cultures produced flies showing the same effect. *per*[L] gave a different pattern of aftereffects: shorter adult periodicities after LL or LD rearings (various photoperiods for the latter) than those displayed by flies whose progenitors had been subjected to DD during one developmental period (Tomioka *et al.*, 1997).

Pausing for a moment from consideration of the constant-light effects, it is important to point out that one generation of rearing in constant *darkness* led to

---

increase in the abundance of PER protein (respectively, these two types of macromolecules peak during approximately the middle of the night and late at night, in 12:12 LD cycles operating at milder temperatures such as 25°C). This temperature-sensitive feature of *per* RNA processing is inferred (by Majercak *et al.*, 1999) to correlate with the preferential daytime locomotor activity that is exhibited by flies at low temperatures (compare the behavior in Figure 1A, which had been monitored at 25°C). Given (a) the likelihood that PER accumulation is dependent on increasing TIM abundance (at least during the risetime, throughout the second half of the night in 12:12 LD at 25°C) and (b) that light depresses TIM levels (Section IX.B.2), long photoperiods partially counteract the cold-induced daily advance of *per* mRNA risetime (lower right). The interplay between temperature effects on intraintronic *per* RNA splicing → mRNA accumulatability and of light effects on TIM abundance are concluded to be important features of "integrating information regarding ambient temperature and day length" (Majercak *et al.*, 1999). One biological consequence of these environmentally modulated molecular dynamics is signified by the following shorthand: gene-controlled *Drosophila* rhythmicity includes regulating a shift of the fly's most active behavioral phase into the daytime when it gets cold.

flies whose free-running periods were basically appropriate to their genotype with reference to the results of developing in more standard LD conditions (Sehgal *et al.*, 1992; Tomioka *et al.*, 1997), whereas several generations's worth of chronic DD leads to per^+, per^S, and per^L flies that give highly variable results (within a genotype) in terms of their adult locomotion: A per^S fly can be very long; a per^L one short; per^+'s, either; and all three types included high proportions of individuals showing anomalous ultradian rhythmicity (Section IV. C.4) compared with the aftereffects of either LD or LL rearing (Dowse and Ringo, 1989; Power *et al.*, 1995a, b). What about the effects of exposing already developed flies to LL (Konopka *et al.*, 1989)? The DD rhythms of per^S were lengthened (compared with DD-stored/DD-tested controls) and those of per^L were shortened (about a 1-hr change in each case). The responses of per^S and per^L response to this kind of photic manipulation of adults (only) is the opposite to what happened in the LL-*rearing*/DD-monitoring tests of Power *et al.* (1995b).

Testing the basic responsiveness of *per* mutants *in* constant light showed both mutant types to get longer and longer as the intensity of (dim) light was raised, but per^S and per^L were more sensitive than wild-type to the arrhythmia-inducing effect of LL, which began to kick in once illumination levels to which these mutants were subjected climbed above about 1 lux (Konopka *et al.*, 1989). Similar hypersensitivity was observed for the *Toki* mutant (Matsumoto *et al.*, 1994). The accentuated sensitivity of per^L alone (among these three mutants) is potentially rationalizable (per^S and *Toki*? who knows?). Thus, we should remember that this long-period *per* mutant generates a product whose effective level of function is (already) lower than normal and that this seems correlated with the mutant's long-period biological phenotypes (Sections II.A.1 and V.B.2.). If constant, very dim light were readily to lower PER^L function even further, behavioral periods could become markedly longer (as they do), and it could be relatively easy to eliminate the action of this protein at a modestly dim level of LL.

The first clues that gave some concrete force to these notions involved subjecting wild-type flies to constant light and monitoring PER levels immuno-chemically: With one such antibody, no tissue staining could be detected in this condition (Zerr *et al.*, 1990); with another, much lower than normal PER abundance was observed in Western blottings (Price *et al.*, 1995). However, PER cycling does not plunge and stay low immediately upon transfer from LD to LL: Oscillation of this protein's abundance continued for at least 2 days, with the only strong effect of staying in L (after the first half of the final LD cycle) being a delay relative to the time of PER's usual decline (Marrus *et al.*, 1996). This is an important and in a way disturbing result, which can be regarded as upsetting the applecart of a scheme that might be used to understand photic effects on he clockworks in terms of daily resets of the circadian pacemaker (see middle of the next subsection).

However, we plunge ahead to deal with additional manipulations of the L versus D portions of daily cycles, made in conjunction with measurements of *per* mRNA abundance. Thus, Qui and Hardin (1996a) performed photoperiod-altering experiments, analogous to those carried out by Majercak *et al.* (1999), but without any temperature variations. In the earlier (LD) study, *per* mRNA abundance peaks got more and more delayed as the L phase was progressively lengthened. Stated another way, the peaks consistently occurred ca. 4 hr after lights-off (e.g., at ZT12 in 8-hr L:16-hr D; at ZT20 in 16:8), such that this form of the *per* product could barely begin accumulating as long as the lights were on (Qui and Hardin, 1996a). In this regard, recall that when PER is knocked way down in another circumstance—by the effects of *per^{01}*—the mRNA emanating from this gene is on average anomalously low, in the sense that it never rises to the usual peak level during any cycle time (Hardin *et al.*, 1990; Qiu and Hardin, 1996a). An almost identical result stemmed from one further photic manipulation to which *per$^+$* was subjected: In 20-hr L:4-hr D tests, transcript levels stayed flat throughout such a cycle, and other wild-type flies tested in parallel were arrhythmic behaviorally, as if they were in thoroughgoing LL (Qui and Hardin, 1996a). The RNA side of this result indicated "intermediate," albeit constitutive abundance (same as the *per^{01}* effect); this should be enough to produce fairly robust levels of PER—unless a posttranscriptional effect of light is operating on this and other components of the clockworks.

2. TIM protein levels rapidly responding to photic stimuli

That there is such an effect of light—actually a posttranslational one—brought *timeless* and its protein product to the fore. Elements of the relevant results permit us to understand, to some extent, why PER is so low when the lights are on, and, more generally, how the effects of incoming light may mediate daily resets of the circadian clock.

A flurry of contemporary experiments was reported which showed that light pulses rapidly cause TIM levels to crash (Hunter-Ensor *et al.*, 1996; Lee *et al.*, 1996; Myers *et al.*, 1996; Zeng *et al.*, 1996). The results came from an (interinvestigation) combination of immunohistochemical and Western-blotting assessments of such photic effects. In a further application of the latter method, TIM was found to go low during the first half-day of LL (contra PER), that is, not rise toward its usual peak level (had these particular flies remained in LD instead of the lights being kept on after a 12-hr L period); the normal nighttime phosphorylation of TIM also did not occur in this LD → LL experiment (Marrus *et al.*, 1996). After such a transfer into constant light, there was only a modest immediate effect on *per* and *tim* mRNA cycling (Marrus *et al.*, 1996), consistent with the *per* results of Qiu and Hardin (1996a) and explainable in part by the ca. 2-day persistence of quasi-normal PER cycling after the LD → LL shift.

Just as there was no immediate effect on PER cycling upon transfer of flies to LL, light pulses caused no acute effect on the apparent abundance of this protein (Hunter-Ensor *et al.*, 1996). In one such study, there was also no noticeable effect of pulsed stimuli on *tim* or *per* mRNA levels (Hunter-Ensor *et al.*, 1996). However, Lee *et al.* (1996) found that PER's nuclear entry (and extent of phosphorylation) got delayed by early-night pulses, and there was an advance of its (normal) early-morning disappearance after an equivalent light stimulus was delivered late in the night. These investigators also reported compensatory shifts of the *per* mRNA rhythm after early- or late-night light pulses, along with an acute, but modest drop in the level of this transcript after the former stimulus (contra Hunter-Ensor *et al.*, 1996). These effects were observed in wild-type flies, but a different influence of light on the system was observed in transgenics carrying the Short-Domain PER deletion (Section V.C.1): Concomitant with the impaired negative feedback inferred to be caused by this PER-stabilizing removal of amino acids located between the PAS domain and the per^S-mutated site, light pulses delivered to $per^{\Delta C2}$ transgenic flies in the middle of the night were found to induce an increase in *per* or *tim* mRNA (Schotland *et al.*, 2000).

Standing back from certain of these complexities, the most consistent and salient effects of light on a "state variable" of the clockworks are the following: The first and fastest effect of photic stimulation is to induce degradation of TIM (this being the first-stage conclusion from the seminal quartet of light-induction studies). Inasmuch as the abundance of that protein normally begins to rise around the time of dusk, light at that phase (or in the early night) delays its accumulation. TIM exhibits its natural decline around the time of dawn, and light reaching the clockworks shortly before then causes an advance in its disappearance (Hunter-Ensor *et al.*, 1996; Lee *et al.*, 1996; Myers *et al.*, 1996; Zeng *et al.*, 1996). These molecular responses of the system can be rationalized to correlate with the dusk delays and dawn advances that are described by the PRCs for the fly's biological rhythms (Section IV.C.1). The system in this context includes the eventual effects of light on PER abundance and the dynamics thereof—given that PER partners with TIM and seems therefore to protect it. Looking at this scenario from a slightly different angle: Monomeric PER is especially subject to degradation, promoted in part by DBT-mediated phosphorylation (Section V.B.4), and the former of these two proteins needs TIM to accumulate to a high enough level to encourage the formation of PER-including dimers. However, if such accumulation is delayed (early-night light) or caused to drop prematurely (late-night light), PER protection is stalled or removed. Thus, the molecular PRC for PER would in effect follow that of TIM. That PER is perhaps more than a passive player in molecular light responsiveness is suggested by the per^S mutant's hypersensitivity to light-induced transcript-cycling arrhythmia, as observed in the aforementioned LD → LL experiment (Marrus *et al.*, 1996). This could relate to the manner by which PER^S interacts with TIM, although this form of the protein is not mutated in a known "interaction domain" (Section V.B.2). In any case, the accentuated molecular response to extended L may explain why per^S cultures or flies gave larger-than-normal light-induced phase shifts in certain PRC assessments (Section IV.C.1). Detailed examination of cycling parameters monitored in per^S extracts (Marrus *et al.*, 1996) rationalized the ability of this mutant to effect 5-hr daily phase delays, such that its locomotor cycles are entrained to 12-hr L:12-hr D (Section IV.C.1).

Recalling an effect of another clock mutation on these features of the system's functioning: The effects of light-induced TIM decrements on PER are consistent with the similar lowering of the latter's abundance that is caused by genetic elimination of TIM (Price *et al.*, 1995). However, rather long-term exposures of wild-type flies to LL are necessary to observe a phenocopy of tim^{01} in terms of PER going down and staying chronically low compared with the much quicker protein-degradative effects that are exerted by "unexpected" light (nighttime pulses or LD → LL) on TIM. How TIM, and PER as well, interact physically with an *additional* factor involved in getting light to the clock will be described in the next subsection.

Lest we get lost once again among some of the complexities just described, it is useful to state a general scheme for these molecular chronobiological phenomena and how they provide a first-order explanation for the phase-shifting effects of light on *Drosophila*'s biological rhythms: Even if the primary effect of light on the key state variable (TIM) is the *same* no matter what the phase of stimulus delivery, the consequences on dynamics of the molecular cycling are *opposite* when light enters the system during the rising phase versus the falling one. Throughout the other half of the cycle, during the daytime, light effects are moot because there is naturally almost none of the state variable around. The problem with this verbal device is that TIM's pretty much immediate light-induced crash does not lead to a rapidly occurring decrement of PER abundance (Marrus *et al.*, 1996), even though the absence of protective effects of TIM on PER are realized in the total absence of the former (as caused by a *tim*-null mutation) and are observed in a chronic LL situation (Price *et al.*, 1995). However, if PER plays a (or *the*) key role in the negative side of the feedback loop (Rothenfluh *et al.*, 2000c; Shafer *et al.*, 2002), and if the pacemaker is to be quickly reset during a given daily cycle by dusk or dawn light, then it seems as if the light → TIM → PER effect should be rapidly realized insofar as a PER decline is concerned. Again, this is belied by the results of Marrus *et al.* (1996).

Even with such a caveat in mind, it is worthwhile to consider the actual degradative processes to which TIM is subjected after a light stimulus (or a high-temperature pulse) reaches the clockworks. The relevant factors can be surmised to involve general features of cellular metabolism. This presumption is partly based on the role of the DBT kinase, which is neither a state variable (*dbt*-product production is constitutive during a daily cycle) nor a dedicated chronobiological entity (the gene acts pleiotropically and causes developmental lethality in its severely mutant forms). As to what mediates this degradation, it is not known what are the particular proteases, let alone whether they are mundane pieces of cellular machinery. That they could be so characterized was suggested by applications of protease inhibitors to extracts of fly heads after exposing them to light (or not), or by applying such agents to dissected-out, incubated larval CNSs, then exposing such specimens to light pulses. The categories of drugs that inhibited light-induced TIM degradation compared with other types that were ineffective suggested an involvement of proteosomal enzymes (Naidoo *et al.*, 1999). These factors play roles in many cellular processes beyond those involving functions dedicated (*tim* itself) or semidedicated (*dbt*, *sgg*) to chronobiology. A further drug experiment indicated that "the ubiquitin-proteosome system" is operating here: TIM, produced in transfected cells, became ubiquitinated after exposing the cultures to light, whereas PER did not have this oligopeptide attached to it with or without light (Naidoo *et al.*, 1999). The meaning of another kind of posttranslational modification, involving the aforementioned phospho-TIM, was also tested in the larval CNS preparation: A tyrosine kinase inhibitor blocked light-induced disappearance of TIM immunoreactivity from the relevant brain neurons (Naidoo *et al.*, 1999; cf. Kaneko *et al.*, 1997). Naturally, SGG is an enzyme in this category (Figure 3).

3. Cryptochrome as a transducer of inputs into extraocular photoreceptors

How does light get to TIM and, as a knock-on consequence, to PER? It would seem that the TIMELESS molecule could not directly respond to such a stimulus in terms of the protein changes that make TIM susceptible to light-induced degradation. First, recall that TIM levels stay high throughout the daytime in certain larval-brain neurons (Kaneko *et al.*, 1997), and this is not because the protein does not in general turn over in such cells (there is strong TIM cycling in these dorsally located neurons, occurring out of phase with oscillation running in other DNs or the larval LNs). Therefore, TIM is not inherently and globally light-sensitive. It is as if a factor that transduces light input into effects on TIM is inoperative or not is present in certain *tim*-expressing locations. Now it was sufficient to transfect *tim* sequences into cultured cells in order to achieve the observed ubiquitination of TIM (Naidoo *et al.*, 1999), but this does not mean that modification was sufficient for the clock protein to be degraded in response to light. Nevertheless, it turns out that the light-to-TIM transducer (to be discussed) is constitutively present in these garden-variety cells (Lin *et al.*, 2001).

What arguably is sufficient to add to a TIM-containing cell in order for an artificial system to be able to degrade that clock protein once light enters it is the subject of the remainder of this section. The background information was presented in Section II.C, which discussed roles played by blue light in the phase shifting and entrainment of *Drosophila* rhythms. Molecular meat was added to these formalistic bones in two ways: (i) Exposing flies to a series of monochromatic light pulses led to maximal degradability of TIM roughly in the "blue range" (Suri *et al.*, 1998); a companion study went deeper than demonstrating the basic existence of extraocular photoreception (Section II.A) by showing light-induced TIM disappearance in the brains of eyeless flies and attenuated such molecular responsiveness in other variants expressing external-eye phototransduction mutations (Yang *et al.*, 1998). (ii) It was found that *D. melanogaster* makes cryptochrome, a blue-light-absorbing substance known for that property and others in a wide variety of organisms; identification of the fly's *cry* gene (e.g., Emery *et al.*, 1998; Selby and Sancar, 1999; Okano *et al.*, 1999) was accompanied by mutating it (Stanewsky *et al.*, 1998) and manipulating it in certain chronobiological contexts (Emery *et al.*, 1998; Egan *et al.*, 1999; Ishikawa *et al.*, 1999).

Could CRY be *the* substance that absorbs light to subserve the extraocular part of *Drosophila's* circadian-photoreceptor system? We will see that the answer is "yes" from certain experimental perspectives (Figure 16; see color insert) or "not the only such substance" from others. Moreover, CRY may play a chronobiological role beyond its photoreceptive one. A key genetic tool for answering these questions and addressing these issues came from a mutant screen aimed at inducing variants that disrupt *period* gene expression: Flies carrying a *per-luc* transgene and homozygous for mutagenized chromosomes led to several lines (less than 1% of the total) in which the daily cycles of reporter activity were anomalous or eliminated. Two such lines harbored proof-of-principle variants (Stempfl *et al.*, 2002; R. Stanewsky and J. C. Hall, unpublished): recessive second-chromosomal *timeless* mutations, which were subsequently shown to eliminate TIM and dubbed tim^{03} and tim^{04} (Table 1). Their isolation phenotype was flattening of *per*-product cycling, as should occur under the influence of a loss-of-function *tim* allele (Sehgal *et al.*, 1994). In another conspicuously aberrant line, a recessive mutation had been induced on the third chromosome that eliminates the normal daily oscillations of *period* products; *tim-luc* cycling was then shown to be wiped out as well (Stanewsky *et al.*, 1998). Strikingly, TIM cycling in adult-head homogenates did not occur either in this mutant, in that the immunochemically determined abundance of this protein stayed high during both day and night phases of an LD cycle. Thus, the protein was insensitive to the degradative effects of light during the L phase. TIM's phosphorylation cycle was also knocked out in the new mutant (Stanewsky *et al.*, 1998).

Genetic mapping of this mutation (Figure 18, color insert), and showing that that position coresponds to a chromosomal location annealing to a cryptochrome-encoding clone found in a database (Emery *et al.*, 1998), led to the demonstration that the mutant harbors an amino-acid substitution within this sequence (Stanewsky *et al.*, 1998). Thus, the variant was called cry^{b}, which leads to a CRY protein substituted in one of the sites that can be predicted to bind a flavin cofactor (Figure 3). For more about this feature of normal fly CRY and other of its photochemical properties, see Selby and Sancar (1999) and Okano et al. (1999). Additionally, the *Drosophila's* CRY has been subjected to light-response analysis by assays more in the molecular biology realm. For this, Froy et al. (2002) subjected *cry* coding sequences to *in vitro* mutageneses, focusing on four prognosticated flavin-binding sites, including that defined by *cry*[b]. Using multiply transfected cells in culture as an assay system, these investigators asked whether normal CRY or a given altered form of it would relieve the PER:TIM inhibition of transcription activated by CLK:CYC (Ceriani et al., 1999), *and* the extent to which these normal or amino-acid substituted CRY molecules would be degraded after light stimulation. Three of the four CRY-mutated proteins, including a mimic of CRY[b], exhibited no detectable light-responsiveness (in terms of light-induced enhancement of transcription or flavoprotein degradation); two of the muant forms were "constitutive high" in terms of their transcriptional-enhancing activity (Froy et al., 2002). These CRYists went on to *in vitro*-mutagenize *cry*-encoded trytophan residues, given that photic stimulation of photolyases leads to eventual regaining of the protein's activity by transfer of an electron through CRY Trp's to flavin moieties. Two of the three Trp → Ala substitutions led to CRY molecules that were not light-responsive in terms of transcriptional derepression or degradation. The results of these *cry* manipulations bring this subsystem (within the larger one that subserves *Drosophila's* light-effected clock resetting) to the cusp of revealing the intramolecular redox reactions revolving round CRY's circadian-photoreceptive functions (as discussed by Froy et al., 2002).

   Application of an antibody raised against CRY sequences showed that the mutated CRY[b] protein is present in an undetectably low concentration in fly-head extracts (Stanewsky et al., 1998). However, transfection of CRY[b]-encoding sequences into cultured cells, or of another *in vitro*-mutated form of the *cry* ORF that deletes the N-terminal 243 amino acids of the protein, led to detection of normal levels for both the altered and the incomplete forms of CRY, which did not drop after light exposure (Lin et al., 2001; Froy et al., 2002). Therefore, extremely low levels of CRY[b] in the mutant fly may not be due to the altered protein's inherent inability to accumulate. Such assays of CRY[b]'s apparent concentration in the actual animal needed to be assessed with a time-of-day consideration, for CRY normally cycles in its abundance (Emery et al., 1998), rising throughout the nighttime and being low during the daytime hours, except for a couple hours after lights-on, but nonetheless correlated with CRY undergoing light-induced degradation (as assessed more directly by Lin et al., 2001). The proteolysis of this molecule that occurs after photic stimulation is blocked by proteosome and electron-transport inhibitors (Lin et al., 2001). The latter effect is probably related to the likelihood that flavin associations with proteins of this type are required for redox-mediated changes in their conformation (Selby and Sancar,

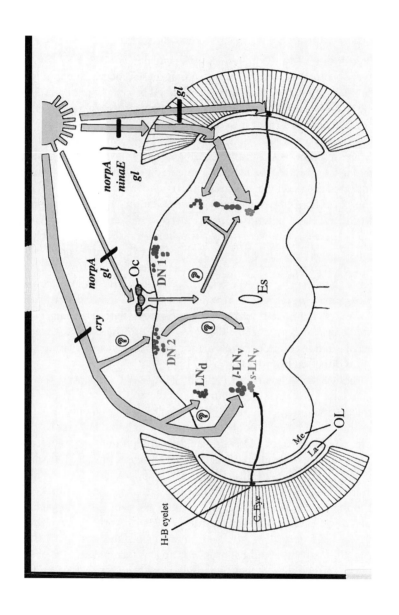

Figure 16. Input routes for light-mediated resetting of *Drosophila*'s behavioral clock, and molecules functioning in these anatomical pathways. This diagram was adapted from Helfrich-Förster *et al.* (2001). It summarizes results obtained by those investigators along with findings previously reported by Helfrich-Förster (1997a), Stanewsky *et al.* (1998), Ohata *et al.* (1998), and Emery *et al.* (2000b). These relatively refined applications of mutations or dietary treatments (see below) stemmed from still earlier ones in which the external eyes were eliminated or inactivated, which nevertheless allowed *D. melanogaster* adults to entrain to light:dark (LD) cycles (e.g., Helfrich, 1986; Dushay *et al.*, 1989). The current picture shows the blocking effects (heavy lines crossing the arrows) of various visual-system mutations, expressed singly or in combination. One element of the scheme is based on the entrainability to LD cycles (and reentrainability to shifted ones) of flies that express *no-retinal-potential-A* (*norpA*) mutations, which cause physiological blindness of the compound eyes and ocelli; or expression of a *neither-inactivation-nor-afterpotential-E* (*ninaE*) mutation (which eliminates rhodopsin from most compound-eye photoreceptors); or of an *eyes-absent* mutation in combination with an *ocelliless* one. Note that an ocellus-to-clock input route is questionable (circled question mark), mainly because it is essentially unknown whether elimination of the simple eyes or their function is sufficient to cause a sensitivity reduction or is necessary to reveal the full effects of *ninaE* or *eya* mutations. Indeed, the sensitivity for LD entrainment was sharply reduced in the mutant types that include effects on compound-eye function or presence, results that are consistent with the effects of rearing *Drosophila* on a vitamin A-deficient medium. That LD entrainability was not eliminated in these situations implied a light-to-clock input role of extraocular photoreception (implicit in the studies reported in the 1980s). The one extant *cryptochrome* mutation—induced in a gene that encodes a blue-light-absorbing protein present in the brain and many other tissues—was found to eliminate behavioral phase shifts that are normally induced by short light pulses in the early or late subjective night (in DD). CRY is found (in part) within LN pacemaker neurons; and transgenically mediated restoration of *cry*$^+$ function to certain LNs only (in flies homozygous for *cry*^{b}) allowed for solid light-pulse responsiveness, as if extraocular photoreception is mediated partly by pacemaker cells themselves: among the LN cell types, at least the LNv's (owing to the *Pdf*-driven *cry*$^+$ expression in these neurons, as alluded to above) if not the LNd's (circled question mark, because whether CRY is in these cells has not been determined). The *cry* gene may be expressed in certain DNs as well, and a DN1-to-LN (axonal) connection (Figure 12) suggests that the latter neuronal types may receive light (circled question marks) in part by way of intrabrain intercellular communication. Locomotor cycles could still be readily reentrained in LD tests of singly mutant *cry*^{b} flies; such responses (to shifted cycles) were poor (mainly meaning substantially reduced sensitivity) when a *norpA* mutation was combined with *cry*^{b}. Retention of some reentrainability by the *norpA*;*cry*^{b} flies implied the existence of one more input route, which would contribute to LD-cycle synchronization (if not to phase shifts induced by short light pulses). It was presumed that the Hofbauer–Buchner (H–B) eyelet could be the input site for this "third" pathway (in addition to the one starting in the external eyes and that defined by certain *cry*-expressing internal structures). One reason for this presumption is that eyelet cells send projections (designated here by mildly curving horizontal lines) to the region of LNv pacemaker cells (e.g., Yasuyama and Meinertzhagen, 1999). A role for the third pathway—which was assumed to remain functional in the face of mutated *norpA*, *cry*, or both (see above)—was tested by combining a loss-of-function *glass* (*gl*^{60j}) mutation with *cry*^{b}, owing to *gl*^{60j}'s elimination of external-eye photoreceptors and of the H-B eyelet as well. This double-mutant type was found to be essentially unentrainable.

1999; Okano *et al.*, 1999; Froy *et al.*, 2002), in the context of a supposition that light would induced such a change in CRY's three-dimensional structure.

Over the course of a DD cycle, CRY levels oddly exhibited a monotonic increase (Emery *et al.*, 1998), as if it might eventually soar off into oblivion. However, *cry* RNA was found to cycle in both LD and DD, exhibiting a peak or plateau during the daytime in the former condition and a weaker maximum during the corresponding phase in DD (Emery *et al.*, 1998; Egan *et al.*, 1999; Ishikawa *et al.*, 1999). Thus, the mRNA peak is displaced by several hours from the CRY maximum in that condition, and the transcript cycling in DD (its dampened attribute notwithstanding) is very different from the steady rise of the protein that was observed in constant darkness (Emery *et al.*, 1998).

Certain of these assessments of *cry*-product levels and their dynamics spoke to the matter of where the gene is expressed. Thus, *cry* mRNA is found in homogenates of both head and body tissue (Egan *et al.*, 1999; Ishikawa *et al.*, 1999; Okano *et al.*, 1999), and it cycles in both parts of the fly (Emery *et al.*, 1998). More refined appraisals of adult tissues, including *in situ* hybridizations, showed *cry* transcripts to be present in putative LN and DN cells within third-instar larval brain (Emery *et al.*, 2000b). Thus, the lack of *cry* mRNA signals in larval homogenates (see subsequent discussion) was a false negative, perhaps because the gene's expression at this stage is very limited spatially. In adults, *cry* mRNA was found in head and thoracic appendages (including the antennae), an anterolateral region of the adult brain (possibly including some of the lateral neurons), and other brain regions as well (Egan *et al.*, 1998; Okano *et al.*, 1999). No signals were detected in the compound eye (Egan *et al.*, 1999), which is curious (see further discussion below). *cry* mRNA was also undetectable in homogenates of embryos or larvae (Egan *et al.*, 1999) and was first spotted in terms of life-cycle stages via Northern blottings of ground-up pupae (Ishikawa *et al.*, 1999). Recalling that phase shifts of the eclosion rhythm induced by light during metamorphosis seem not to rely on opsin-mediated photoreception and that the peak sensitivity for such shifts is in the blue (Section II.C), one expects a cryptochrome-encoding gene to be turned on (at the latest) during the pupal period. Incidentally, transcriptional activation is pertinent to the gene's expression during adulthood: The daily mRNA abundance cycle is underpinned by oscillations of transcription rate *per* se (Emery *et al.*, 1998).

This brings us to clock regulation of *cry*-product cycling. For this, adult *cry* mRNA was shown to stay near its trough levels in per^{01} or tim^{01} head homogenates (Emery *et al.*, 1998; Ishikawa *et al.*, 1999). However, in the Clk^{Jrk} or cyc^{01} mutants, the transcript was constitutively high, and these effects were still observed by Emery *et al.* (1998) when either of these mutations was combined with per^{01} (which by itself causes *lower*-than-normal levels of *cry* transcript). CRY protein as extracted from any of the four mutants just named was forced to be at relatively low abundance during the daytime in LD (Lin *et al.*, 2001), even though *cry* mRNA is pinned at a high level in these genetic backgrounds in this condition (Emery *et al.*, 1998).

Further application of chronogenetic variants spoke to the biological meaning of CRY in *Drosophila*. Overexpressions of *cry* sequences, fused downstream of UAS ones, were mediated by heterologous transgenes carrying the yeast *gal4* factor. Driving CRY to higher-than-normal levels by 5′-flanking DNA from the *timeless* gene led to enhanced phase delays and advances of the fly's locomotor rhythm (Emery *et al.*, 1998, 2000b). These results were consistent with the effects of blue-light pulses delivered to flies heterozygous for the normal *cry* allele and a deletion of the gene (Egan *et al.*, 1999). Here, diminished phase delays were induced when only one dose of *cry*$^+$ was present (the deletion used by Egan and co-workers also happened to be heterozygous for removal of three opsin-encoding genes). However, another kind of overexpressing double transgenic—in which CRY production was brought under the indirect control of an actin promoter—responded in a seemingly anomalous manner: The delay portion of the behavioral PRC revealed *diminished* phase shifts, but the magnitudes of locomotor advances were normal after the white-light-pulsing of *Act5C–gal4*;UAS–*cry* flies (Ishikawa *et al.*, 1999). Perhaps the regulatory sequences of the *Act5C* gene led to *too* much enhancement of *cry*$^+$ expression compared with that mediated by the *tim* promoter. This possibility was suggested by Lin *et al.* (2001), who found that molecularly engineered overexpression of CRY in transfected cells attenuated the ubiquitination of TIM, which is related to its degradatability (Naidoo *et al.*, 1999; also see subsequent discussion). These investigators went on to show that, in flies carrying the *Act5C–gal4*;UAS–*cry* combination, TIM levels were quite a bit higher than in wild-type flies during the first third of a 12-hr photoperiod (Lin *et al.*, 2001), a phase within the daily cycle when TIM is usually subjected to light-induced disappearance.

To understand how *cry* functions may be involved in transducing light stimuli into the clockworks, it is arguably insufficient solely to manipulate CRY levels in the face of the normal gene's expression. Therefore, applications of the nearly null *cry*^{b} mutant were informative. In certain circumstances, these flies were found to be circadian-blind: Little or no phase shifts occurred after brief pulses of bright white light were delivered in the early or the late night (Stanewsky *et al.*, 1998). However, *cry*^{b} flies entrained well to LD cycles and re-entrained to shifted-over ones, in which the L portions involved an experimental series of dim blue light ranging down to a fraction of 1 lux (Stanewsky *et al.*, 1998). Thus, brief light exposures are far less efficacious in the phase-shifting effects compared with a full half-cycle worth of stimulation. To apprehend the latter circumstances, the apparent involvement of *external* photoreceptors must be remembered (Section II). Whereas the compound eyes and ocelli need not be present or functioning for light effects to influence locomotor rhythmicity (Figure 16), the defects in externally mediated photoreception just implied make the system far less sensitive than normal insofar as entrainment is concerned. Analogous low sensitivities were observed when presumptive *trp* and *trpl* "eye-only" mutations (not really: Störtkuhl *et al.*, 1999) were applied in light-pulse experiments in conjunction with H. Q. Yang and colleagues's assessments of effects of such mutations on light-induced TIM degradability.

When another phototransduction mutation (*norpA*) was combined with *cry^b*, the effects on entrainability were relatively severe: Only about one fourth (16-lux blue) down to none (0.16-lux blue) of the doubly mutant individuals could re-entrain to shifted LD cycles, whereas wild-type flies and both singly mutant types responded reasonably well even at the lowest intensity (Stanewsky *et al.*, 1998), although subsequent testing of these mutants showed that *norpA* or *cry^b* flies each took longer than normal to shift over (Emery *et al.*, 2000b; cf. Wheeler *et al.*, 1993). In any case, residual responsiveness of the *norpA*; *cry^b* flies at the *behavioral* level is correlated with the synchronizability of clock-protein cycling *among LNs* within the adult brain (Helfrich-Förster *et al.*, 2001). An additional doubly mutant type—*per^S*;*cry^b*—was asked to entrain to 24-hr LD cycles, but about 40% of these behaviorally tested flies exhibited free-running 19-hr components (Stanewsky *et al.*, 1998; Emery *et al.*, 2000b; Rosato *et al.*, 2001), whereas all *per^S* (*cry$^+$*) individuals entrain in 12-hr L:12-hr D conditions (Hamblen-Coyle *et al.*, 1992).

Even in these genetically stringent circumstances, *Drosophila* with no known external photoreception and little or no CRY were not circadian-blind, nor were flies whose CRY decrement impinged on their ability to achieve several hours's worth of daily reset. Perhaps this is because *cry^b* is not a fully null mutant. However, an additional entity needs to be considered (as introduced in Section II): the extraocular H–B eyelet (Figure 16). If it can bring in light and send it to the brain neurons subserving locomotor rhythmicity, combining *norpA* with *cry^b* would leave part of the response system intact—if, in turn, the *norpA* mutation does not affect phototransduction within the H–B cells, notwithstanding the presence of at least one rhodopsin subtype (called Rh6) within them (Yasuyama and Meinertzhagen, 1999). Further studies of H–B rhodopsins confirmed that Rh6 molecules are present and that Rh5 ones may be as well (Malpel *et al.*, 2002). It is not known whether *cry* is expressed in H–B cells; this is worth looking into, given the CRY-like immunoreactivity reported for supposed extraocular photoreceptors in cockroaches (Fleissner *et al.*, 2001).

Neither the Rh5 subtype nor the Rh6 one is "classically" known to be involved in phototransduction in *Drosophila* (these forms of rhodopsin are not present within compound-eye or ocellar photoreceptors). Intriguingly, the *norpA*-encoded phospholipase C (PLC) is also found within the H–B eyelet (Malpel *et al.*, 2002), just as this PLC is present within external photoreceptors (naturally). Therefore, it seems as if a *norpA* mutation, when combined with *cry^b*, should come close to eliminating the adult fly's circadian photoreception. Not so (see earlier discussion). Assuming that circadian photoreception mediated by rhodopsin and PLC actually *functions* in H–B cells, and that the light input thereby transduced is then sent to the LNs via the afferent pathway connecting the former to the latter structures (described in most detail by Malpel *et al.*, 2002), a *norpA*-null mutant would eliminate all transmission of light from peripheral photorececptors to the brain. Again, all that should be required to knock out all circadian photoreception would be to combine *cry^b* with a *norpA* mutation. That that double mutant did not do the job, and tacitly taking to heart the assumptions just cited, Malpel and co-workers take their results as supporting "the existence of as yet uncharacterized photoreceptive structures in *Drosophila*."

Potentially to eliminate more inputs to the *Drosophila* (behavioral) clock than the *norpA;cry^b* combination is able to do, the latter mutation was combined with a null allele of the *glass* eye gene. This *gl^{60j}* mutation anatomically ruins compound-eye, ocellar, and H–B photoreceptors (Helfrich-Förster *et al.*, 2001). By behaviorally testing *gl^{60j} cry^b* flies for re-entrainment to shifted-over LD cycles, at last a multiply defective type of *Drosophila* (Figure 16) was essentially "circadian-blind." In addition, LNs could not be synchronized for cycling of clock proteins in the brains of this double mutant, whereas such histochemically monitored neuronal rhythmicity is robust in *cry^b* (Stanewsky *et al.*, 1998) or *norpA;cry^b* flies (Helfrich-Förster *et al.*, 2001). If Malpel *et al.* (2002) are correct (as bottom-lined in the preceding paragraph), the effects of *gl^{60j}* in the Helfrich-Förster *et al.* experiments were more severe than removal of the H–B eyelet (newly reported for this mutant), along with compound-eye and ocellar photoreceptors (these being the externally abnormal phenotypes by which the classic *glass*-eye mutants were recocognized). In other words, *gl^{60j}* would also take out the "uncharacterized [circadian] photoreceptor."

A light-off-induced bump in locomotion, observed upon subjecting the *gl^{60j} cry^b* double mutant to high-intensity LD cycles (Helfrich-Förster *et al.*, 2001), was interpreted as a "masking" effect in which the stimulus would not be operating through the circadian system (i.e., accentuated activity in the early part of the dark period may not be related to clock-mediated entraining of behavior, which, in wild-type or singly mutant flies, controls their synchronization with environmental cycles). Given that a changes in light levels *do* cause *gl cry^b* flies to alter their behavior, this explanation is gratuitous, but it leads to consideration of a putative masking phenomenon that may be involved in another circumstance for which light affects *Drosophila*'s locomotor rhythmicity: constant light, which abolishes it.

Remarkably, *singly* mutant *cry^b* flies were found to be insensitive to LL in that they remained rather robustly rhythmic in that condition (Emery *et al.*, 2000a). However, if external photoreceptors, the H–B eyelet, or the "uncharacterized photoreceptor" are each sufficient to "get light to the clock," how would *cry^b*'s LL insensitivity occur? Recall in this regard that that mutation alone also causes flies to be insensitive to the phase-shifting effects of short light pulses (Stanewsky et al., 1998). Perhaps the effects of light, as its enters the animal via external locations or the H–B eyelet's relatively peripheral one, reach the neural substrates of adult locomotion in a masking-like manner (proposed by Emery et al., 2000a). Rest–activity cycles would be affected, but not by way of the circadian system functioning as an intermediary. That system may use CRY as the sole substance that is a circadian photoreceptor per se. Thus, cry^b would wipe out this functionality in terms of the efficacy of brief light stimuli: Light effects entering by way of the eyes would not summate sufficiently to cause phase shifts, or this input route would not even reach the pacemaker. In the latter scenario, and now referring to an LL setting, there would be a cry^b-induced blockade of light to the clock: Light constantly entering the system otherwise would go from eyes to who-knows-where—brain regions whose interneuronal responsiveness and outputs are unable to knock out cyclical locomotion. Actually, it could be that some light effects leak into the circadian system in cry^b, perhaps by way of photoreceptors unaffected by that mutation (such as external eyes or the H–B structure): A subsequent set of LL tests showed such singly mutant flies to be rhythmic in this condition, but with longer-than-normal periods (Helfrich-Förster et al., 2001). It was as if the mutant were interpreting constant light, which was reasonably intense in the experiment just cited, as dim LL (Konopka et al., 1989).

The foregoing discussion implies that, if CRY is at least an important circadian photoreceptive substance, it is likely to function extraocularly. Where? It seems as if the H–B eyelet is not involved, else $norpA;cry^b$ flies would be as circadian-blind as are $gl\ cry^b$ ones. Thus, CRY cells in the brain proper should be considered (Egan et al., 1999). Presumably, the accentuated phase-shiftability of tim-gal4;UAS-cry flies (Emery et al., 1998) is caused by excess CRY in the CNS, although the tim promoter region drives expression of the GAL4 regulator pretty much all over (Kaneko and Hall, 2000). Therefore, the alleviation of cry^b's severe decrement in light-pulse-induced phase shifts by the transgene combination just noted (Emery et al., 2000b) is not particularly informative. An additional transgenic type was applied in which very limited spatial expression of cry^+ was effected by applying a fusion construct containing regulatory sequences of a gene that encodes the PDF neuropeptide (whose chronobiological function keeps getting previewed; see Section X.B.1). This substance is normally expressed only in ventro-lateral brain neurons, both small and large LN_v's (Figures 11–13). When gal4 and UAS-cry were driven by these sequences, homozygous cry^b flies were partly rescued in terms of behavioral phase shifts caused by short light pulses and the 5-hr daily shifts necessary to entrain in the face of per^S (Emery et al., 2000b). In constant-light tests of cry^b's (usual) nonresponsiveness, the tim-gal4;UAS-cry combination made the flies arrhythmic, and cry^+ expression that was transgenically limited only to the aforementioned lateral neurons led to a partial restoration of the arrhythmia-inducing effects of LL (Emery et al., 2000a,b). It was concluded that these rescues occurred because CRY can function as "deep-brain circadian photoreceptor." Are the CNS cells that have this capacity the clock-gene-expressing neurons? That this is possible was examined by relatively high resolution determinations of reported *cry* expression: A *cry*(5′-flanking)-*gal4* construct was combined with a UAS-GFP one, and the fluorescent marker was found to be colocalized with the neuropeptide in LN$_v$'s of adults; singly labeled (GFP-expressing) cells were also observed in the dorsal brain (Emery *et al.*, 2000b).

Therefore, it seems as if light-induced decrements of TIM in LN pacemakers (e.g., Yang *et al.*, 1998) whose function underlies behavioral rhythmicity could be mediated through CRY's light reception in the same cells. By the way, several of the tests in which the light-induced TIM crash were measured involved whole-head homogenates (Section IX.B.2). Most of this protein in the anterior parts of the fly appears to be in compound-eye photoreceptors (e.g., Hunter-Ensor *et al.*, 1996; Myers *et al.*, 1996). Inasmuch as blue-light exposure, presumably acting through CRY, causes TIM to go way down (Suri *et al.*, 1998) one would expect the former protein to be present in the eye. Moreover, eye-only extracts showed that a *Rhodopsin-gal4* construct set up to drive *cry*$^+$ sequences solely in the outer photoreceptors in each facet allowed for TIM cycling in a *cry*^{b} genetic background. However, no specific *cry* mRNA signals were observed in sections through the compound eye (Egan *et al.*, 1999). Presumably, CRY-mediated light reception does get to TIM in that tissue, but it is puzzling that *cry* expression is below the level of detectability in the part of the head that contains the majority of the *timeless* gene products.

In any case, *how* the phototransduction process just implied may operate has been examined so far only in cultured cells (as introduced earlier in this section via Lin *et al.*, 2001), including nonfly ones. Thus, transformation of expressible *cry* and *tim* genes into yeast cells in the context of two-hybrid testing for protein interactions led to the demonstration that CRY and TIM can physically interact in a light-dependent manner (Ceriani *et al.*, 1999). In further experiments that introduced *Drosophila* transgenes into this fungal assay system, light was shown to alleviate the usual PER–TIM inhibition of *tim* transcription. Presumably this stimulus caused TIM to be tied up by CRY such that the former protein gets degraded; PER, being released from protective dimerization with TIM, could become similarly vulnerable (see later). However, transfection of *cry*$^+$ and *tim*$^+$ sequences into cultured *Drosophila* cells did not lead to demonstrable light-sensitive interactions between the two proteins in extracts (Ceriani *et al.*, 1999).

Nevertheless, if CRY can be induced by light to interact with TIM in the actual fly, the transduction pathway from environmental stimulus to the clockworks would be a very short one: Light-induced changes in CRY protein would permit it to interact physically with TIM and somehow trigger the latter's degradation. In the context of that process involving ubiquitination of TIM (discussed earlier), this posttranslational modification is likely to be dependent upon a light-induced

CRY–TIM interaction. However, it does not require that CRY subsequently disappear from the heterodimer, because TIM ubiquitination precedes the proteolysis of CRY that is stimulated by light (Lin *et al.*, 2001; cf. Emery *et al.*, 1998). Leaving this biochemical detail aside, CRY's interaction with TIM should influence PER's concentration or function because, within fly cells and tissues, that protein seems sufficient for operation of the negative side of the feedback loop (Section V.B). If a light → TIM → PER sequence of events is operating when both proteins are still cytoplasmic in LNs during the relatively early nighttime (Shafer *et al.*, 2002), PER's concentration should somehow fall (though see Marrus *et al.*, 1996), such that the nuclear presence of this protein gets diminished at a phase when it alone seems to cause a substantial turndown of *per* and *tim* transcription during the operation of an undisturbed cycle. As dawn approaches, PER and TIM are both in the nucleus and presumably heterodimerized (e.g., Zeng *et al.*, 1996), which may provide extra protection for PER.

Viewing the TIM–PER association from a different perspective, the existence of this dimer during the dawn hours may accentuate the extent to which CRY-mediated degradative effect on TIM can influence PER, such that proteolyzing the latter molecule eliminates its negative effects on *per* and *tim* transcription. In other words, it would not be the mere existence of monomeric PER that causes its susceptibility to proteolysis; the process per se of freeing up PER from its binding with TIM could promote the former's degradation at that moment.

These notions about interactions among these three molecules needs embellishment owing to further analyses of such processes that concentrated on *cry*, *per*, their protein products, and fragments thereof. That the cry^b and per^S mutations interact (Stanewsky *et al.*, 1998) was initially interpreted from the perspective of applying this *period* mutant as a tool (i.e., forcing flies with a cryptochrome deficit to effect several hours of daily phase shift in order to entrain). However, CRY^b:PER^S interactions were inferred to have more substantive meaning when doubly mutant flies were found by Rosato *et al.* (2001) to exhibit accentuated percentages of 19-hr locomotor components in 12:12 LD at increasing temperatures (*no* such components at 18°C; up to about 80% of the behaviorally tested flies exhibiting principally 19-hr periodicities at 28°C). These results could be reflective of normal interactions between the actions of the two genes, including direct physical associations between CRY and PER. The yeast two-hybrid system was exploited further (cf. Ceriani *et al.*, 1999) to examine this possibility. The following experiments and results are dizzyingly complex, but please bear with me (on behalf of Rosato and co-workers).

A PER fragment starting with the PAS domain at its N-terminus and extending ca. 450 aa downstream (here called PER^{P450}; cf. Figure 14) was shown to interact with full-length CRY (CRY^F) in a light-dependent manner within the nuclei of appropriately transformed yeast cells (Rosato *et al.*, 2001). A fragment consisting of approximately the middle-third of TIM (TIM^M) did not show such an interaction with CRY^F. However, light-dependent association between CRY and TIM was detected when full-length versions of each protein were produced in the fungal cells, but a CRY^F:PER^F interaction was not observed even in the light (Rosato et al., 2001). Both of these results are in agreement with those of Ceriani et al. (1999), who inferred a primacy for CRY:TIM interactions. Yet a hypothetical proviso should accompany these inferences: Rosato et al. (2001) isolated two mutants of *S. cerevisiae*, in which nuclear CRY:PER interactions occur *in the dark*. Therefore, associations between these clock proteins, possibly extending to CRY:TIM contacts, are not per se light-dependent (ratified by elements of the results from studying these interactions in cultured *Drosophila* cells, as described both earlier and subsequently). Therefore, as if the overall scenario involving "light to CRY, then onto the clockworks" were not complicated enough (read on), it is possible also to invoke a *nuclear factor that normally inhibits* CRY:PER (and possibly CRY:TIM) interactions; this conceptual entity might therefore be inactivated by light and could be just as much of a "light sensor" as the CRY molecule; or, perhaps a further role played by light-activated CRY is somehow to restrain the operation of this "nuclear factor" (if any).

The association of CRY^F with PER^{P450} observed by Rosato and coworkers in yeast cells prompted these investigators to piggyback onto one of the experiments performed by Ceriani et al. (1999) and thus transfect sequences encoding full-length versions of both molecules into cultured *Drosophila* cells (which are naturally TIM-less). Within them CRY^F:PER^F binding was detected in the dark by immunoprecipitation (Rosata et al., 2001). Moving back to the yeast system, a deletion of 20 residues from CRY's C-terminus was engineered by these investigators, by analogy to manipulating the carboxy terminals of *Arabidopsis* cryptochromes to demonstrate that these regions of the plant proteins are involved in mediating their light-responsiveness (Yang et al., 2000). The similarly truncated form of *Drosophila* cryptochrome ($CRY^{\Delta C}$) was shown to interact with PER^{P450} or with TIM^F in *both* light and dark conditions (Rosato et al., 2001). Such intrayeast associations did not occur when PER fragments consisting of the N-terminal (ca.) 170 or 250 aa of the P450 region (i.e., PER^{NP170} and PER^{NP250} subsets of PAS) were challenged with $CRY^{\Delta C}$, but a C-terminal, PAS-less subset of the P450 residues (PER^{C160}) did interact with the C-terminal-truncated form of CRY (Rosato et al., 2001). Owing to an intrayeast TIM^M-with-PER^{NP170} interaction that was demonstrated in parallel (as in the transfected-cell studies of Saez and Young, 1996), Rosato and colleagues speculated "that PER, TIM, and CRY can be found in the same complex." Finally, interactions between $CRY^{\Delta C}$ and either PER^{P450} or PER^{C160} were shown to be attenuated in yeast cells grown at a relatively high temperature, whereas PER^{P450}:TIM^M dimerization was not temperature-sensitive (Rosato et al., 2001).

This molecular-genetic experiment brought these investigators back to their seminal chronogenetic one involving the heat-sensitive per^S;cry^b double mutant, which shows such poor light-responsiveness (minimal entrainment) at high temperatures. Rosato et al. (2001) remembered in this regard that the site of the former mutation is contained within both P450 and C160 fragments of the PER protein (Figure 3). It also should be recalled that per^S flies are hypersensitive to the circadian phase-shifting effects of light (Section IV.C.1), and that those whose only functional *period* gene is deleted of the so-called Short Region (corresponding approximately to the N-terminal 50 residues of PER^{C160}) exhibit anomalous behavioral responses to light pulses (Schotland *et al.*, 2000). With regard to light leading to decrements in state variables of the pacemaker—and the role played by DBT in destabilizing PER—genetic interactions between these two factors are likely to be relevant, in that per^S and per^T, both mutated near each other within the C160 region of PER, partly suppress behavioral arrhythmicity caused by the dbt^{ar} mutation (Rothenfluh *et al.*, 2000b). We also want to bear in mind that DBT can be thought of as a further member of the putative PER:TIM:CRY complex (Kloss *et al.*, 2001).

However, an assertion of this multimer's existence leads mainly to musings about the potential meaning of PER's direct interactions with CRY within the complex in question. Thus, imagine that light alleviates the normal blockade of CRY:PER interaction (which wants to occur in general terms, i.e., absent CRY's C-terminal residues). When light comes in, CRY:PER associations increase or form anew, pulling PER relatively away from TIM—which naturally interacts with PER, notably in the dark. Thus, protection by PER of TIM's stability decreases. Moreover, CRY—which minimally touches TIM in the dark—now does do (given light), owing in part to TIM being relatively *pulled away from its PER interaction* by the induction of a CRY:PER association. CRY:TIM, but not PER:TIM, is relatively vulnerable to degradation, so TIM crashes with light.

These suppositions, fanciful as they are, would seem to be overly mesmerized by the primacy of TIM as the clock factor responding to light—that the "disappearance" of this protein upon such stimulation alleviates TIM's negative effects on feedback-loop functions, which results in abrupt phase shifts of the daily molecular cycle. However, if PER actions are equally important or more so in this aspect of the clock workings (Rothenfluh *et al.*, 2000c; Shafer *et al.*, 2002), then the results of Marrus *et al.* (1996) will not quit: Light at dawn does not make PER's abundance decrease (suddenly) to anywhere near the extent that TIM's concentration falls. To cope with this apparent conundrum, let us bring the CRY–PER interaction into play. Thus, light-induced binding of these two proteins could *tie up* PER such that it cannot carry out its negative effects on the CYC:CLK dimer optimally (Figure 7). In this feature of the hypothesis about CRY–PER binding, it is not that light causes PER levels as such rapidly to decrease. Instead, the protein is put into an effectively inactive form by CRY undergoing a light-induced conformational change that allows it to sequester PER.

The actions of the *cryptochrome* gene just summarized were dealt with at the level of the protein it encodes and how that molecule interacts with TIM

(Ceriani *et al.*, 1999) and PER (Rosato *et al.*, 2001). However, we need to remember that *cry* also influences cyclical expression of the clock genes that encode the other two proteins. In this regard, the initial consideration of *per* and *tim* cycling in the *cry*[b] mutant focused largely on molecules taken from or observed in tissues of the adult head. However, *cry* is expressed in many head and body parts of the fly (e.g., Emery *et al.*, 1998), and the mutation in this gene apparently flattens the *per* and *tim* timecourses all over the animal (Stanewsky *et al.*, 1998).

Such pleiotropic expression of *cry* can be surmised to connect with the circadian light responsiveness of far-flung tissues (Section VI.D). These light effects are now known to involve not only the phase-shiftability of clock-gene cyclings, but also an acute TIM response in one posterior organ: Light pulses caused a sharp drop in that protein's immunoreactivity within cells of the Malpighian tubule (Giebultowicz *et al.*, 2000). However, this does not mean that CRY is responsible for this particular effect of light on TIM. Indeed, the notion that that blue-light receptor is the only extraocular photoreceptive molecule functioning in posteriorly located nonneural tissues is not yet compelling: More "standard" substances that mediate light reception and phototransduction are found in at least one posterior tissue—the male gonad, which was reported to contain a rhodopsin, a G-protein that binds it, and one of the fly's two arrestin subtypes (Alvarez *et al.*, 1996). Perhaps a clock is running in these reproductive structures, analogous to that found in other insects (reviewed by Giebultowicz, 1999, 2001), and its autonomy would include photoreceptive capacities unrelated to cryptochrome. Malpighian tubules, and their inherent light-responsive with regard to clock-molecule cyclings, may also be subserved by non-CRY photoreception because RNA encoded by the *norpA* gene (introduced in Section II) is found in MTs (Ivanchenko *et al.*, 2001).

In *Drosophila* the meaning of molecules functioning in light-responsiveness with respect to non-eye and non-CNS tissues has been examined mainly for CRY (the presence of gonadal rhodopsin, etc., continues to fester as solely a descriptive finding). In this regard, bear in mind that the *cry*[b] mutation was identified owing to its ostensible elimination of cyclical *per* expression (Stanewsky *et al.*, 1998), the detection of which by luciferase reporting involves mainly body tissues (Stanewsky *et al.*, 1997b), certainly not brain cells only. *cry*[b] was soon found also to eliminate the normal daily oscillations of *tim-luc* expression (Stanewsky *et al.*, 1998), presumably that which occurs throughout the adult animal (Table 4). Indeed, TIM abundance and phosphorylation cyclings were found to be absent in extracts of heads or bodies separated from *cry*[b] flies; adding the *tim-gal4*; UAS-*cry* combination to the mutant rescued these biochemical defects (Emery *et al.*, 2000b).

Given *cry*[b]'s isolation phenotype and the molecular-timecourse defects determined subsequently—strictly speaking, the mutant exhibits faulty clock functioning, not necessarily or solely circadian blindness—we will now entertain the possibility that CRY can act as a state variable in the circadian pacemaker. However, the ability of *tim* and *per* products to oscillate is not globally perturbed in *cry*[b]. In extracts of fly heads, the concentration of TIM and PER proteins cycled when the temperature went up and down systematically (in DD), and such en-trainment of the molecular oscillations continued to manifest themselves when the mutant proceeded into thoroughly constant conditions (Stanewsky et al., 1998). Yet *cry*[b] caused an overall 50% reduction of TIM and PER in this temper-ature experiment, as if the molecular pacemaker is not fully operational. Perhaps it is at least partly functioning in LD cycling conditions as well, despite the flat-ness of *tim* and *per* gene products in whole-fly monitorings or gross tissue extracts. This supposition is based on the possibility that, in LD cycles, the *initial stimulus* to start the clock does not work in *cry*[b]. Light input would get the clockworks moving to begin with (in *cry*[+], but not in *cry*[b]), although the required stimula-tion can be delivered by temperature changes (in animals of either genotype). A related hypothesis is that TIM and PER do oscillate on a cell-by-cell basis in *cry*[b] peripheral tissues, but such cycling could be asynchronous among cells. The over-all result would be flattened levels of these clock-gene products in gross tissue assessments. However, this hypothesis, in which CRY is apprehended as a clock-*input* molecule *only*, was undermined by asking whether there is *heterogeneity* for TIM and PER immunostaining among cells in *cry*[b] compound eyes at a given time point (Stanewsky et al., 1998). That *cry*[b] caused PER and TIM levels to be apparently *flat as such* throughout the retina indicated that the answer is "no."

Elimination of *per*- and *tim*-product cyclings by this mutation is not spatially ubiquitous: In LD, PER and TIM still went up and down within the small-LN$_v$ cells of the *cry*[b] brains (Stanewsky et al., 1998); adding the effects of *norpA* to that of the cryptochrome mutation left the s-LN$_v$'s molecularly entrainable in terms of PER cycling (Helfrich-Förster et al., 2001). These cells are such a small proportion of the head that abundance fluctuations of clock-gene products, as affected by *cry*[b], would be swamped in homogenates (Stanewsky et al., 1998; Emery et al., 2000b). However, the neuronal oscillations could be sufficient to entrain the system such that these brain neurons would sustain their synchrony in DD, allowing for the cryptochrome mutant to exhibit free-running locomotor rhythmicity, which it does (Stanewsky et al., 1998).

A different cytomolecular result was obtained for the s-LN$_v$ precursors in the larva. When its peripheral photoreceptive organ was presumably rendered nonfunctional by *norpA* (Section II.B), and putative intrabrain reception was at-tenuated by *cry*[b], larval LNs could *not be light-entrained* for PER or TIM cycling (Kaneko et al., 2000a). The reason for this nonresponsiveness is presumably be-cause there is no backup photoreceptive structure present in larvae, and the H–B eyelet has not yet formed during this stage (Yasuyama and Meinertzhagen, 1999). However, one predicts that an eye- and eyelet-eliminating *gl*[60j] mutation, when combined with *cry*[b], should knock out s-LN$_v$ cycles in adult brains; indeed (as first noted earlier in this section), PER immunoreactivity was equivalent at the usual trough and peak times (Helfrich-Förster *et al.*, 2001).

Further analyses of these neuronal signals led to ostensible correlations with the free-running behavior of $gl^{60j} cry^b$ flies: They exhibited relatively high proportions of arrhythmicity or complex periodic locomotor components. The neural substrate of these anomalies could be the frequent *intraindividual asynchrony* of PER and TIM cycling that was observed for the LNs within the brains of this doubly mutant type (Helfrich-Förster *et al.*, 2001). Therefore, the absence of external photoreceptors and the H–B eyelet, combined with a near elimination of CRY functioning, creates a genocopy of constant darkness. This genetically effected mimic of an environmental manipulation would go as follows: Remember that several generations of DD maintenance results in flies that exhibit phenocopies of genetic clocklessness, along with other complex locomotor patterns (Dowse and Ringo, 1989; Power *et al.*, 1995b). One imagines that light-depriving these animals (and their ancestors) would cause substantial *intrabrain asynchrony* in the functions of neuronal pacemakers. Yet Zeitgeber deprivation in which the behaviorally monitored adults stemmed from DD development occurring only for their own embryo-to-adult stages resulted in flies that were nearly always rhythmic (Sehgal *et al.*, 1992; Tomioka *et al.*, 1997; Kaneko *et al.*, 2000a). There was *interindividual* asynchrony among their free-running locomotor cycles, but apparently a one-time DD rearing permits clock-gene expressions to operate such that a given fly possesses *intrabrain* synchronies of periodic functions. In this circumstance, *per* and *tim* (possibly *Clk*, *cyc*, etc.) would somehow be activated by virtue of internal gene-regulatory circuitry; intercellular synchrony of such expression would then be sustained during the 1.5–2 weeks of maintenance in the absence of environmental cues. In natural LD conditions, photic signals would accentuate such synchronous cellular functions, causing all flies within a group to be in phase with one another. However, initial establishment of the molecular events does not as a matter of course occur at the exact same developmental stage in all animals, such that adults resulting from DD rearing are unphased behaviorally (e.g., Sehgal *et al.*, 1992). In contrast, chronic DD maintenance of wild-type cultures and flies, or the creation and multigeneration sustenance of a circadian-blind, doubly mutant stock in LD, leads eventually to intra-animal drift among the gene expressions operating among behaviorally relevant pacemaker structures.

These concerns bring us back to considerations of the more global effects of cry^b on *Drosophila* tissues. Are the dearths of *per*- and *tim*-product cyclings caused by a mutationally induced absence of light-influenced activation of circadian clocks during the times when all these nonneural tissues are developing? Alternatively, the clock could start normally within a given structure or cell as a cry^b animal develops, but these entities would never get properly synchronized. The result would be a severe temporal smearing of *per* and *tim* expressions as determined by gross monitorings of whole flies (Stanewsky *et al.*,

1998) or product-abundance measurements based on homogenates of major portions thereof (e.g., Emery *et al.*, 2000b). This explanation of the overall flatness exhibited by *per* or *tim* mRNAs, proteins, or reported gene activities will not wash for one peripheral tissue: the adult's compound eye (Stanewsky *et al.*, 1998). In it, and perhaps in other locations outside the CNS, CRY functions may participate in those of the pacemaker. Thus, *cry^b* would knock out clock functioning itself in certain peripheral tissues, but not in all of them, else extracts of mutant tissues would not exhibit entrainability of PER and TIM cycling to temperature cycles (Stanewsky *et al.*, 1998).

There is one peripheral structure that was shown to be thoroughly unentrainable by cyclical Zietgebers when it was homozygous for the cryptochrome mutation. This is the antenna. First, it needs to be registered that that structure represents the only known *biological* rhythm in adult *Drosophila* other than its behavioral cycles. The odor-sensitivity rhythm peaks during the middle of the night, free-runs in DD, is flattened by the effects of *per^{01}* or *tim^{01}* (Krishnan *et al.*, 1999), and operates autonomously within this appendage (Section VI.D). Culturing of antennal specimens carrying the *per* or *tim* promoter regions fused to *luc* showed that the reporter autonomously cycled in LD; this rhythmicity dampened in DD (Plautz *et al.*, 1997a; Krishnan *et al.*, 2001). Therefore, the antenna's molecular rhythmicity is inherently light-responsive (including that it was re-entrainable in the LD → DD → LD experiment of Plautz *et al.*, 1997a). A corollary, which seems unlikely to be coincidental, is that *cry* mRNA was readily detectable in antennal homogenates (Okano *et al.*, 1999). These results poise this peripheral pacemaker to be apprehended as functioning by virtue of molecular cycles (at least that of *per*) that are entrained by CRY-mediated input. However, cultured *cry^b* antennae were found to be flat, or they exhibited extremely weak *per-luc* and *tim-luc* cycles—depending on the specimen—in both LD conditions *and* in those for which the temperature systematically fluctuated (Krishnan *et al.*, 2001). Measurements of *cry^b*'s odor-sensitivity rhythm gave the same apparent clocklessness from a series of recordings made in LD, DD, or high–low temperature.

Tests of this mutation's effects partly analogous to those just described were performed with respect to molecular cyclings in the Malpighian tubules. Degradation of TIM in this tissue, which occurs after brief light exposures (Giebultowicz *et al.*, 2000), did not occur in the MTs of *cry^b* flies until long, 5-hr pulses were delivered (Ivanchecko *et al.*, 2001). In LD, oscillations of TIM and of PER as well were observed in the mutant, although the usual TIM decline after lights-on was anomalously gradual; adding a *norpA* mutation was concluded not to cause a further decrement in such responsiveness (despite the aforementioned expression of this gene in the MTs). Thus, the nearly null *cry* mutation was sufficient severely to attenuate light responsiveness; however, neither it nor the *norpA;cry^b*, double mutation made the MTs completely blind. In DD, the usual free-running rhythmicity of PER and TIM in these excretory structures was abolished by *cry^b*, in that the immunoreactivity of either protein stayed near trough levels throughout the subjective day and night (Ivanchenko *et al.*, 2001). That these effects of the mutation may reflect *pacemaking* problems was examined further by testing dissected MTs in cultures. Reported molecular rhythmicity of *tim-luc* specimens homozygous for cry^b was very weak in LD (with peaklets that were also phase-delayed compared with the cry^+ luciferase peaks) and not discernible in DD (Ivanchenko *et al.*, 2001).

So, the antennae of the cryptochrome mutant are non-entrainable, as well as being arrhythmic in constant conditions, and there was no free-running molecular cycling in the Malpighian tubules. These results make it seem as if CRY can function closer to the clock outside of the CNS compared with how it acts within LN cells of the brain. Appreciation of a free-running clock function for CRY in the antennae or MTs is not difficult. For simplicity (cf. Rosato *et al.*, 2001), let us consider CRY interacting with TIM alone in free-running conditions. Thus, when *cry* mRNA rises at subjective dawn, newly manufactured CRY may be required such that these molecules can interact with TIM polypeptides and promote the latters' disappearance at this phase of the cycle. Such CRY–TIM interactions are likely to occur only after light effects enter the system in some fly tissues; however, in others, the two molecules could dimerize in the dark (around the time of subjective dawn), such that CRY is functioning squarely within the cogs of the pacemaker. How could this molecule be part of the clock in one cellular circumstance and only feed environmental information to it in another? Presumably, this would depend on the manner by which CRY is posttranslationally modified, or on the quantity and quality of other rhythm-related molecules with which it might interact, in one cell type versus another.

X. GENE-DEFINED FUNCTIONS CONNECTING CENTRAL PACEMAKING TO CIRCADIAN CHRONOBIOLOGY

A. Genes encoding factors that control eclosion

1. Neuropeptides

Several small molecules are known to function as coordinators of tissue events underlying the gated emergence of metamorphosing insects into adulthood (reviewed by Gäde *et al.*, 1997). Most manipulations of hormones and other small molecules that have contributed to understanding of these processes have been performed in large insects, notably moths (e.g., Truman, and Morton, 1990; Truman, 1992a,b). Such studies have been unable to draw on eclosion mutants, although the normal genes encoding several of the relevant peptide hormones (Table 6) have been isolated from insects representing these nongenetic systems (e.g., Horodyski *et al.*, 1989; Kawakami *et al.*, 1990; Kamito *et al.*, 1992; Noguti *et al.*, 1995; Horodyski, 1996; Sauman and Reppert, 1996b; Zitnan *et al.*, 1999).

Table 6. Genes and Encoded Products Known or Suspected to Mediate Outputs from Circadian-Clock Control of Insect Rhythms

Gene and/or peptide	Molecularly characterized	Manipulated	Mutated	Remarks
CCAP	+	—	—	Sequence encoding this small molecule, called (for historical reasons) crustacean cardioactive peptide, cloned in *D. melanogaster*; exerts proximate regulation of ecdysis behavior
Corazonin	+	—	—	cDNA encoding this 11-mer (along with signal peptide and a 39-aa corazonin- related peptide) cloned in *D. melanogaster*, after similar material isolated from other insects [not mentioned in text, but see Veenstra (1994) and references therein]; in most species, the peptide or mRNA encoding it found limited to certain cells within CC; possible chronobiological role: to increase heartbeat rate as prelude to eclosion (and accompanying "pumping up" of hemolymph); this peptide coexpressed with PER in a small subset of PER-containing neurons within the *Manduca* brain
Crg-1	+	—	—	Identified by virtue of daily oscillating mRNA; such cycling transcripts (three isoforms) encode forked-head (a.k.a. HNF3)-type transcription factor; mRNA oscillations eliminated by per^{01} or tim^{01}; this *Clock- regulated-gene-1* likely to be coexpressed with *per* in part of latter's expression domain in adult brain
Dat	+	—	+	Mutant exhibits sleep-deprivation rebound defect, which seems to be in part under clock control (Table 2); one isoform of the encoded aryl alkylyamine *N*-acetyltransferase likely to be coexpressed with *per* in part of latter's expression domain
Adh	+	—	—	Identified chronobiologically as 350-bp cDNA called *Drosophila rhythmically-expressed gene-1*, which is complementary to LD-oscillating mRNA; such cycling f-r in DD; eliminated by per^{01} and in both conditions
Dreg-3	+	—	—	310-bp cDNA complementary to 2.7-kb LD-oscillating mRNA encoding protein of unknown function; mRNA oscillation f-r in DD and is eliminated by per^{01} in that condition, but the mutation has no effect on LD cycling
Dreg-5	+	—	—	385-bp cDNA complementary to 1.8-kb LD-oscillating mRNA; such cycling eliminated by per^{01}; mRNA oscillation f-r in DD; the gene encodes a ca. 300-aa protein which oscillates at least in LD

Table 6. *continued*

Gene and/or peptide	Molecularly characterized	Manipulated	Mutated	Remarks
Dreg-6	+	—	—	310-bp cDNA complementary to 7-kb oscillating mRNA, whose cycling f-r in DD
Dreg-10	+	—	—	460-bp cDNA complementary to three oscillating mRNAs, whose cycling f-r in DD
Dreg-15	+	—	—	650-bp DNA complementary to 3.5-kb oscillating mRNA, whose cycling f-r in DD
Eh	+	+	—	Driving of cell killers by regulatory region of this eclosion-hormone-encoding gene leads to ecdysis and eclosion anomalies, but permits adult emergence still to occur
ETH	+	—	—	Ecdysis-triggering hormone, sequences encoding which were cloned in *Bombyx mori*, *Manduca sexta*, and *D. melanogaster*; injection of this oligopeptide into pharate adults of the latter causes premature onset of eclosion behavioral sequence; in *M. sexta*, natural ecdysteroid rise prior to onset of ecdysis behaviors induces expression of ETH-encoding gene in certain endocrine (Inka) cells
lark	+	+	+	Dominant mutation at this vital locus (or deletion of it) causes earlier-than-normal eclosion (heterozygous effects); "manipulated" refers to further gene-dosage alterations, which showed that extra copy of *lark*$^+$ causes delayed eclosion; the gene encodes an RNA-binding protein, which exhibits a daily cycle of abundance and is widely expressed (life-cycle stages; neural and nonneural tissues); LARK coexpressed with CCAP in VNC; *per*^{01} causes the protein to be constitutively high in these neurons
Nf1 and Ras/MAPK	+	+	+	*Nf1* mutations cause largely arrhythmic locomotor behavior, which is partially suppressed by Ras/MAPK-affecting mutations; *Nf1* mutations do not affect the clockworks, but anomalously upregulate phospho-MAPK levels; this activated form of MAPK has its production attenuated by a PDF-null mutation (see below) and by *tim*^{01} (cf. Table 1), which apparently eliminates PDF cycling in the termini of axons projecting from certain clock-gene-expressing, PDF-containing neurons (see below)
Pdf	+	+	+	Sequences encoding insect pigment-dispersing factors (historically named for its role in crustacea) cloned in *D. melanogaster*, in the cricket *Gryllus*, and in the grasshopper *Romalea*;

continues

Table 6. *continued*

Gene and/or peptide	Molecularly characterized	Manipulated	Mutated	Remarks
				see the footnote for information on how this gene is currently abbreviated; PDF-null mutant in *Drosophila* causes poor or absent f-r B rhythmicity, as does driving of cell killers by *Pdf* regulatory sequences (which ablate PDF-containing s-LN$_v$'s and l-LN$_v$'s; cf. Figure 13); transgenically mediated ectopic expression of *Romalea* PDF causes E and B anomalies; *Pdf* expression absent or very low in s-LN$_v$ neurons of Clk^{Jrk} or cyc^0 mutants; per^{01} or tim^{01} eliminates cycling of PDF in dorsal-brain nerve terminals of adult s-LN$_v$'s
PETH	—	—	—	Pre-ecdysis triggering hormone, sequences encoding which were cloned in M. *sexta* and D. *melanogaster*; in both species, one gene produces an mRNA encoding both PETH and ETH; see entry about the latter for biological effects of PETH injection into flies and steroid induction of this gene in the moth
PTTH	+/—	—	—	Prothoracico tropic hormone, which (when released from terminals of certain brain neurons) stimulates prothoracic gland to synthesize and release ecdysteroid; sequences encoding ca. 100-aa PTTH polypeptide (which forms a homodimer) have been cloned from *Anthereae pernyi* and B. *mori*; PTTH not coexpressed with A. *pernyi per* gene in this silkmoth's brain; 45-kD native PTTH protein (but not the gene or cDNA) isolated from D. *melanogaster*, partially sequenced, and shown to induce ecdysteroid secretion from isolated ring gland
to	+	—	+	Identified by virtue of mRNA whose abundance is lowered by the effect of cyc^{01} mutation; *takeout* mRNA exhibits abundance oscillations in LD and DD; transcript undetectable in cyc^{01} or Clk^{Jrk}; lower-than-normal abundance in per^{01} or tim^{01}; TO protein (a medium-sized one) exhibits abundance oscillation; tissue expression not especially broad, but includes adult CNS and portions of thoracic alimentary system; *to* expression induced by starvation; a rough corollary is that hypomorphic *to* mutant exhibits aberrant rhythmic B and dies prematurely in starvation conditions

Table 6. *continued*

Gene and/or peptide	Molecularly characterized	Manipulated	Mutated	Remarks
vri	+	+	+	Identified chronobiologically by virtue of intra-nighttime oscillating mRNA; such cycling revealed by subsequent study to be a daily one, which f-r in DD; LD cycling attenuated in per^{01} or tim^{01}, and the mRNA is constitutively very low in Clk^{Jrk} or cyc^0; sequence complementary to this transcript reidentified the developmentally vital *vrille* gene, which encodes a bZIP-type transcription factor; "mutated" refers to B tests of *vri* or *Df-vri* heterozygotes, which exhibit shortened f-r period; transgenically effected overexpression leads to long-period B or arrh. and lowered expression of *per* and *tim* products and of PDF—as if *vri* is in part a clock gene
dfmr1	+	+	+	Ambiguous and controversial as to whether this gene is in part clock-involved or instead only an output factor: Are eclosion as well as locomotor rhythms disrupted by *dfmr1* (a.k.a. *dfxr*) mutations? Are the temporal dynamics of PER and TIM expressions altered? If the answer to the latter question is "no," then bear in mind that there are dosage effects of this gene on circadian behavioral periods (which imparts to *dfmr1* the whiff of a clock-gene property); incidentally, among the anatomical anomalies caused by *dxfr* variants are dosage effects on certain neurite patterns (not discussed in Section IV.C.6, but see Morales *et al.*, 2002)
Several and various (enhancer trapping)	+	—	+	Ca. 5% of newly created *luc*-transposon-containing strains exhibited daily cycling of the reporter-enzyme activity; this "temporal enhancer trapping" allowed cloning and analysis of the trapped loci (a subset of the ca. 70 identified on one study; the 1 gene, out of 20 screened strains, in another); some (but not all) of these genes exhibited mRNA cycling in direct measurements; the analyzed genes encode a variety of enzymes, transcriptional regulators, and/or genes known to be involved in certain features of *Drosophila* development; see text for discussion of the two most heavily analyzed cases: *numb* in one study, *regular* in the other (the latter being one example of certain "pioneer" genes that were tapped into by this strategy for identifying rhythm-related genes)

continues

Table 6. *continued*

Gene and/or peptide	Molecularly characterized	Manipulated	Mutated	Remarks
Many and various (genome scans)	+	—	—	Drawing on *Drosophila* genomics and DNA-microarry technology, many known sequences (including scores of comprehensible genes) were identified anew by virtue of daily cycling (in DD) of their encoded mRNAs; the "unknown" genes alluded to in previous entry as well are open-reading frames presumed to encode proteins that have no established functions; the remainder involve known or suspected functions, most of which could be categorized as involved in: transcriptional control, neuropeptide actions, olfactory functions, nutritional phenomena, cuticle formation, potassium-channel regulation, signal transduction, detoxification, and immunity; certain of these genes and others (e.g., *to* family members, serine proteases, some of the unknowns) were quickly revealed (via the genomics feature of the screen) to be clustered at particular chromosomal loci; these exemplary results are drawn mainly from McDonald and Rosbash (2001); see footnote for citations to additional, analogous studies

[a] The putative clock-output genes are in italics, and the small or oligopeptides encoded by other genes are in upper case. Most of these factors have been identified in *D. melanogaster* only, although some of the peptide-encoding genes have also (and in one case instead) been cloned in a few additional insect species (see Remarks). One gene in the latter category, *Pdf*, was originally called *pdf* (Park and Hall, 1998); but this symbol had been preempted long ago by the abbreviation for a mutant in *D. melanogaster* called *podfoot* (http://flybase.bio.indiana.edu/), so the PDF peptide-encoding gene is renamed here (also see Toma *et al.*, 2002). "Molecularly characterized" refers to the fact that all these factors have been cloned and sequenced (usually for the gene's open-reading frame only); + in this column indicates that expression patterns of the mRNA, protein, or both have been examined with at least modest intensity. Such patterns include temporal ones in some cases—life-cycle stages and/or daily fluctuations in product abundances (the latter being how some of these factors were originally identified). The Manipulated and the Mutated columns refer to whether a given output factor has been manipulated (usually at the genic or cDNA level, possibly at that of the translated product) or studied in at least one mutant form for effects on some aspect of the fly's chronobiology. Not all the *Dreg*'s that were identified by mRNA fluctuations in LD (Van Gelder *et al.*, 1995) are listed because such cycling did not occur in DD and/or in certain culture conditions, and most *Dreg*'s were not tested for effects of *per^{01}* on transcript cyclings (the other *Dreg*'s are numbered 2, 7–9, 11–14, 16–20). None of the genes highlighted (under Remarks in the final tabular entry) as rhythm-regulated, based on a microarray-based "genome scan" (McDonald and Rosbash, 2001), are specified because they are too numerous to list as such; the same incipiently overwhelming situation obtains for genes detected in additional screens that involved chronobiological applications of "chip technology," which to a significant extent turned up only partially overlapping sets of cyclically expressed factors (Claridge-Chang *et al.*, 2001; Lin *et al.*, 2002; Ueda *et al.*, 2002). Abbreviations (e.g., B, E, f-r, arrh.) are as in previous tables, except for CC: corpus cardiacum.

One way that eclosion is known to be clock-controlled derives from the effects of mutations in *Drosophila* that almost certainly affect pacemaker functions. Several of these mutants were isolated as defective in eclosion rhythmicity (Konopka and Benzer, 1971; Sehgal *et al.*, 1991, 1994) or turned out to exhibit eclosion abnormalities after being identified initially as adult behavioral variants (Konopka *et al.*, 1994; Hamblen *et al.*, 1998; Allada *et al.*, 1998; Price *et al.*, 1998; Rutila *et al.*, 1998b). The links between the actions of these *period, timeless, Clock,* and *cycle* genes to downstream factors that exert proximate control of eclosion are unknown. However, it may be possible to fill in these regulatory gaps, in part because *Drosophila* genes encoding certain eclosion-influencing hormones (Table 6) have been identified. Isolation of these nucleotide sequences established the (expected) existence of these peptide factors in *D. melanogaster* and set up the wherewithal to manipulate certain components of the fly's eclosion system.

Thus, cloning the eclosion hormone gene from *D. melanogaster* (Table 6) led to description of a pair of medially located brain cells that contain *Eh* RNA in the brains of larva, pupae, and young adults (Horodyski *et al.*, 1993). This dipteran cDNA was isolated based on the *Eh* coding sequence in *Manduca sexta*, and an antibody against the moth peptide apparently stained the same *Drosophila* cells containing the transcript. These neurons are not particularly near any of those that express clock genes in larvae (Sections VI.A and VI.B); the pharate-adult brain cells are somewhat close to the locations of clock-protein-containing DN cells (Figure 11), but the EH neurons are probably too ventrally positioned (i.e., they would not coexpress the *per* or *tim* genes). The labeled medial cells in *Drosophila* project EH-immunoreactive fibers to the corpora cardiaca within the ring gland of larvae (second-instar ones being the earliest stage examined) and pharate adults; another projection runs posteriorly into the ventral cord of larvae and young adults, forming a ventomedial tract that terminates in the abdominal ganglion (Horodyski *et al.*, 1993). EH is released from the analogous neurites in *M. sexta* just prior to the moth's ecdyses; the same sharp decrement in EH immunoreactivity was observed in *Drosophila*, although this was not examined in conjunction with eclosion, but in ecdysing second-instar larvae (Horodyski *et al.*, 1993).

To take this case beyond descriptive studies, 5'-flanking sequences cloned from the *Eh* gene of *Drosophila* were fused to those encoding yeast GAL4. Combining this fusion gene with a GAL4-drivable marker (tau-GFP) led to cell-body and neurite expressions that mimicked the endogenous EH pattern (McNabb *et al.*, 1997). This permitted interpretability of experimental transgenics in which *Eh-gal4* was combined with UAS-*reaper* (*rpr*) or UAS-*head-involution-defective* (*hid*) constructs. The latter two factors are natural cell killers (e.g., Bangs *et al.*, 2000), which can ablate developing cells in which they are ectopically expressed. Elimination of EH cells effected in this manner led to low viability; but this meant that fully one-third of the developing animals progressed through the larval stages and emerged as viable adults, although most of the latter failed to inflate and expand their wings (McNabb et al., 1997). The "eclosion program" was defective in a stochastic manner in the cell-killed transgenics: Tracheal inflation was delayed; the latency between extension of an anterior structure, which ruptures that end of the pupal cuticle, and the time of eclosion itself was substantially extended; eclosion behaviors, categorized into four stages, did not exhibit the usual periods of quiescence between a given pair of stages; and the times required to progress from one behavioral "landmark" to the next were extended (McNabb et al., 1997). Eclosion profiles determined in light:dark cycles typically show a rather sharp peak of emergence after lights-on followed by a broader plateau of emerging flies distributed over most of the photoperiod. Cultures expressing an EH-cell-ablating transgene combination gave no lights-on peak, but otherwise eclosed (in a quasi-normal manner) within the L portion of an LD cycle; these variant cultures exhibited free-running eclosion profiles indistinguishable from controls (McNabb et al., 1997).

EH as and eclosion mediator—a contributing but inessential one, as just described—is only the tip of the hormonal-regulatory iceberg (Table 6). For example, a peptide called ecdysis-triggering hormone (ETH) appears to function with EH in a positive feedback loop in which small amounts of ETH stimulate the release of EH; this stimulates the complete release of ETH, which causes the remainder of the EH stores to be released (Hesterlee and Morton, 1996). The effect of ETH in this system on flies with no EH cells was tested: Compared with controls for which injection of M. sexta ETH (MasETH) caused precocious eclosion, the EH-cell "knockouts" were refractory to this effect, indicating that the EH neurons are necessary for ETH's eclosion-stimulating consequences (McNabb et al., 1997). Further manipulations of ETH were performed in conjunction with the (previously determined) role played by a messenger factor that functions downstream of EH's action: cyclic GMP. In conjunction with the final release of EH (during the final preecdysis hour), this small molecule is produced in certain VNC neurons and in trachea, which is correlated with heart-rate elevation and tracheal air-filling. Injecting Drosophila with MasETH during a late pupal stage caused all these events to occur prematurely, but no such changes were induced when the EH-cell ablatees were injected; yet, the transgenic's response failure was overcome by injecting Eh-gal4;UAS-rpr pupae with a membrane-permeable analog of cGMP (Baker et al., 1999). As alluded to above, there is normally a delay (ca. 40 min) between the last release of EH and ecdysis; however, if after the former event, late-developing flies are decapitated, ecdysis ensues almost immediately. The premature occurrence of this event (in the experimental circumstance just described) did not occur in the EH-cell knockouts (Baker et al., 1999). It was inferred that one element of the roles played by EH cells is to activate descending inhibitory neurons, which function to delay the onset of ecdysis; this would allow for EH-activated processes such as tracheal filling to occur over the course of a 30- to 40-min timespan during which inhibition slowly wanes.

Elements of the effects on eclosion of moth ETH were reproduced by injecting the *Drosophila* form of this peptide. For this, sequences putatively encoding this material were found in a genomic database. Cloning and characterization of the full open-reading frame revealed a transcript encoding two ETH-like peptides, dubbed DrmETH1 and 2, along with a third novel peptide (Park *et al.*, 1999). Synthetic forms of these 14- to 22-mers (including two alternative forms of DrmETH1 encodable by a portion of the ORF) were produced by chemical synthesis, and injection of any one of them into pharate *Drosophila* adults led to premature eclosion and earlier-than-normal onsets of the associated behaviors (Park *et al.*, 1999). The possibility of manipulating the corresponding gene (*eth*), perhaps by analogy to the manner by which sequences from the *Eh* one were handled, was established by characterizing the 5′-flanking region of *eth*, which was found to contain a putative ecysteroid-response element. Indeed, the *Manduca* form of this gene is regulated by fluctuating ecdysteroid levels that precede each ecdysis during moth development (Zitnan *et al.*, 1999). Recall in this regard that the ecdysteroid-synthesizing prothoracic gland in *Drosophila* expresses the *period* gene (Emery *et al.*, 1997).

Acting upstream of the production and release of this steroid hormone is prothoracicotropic hormone (PTTH). This material is produced by neurosecretory cells in insect brains. In *A. pernyi*, PTTH release is clock-controlled, raising the possibility that known clock-gene-expressing cells in the silkmoth's brain might containing this (relatively large) peptide hormone (Table 6). However, cloning of PTTH-encoding sequences from *A. pernyi* (based on portions of the relevant amino-acid sequence determined for other insects) led to demonstrations of noncongruent brain expression of PER and PTTH. At various developmental stages and in adults, cells containing mRNA that encodes the hormone or PTTH immunoreactivity (applying an antibody generated against a synthetic peptide deduced from the PTTH cDNA sequence) were shown to be in the same brain region as PER-containing cells, but the two proteins are not colocalized (Sauman and Reppert, 1996b). Recombinantly produced silkmoth PTTH was shown to be biologically active by virtue of inducing adult eclosion 3–4 weeks after injecting this material into debrained pupae.

Bioassaying protein fractions in extracts of *Drosophila* larvae—by monitoring ecdysteroid secreted by cultured ring glands—led to the purification of putative PTTH from this insect (Kim *et al.*, 1997). Antibodies generated against the purified material led to labeling of dorsomedial neurosecretory cells in the brain of third-instar larvae, along with fiber tracts projecting near the ring gland. A partial (N-terminal) amino-acid sequence of *Drosophila* PTTH revealed no similarity to the *Bombyx mori* form of the hormone (Kim *et al.*, 1997), but these data for the fly material establish the possibility of identifying PTTH-encoding sequences and eventual manipulation of the *Drosophila* gene that encodes it.

A downstream neuropeptide bioactive in insect ecdyses in crustacean cardioactive peptide (CCAP). EH, ETH, and a peptide relative of the latter called pre-ecdysis-triggering hormone (e.g., Zitnan *et al.*, 1999) control the timing of ecdysis (PETH in *Drosophila* is likely to correspond to DrmETH2), whereas CCAP, acting within the brain and ventral cord (Ewer and Truman, 1996; Draizen *et al.*, 1999), turns on ecdysis behaviors themselves. To permit experiments that interrogate the roles played by the small CCAP molecule, cDNA encoding it was isolated from *Drosophila* by an RNA extraction → reverse transcription → PCR cDNA-amplification strategy commonly applied (Table 6) in the isolation of sequences encoding chronobiologically relevant proteins from one organism, using oligonucleotide primers based on amino-acid sequence data known for another organism's version of the substance (Park and Hall, 1998). After isolating a CCAP-encoding cDNA from *Drosophila*, sequences 5′ to the ORF were then cloned (Ewer *et al.*, 2001); this will allow for manipulation of neurons containing this peptide by the production of 5′-flanking *gal4* transgenes (once more referring to the object lessons created by McNabb *et al.*, 1997).

2. *lark*

Taking an entirely different genetic tack led to molecular identification of a factor that is likely to be involved in CCAP regulation. A mutant was isolated in a screen for *Drosophila* eclosion variants caused by transposons mobilized to land in novel genomic locations (Table 1). One such P-element insert (out of approximately 1,000 lines screened) was a dominant early-eclosing mutant; it was baptized *lark* (Newby and Jackson, 1993). Only the phase of eclosion was thus affected (either in LD or high–low temperature cycles), not its free-running period. Moreover, adult locomotor rhythms were normal in their phase and period. Reminiscent of the late-eclosing *psi-3* mutant, isolated after chemical mutagenesis as a mutation that essentially affects only the phase of one kind of circadian rhythm (Jackson, 1983), the potentially transposon-tagged variant can be thought of as identifying a *clock-output* factor (Table 6). A similar argument would be made about the *ebony* body-color mutants (Section IV.A.2), which are not defective for clock functions in general (*e* flies exhibit arrhythmicity for adult behavior, but are eclosion-normal).

The early-eclosing *lark* mutant defines an essential gene: Homozygosity for the P-element, or derivatives of this mutation resulting from remobilization of the transposon and its imprecise excision, cause embryonic lethality (Newby and Jackson, 1993). Note that this strain would have been discarded if the screen had been based on homozygosity for autosomes carrying P-elements newly landing on these chromosomes. DNA putatively corresponding to the *lark*-mutated transcription unit was readily isolable owing to the transposon tagging implied in these passages. Indeed, the etiology of the mutant's anomalous eclosion comapped with the location of the P-element insert (in contrast to the induction and eventual identification of the first *timeless* mutant), and the third-chromosomal *lark* locus is distinct from the sites defined by other rhythm mutations on that chromosome (Figure 18, color insert).

Assessing the spatial distribution of presumptive *lark* mRNA, bolstered by staining for β-GAL activity emanating from the *lacZ*-containing P-element mutant, revealed widespread CNS expression in embryos and more limited labeling of brain locations in pharate adults (Newby and Jackson, 1993). Temporal analysis of *lark* expression accompanied the demonstration that this transcription unit encodes a protein similar to a certain class of RNA-binding factors (Newby and Jackson, 1996). Polypeptides of this sort contain RNA-recognition motifs, and the LARK protein possesses two such RRMs along with a retroviral-type zinc finger (RTZF). *In vitro* mutagenesis of the RRMs and the RTZF domains led to transgenic flies with reduced viability, abnormal wing and bristle morphology, female sterility, and flightlessness; however, mutations engineered within *none* of the three protein motifs just indicated led to anomalous eclosion of the type observed in straight *lark*/+ mutants (McNeil *et al.*, 2001).

Potentially to ratify the chronobiological significance of this gene, *lark*-gene products were assessed for abundance fluctuations. However, mRNA from which the LARK RNA-binding protein is translated was found not to oscillate in its abundance in late pupae (McNeil *et al.*, 1998). However, the meaning of *lark*-product amounts with respect to eclosion was inferred from effects of gene-dosage manipulations. Flies carrying one copy of *lark*+ (formally similar or equivalent to the mutant as originally isolated) exhibited anomalously early eclosion, whereas three copies of the normal gene caused delays of eclosion peak times (Newby and Jackson, 1996). It is as if LARK protein exerts a repressive effect on adult emergence via regulation of other chronobiological factors acting at this stage of the life cycle. These inferences were supported by subsequent characterization of LARK protein expression, analyzed both temporally and spatially. Generation of anti-LARK led to demonstrations that the protein *does* oscillate in its abundance in late pupae, even though the encoding mRNA does not (McNeil *et al.*, 1998), and LARK abundance is temporally constant in larval extracts (Zhang *et al.*, 2000). Nevertheless, production of this protein is under temporally varying posttranscriptional control during the next major developmental stage. During this stage, this RNA-binding protein may influence the products of eclosion-related genes after they have been transcribed. It seems noncoincidental that the LARK trough occurs a few hours before the time of adult ecdysis (McNeil *et al.*, 1998; Zhang *et al.*, 2000). Its normal nadir might therefore alleviate the aforementioned repressive affect on eclosion. However, a low, eclosion-promoting level could be attained prematurely when the overall level of product is halved in $lark^-/+$ cultures, and three copies of $lark^+$ could result in an inappropriately high degree of repression during the late-ecdysis run-up to eclosion.

Consistent with the results of monitoring $lark$ mRNA's presence at various life-cycle stages, LARK was detected throughout development and is even maternally inherited (McNeil et al., 1998, 1999; Zhang et al., 2000). By scrutinizing immunohistochemically stained tissues in third-instar larvae through adults (McNeil et al., 1998; Zhang et al., 2000), the protein was found to exhibit broad spatial patterns (e.g., CNS, ring gland, gut, Malpighian tubules, eye-imaginal disc, adult eye). Within most developing neurons (LARK was not found in glia), the protein is nuclear, but is located within cytoplasm, nucleus, or both in certain nonneural tissues (Zhang et al., 2000).

Particular attention was paid to LARK expression in CCAP-containing neurons within the subesophageal ganglion and the ventral cord of third-instar larvae and pharate adults. This putative RNA-binding protein was observed to be colocalized within several such brain and VNC neurons; the intracellular signals were cytoplasmic, in contrast to other LARK-expressing CNS neurons (McNeil et al., 1998; Zhang et al., 2000). LARK/CCAP brain neurons were shown (by application of a heterologous antibody against the latter substance) to project from larval and pharate-adult brains into the VNC; other neurites terminated on esophageal muscles and in the prothoracic gland of late pupae; VNC neurons containing both LARK and the peptide sent projections posteriorly to gut muscles (Zhang et al., 2000). Temporally based in situ studies concentrated on these coexpressing ventral-cord neurons: LARK cytoplasmic immunoreactivity exhibited a daily oscillation in pharate adults (Zhang et al., 2000). The nadir occurred toward the end of the nighttime, roughly consistent with previous Western-blotting results of McNeil et al. (1998), although the trough in that biochemical timecourse was earlier in the D phase. There was also an immunohistochemically observed dip in the early morning (Zhang et al., 2000).

Other LARK-expressing pupal tissues did not show systematic immunoreactivity fluctuations. More interesting, probably, is that the staining cycle in CCAP neurons was eliminated in per^{01} pupae (Zhang et al., 2000). This result should be considered in the context of $period$-gene expression being undetectable in any VNC neurons of adult flies; anti-PER staining was found only in (many) glia within these ventroposterior ganglia (Ewer et al., 1992). Could it be that a knock-on consequence of PER's action within this part of the CNS influences posttranscriptionally regulated LARK cycling by an intercellular mechanism? It may not be necessary to invoke such communication between cells if PER and LARK are coexpressed in certain pupal VNC neurons (an unexamined possibility) with regard to eclosion control toward the end of this developmental stage.

However it is that the clockworks influence LARK cycling, what does this RNA-binding protein do to its potential regulatees? One candidate factor is of course CCAP, but that peptide's immunoreactivity was found not to fluctuate in the VNC neurons that coexpress oscillating LARK (Zhang *et al.*, 2000). Perhaps CCAP is rhythmically secreted under pacemaker control. An analogous possibility comes into play in the next neuropeptide substory.

B. Genetic variants and clock outputs influencing adult behavior

1. Pigment-dispersing factor

Indirect regulatory effects of clock-gene actions, vaguely similar to the *per* effect just discussed, are operative for a so-called neurohormone that influences the locomotor rhythmicity of adult *Drosophila*. It has long been considered that brain-hormonal effects on this behavior may be relevant because of the aforementioned experiment (Section VI.A.2) in which transplantation of a *per*^{S} mutant brain into the abdomen of *per*^{01} adult hosts caused a fraction of the latter to exhibit short-period free-running rhythms (Handler and Konopka, 1979). This does not, however, mean that a blood-borne clock-output factor functions in *intact* flies—let alone what might *be* the circulating factor(s) that necessarily moved from the implanted brain to the anteriorly located CNS in the Handler–Konopka experiment.

   One candidate factor is a small neuropeptide known as pigment-dispersing hormone (PDH), a. k. a. PD-factor (PDF). This is an 18-mer, discovered in crustacea, but inferred also to be produced in the CNS of insects. The initial such demonstrations were performed in cockroach by M. Stengl and co-workers, who were the first to apply antibodies against crustacean PDH to insect brains (reviewed by Nässel, 2000). Not long afterward, this same reagent was found to mediate staining of CNS cells in *Drosophila*: anterolaterally and dorsally located brain neurons, plus others within the abdominal ganglion (Helfrich-Förster and Homberg, 1993; Nässel *et al.*, 1993). The PDH immunoreactivities elicited by anti-crab-PDH included neurite signals (Figure 12). Those that were scrutinized principally involved projections from the lateral-brain neurons, some of which sent fibers out into the optic lobes (as far as the medulla) or across the midline (terminating near a contralateral cluster of PDH cells); another cluster of cell bodies projected axons into a dorsal (and quasi-medial) region of the adult brain (Helfrich-Förster and Homberg, 1993; Nässel *et al.*, 1993). Developmentally, PDH-immunoreactive brain cells first appear in L1 larvae; the signals persist in all later stages, such that the dorsally projecting imaginal neurons could be revealed to arise from the larval/pupal cells; the optic-lobe/midline-crossing neurites project from adult neurons that either arise during metamorphosis or begin to produce a PDH-like substance then; in pupal and young-adult brains, there is

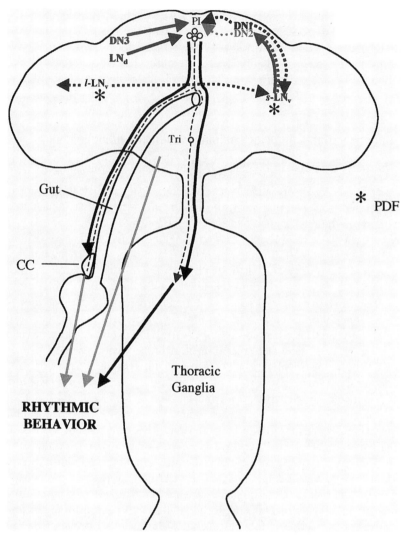

Figure 17. Anatomical and physiological outputs from neurons expressing clock genes in the brain of *Drosophila* adults. The former kinds of outputs are descriptive. Physiological factors known to accompany certain components of these neurite pathways, and chronobiological meanings of structures "targeted" by clock-gene-expressing neurons, are diagrammed here in a largely speculative manner (adapted from Kaneko and Hall, 2000). Major elements of the neuronal projection patterns shown in the adult-brain diagram of Figure 12 are reproduced, augmented by designations of other (generally known) brain and more posteriorly located structures: PI and Tri, brain cells in the pars intercerbralis and tritocerebrum, respectively; CC, corpus cardiacum; Thoracic Ganglia, the fused prothoracic, mesothoracic, and metathoracic neuromeres in the adult VNC; Gut, thoracic portion of

another group of anti-PDH-stained cells in a ventral brain region near the midline, whose immunoreactivity wanes shortly after eclosion (Helfrich-Förster, 1997b).

Such spatial and developmental patterns should seem familiar (Figures 10–13, 16), in particular for the laterally located PDH-immunoreactive neurons (Figures 13, 17; see color insert). Indeed, and as described in Section VI.A.1, these cells are a subset of those that express the *period* gene. This was revealed by coexpression determinations, which took advantage of *per-lacZ* fusion transgenics (Helfrich-Förster, 1995; Kaneko *et al.*, 1997) and showed that the larval/pupal PDH cells are precursors of the PER-containing small ventrolateral neurons (s-LN$_v$'s) in the adult brain. These lateral neurons along with the l-LN$_v$'s (Figures 13 and 17) express both *per* and the gene encoding a PDH-like substance in adults (Helfrich-Förster, 1995). It was proposed that dorsally located PDH-immunoreactive brain neurons in the adult could be equivalent to some of the PER-containing DN cells. Moreover, the anti-PDH-stained neurons are near the perikarya of the mushroom bodies (MBs), and the s-LN$_v$ nerves terminate near this dorsal-brain structure. Such observations prompted circadian locomotor tests of MB-mutated or MB-ablated flies (Section IV. A.1), but to little or no avail (Table 2).

---

alimentary system. The following "output-meaning" speculations seem warranted, based in part on the behavioral/neurogenetic experiments summarized in Figure 13 and the anatomical descriptions presented and discussed (for example) by Helfrich-Förster (1996), Kaneko (1998), and Kaneko and Hall (2000): (a) The s-LN$_v$'s, which contain the PDF neuropeptide as indicated, are key circadian-pacemaker cells that may anatomically connect with DN1s, DN2s, or both; such dorsal (clock-gene-expressing) neurons send signals to the PI, which in turn communicates with the thoracic ganglia—to help mediate daily cycles of rest versus activity; this PI → VNC communication may go through the Tri or CC (and involve humoral factors emanating from the latter), owing to certain PI neurosecretory cells whose projections arborize in the latter two structures (Rajashekhar and Singh, 1994); that the PI is involved in circadian-output systems is considered more generally by Helfrich-Förster *et al.* (1998). Brain cells that are anatomically downstream of s-LN$_v$ dorsal-brain neurites are inferred to possess PDF receptors that activate a Ras/MAPK signal-transduction pathway, which is regulated by neurofribromin (Williams *et al.*, 2001); however, the PDF-target cells are known only to be in the vicinity of s-LN$_v$ axon terminals, that is, are not identified as being among the structures surmised above to be farther along the anatomical output pathway that starts with clock-gene-expressing neurons. (b) LN$_d$'s and DN3s may also interact with the other DNs or the PI and thus contribute to behaviorally meaningful output (see details related to these LNs and DNs that are presented within both panels of Figure 13). (c) Intercellular coupling signals between the LNs, carried by PDF-containing neurites projecting from l-LN$_v$ perikarya in the posterior optic tract (cf. Figure 12), may mediate synchronization between these bilaterally symmetrical pacemaker structures (see Renn *et al.*, 1999, Helfrich-Förster *et al.*, 2001; cf. Petri and Stengl, 2001). (d) Another coupling pathway operating between s-LN$_v$'s and DN1s could subserve a similar function (along with the basics of LN → DN → VNC output, with respect to that direction of this ventral → dorsal pathway); alternatively, or in addition, DN1 → s-LN$_v$ communication may be part of the light-input system (see Figure 16). (e) Possible humoral pathways from the brain and the CC, which could involve rather direct outputs from these relatively anterior structures into the VNC's proximate regulation of cyclical behavior, are indicated by the two downward-pointing arrows on the left.

Another inferential connection of this neuropeptide to the fly's rhythm system came from the demonstration that a *disco* mutant largely lacked anti-PDH-mediated staining (Helfrich-Förster and Homberg, 1993), just as these brain-damaged flies are in the main devoid of lateral-neuron PER immunoreactivity (Zerr et al., 1990). However, a *disco* mutation could exert pleiotropic effects on the developmental appearance (or persistence) of neurons—or expressions of genes within them—in this anterolateral region of the brain. Thus, PDH and PER cells could be separately eliminated by the mutation or silent for activity of the genes encoding these proteins. The coexpression determinations make this unlikely, which therefore gave a boost to the interpretability of a study in which *disco*-induced arrhythmicity was correlated (after behaviorally testing a large series of mutant adults) with PDH immunoreactivity (Helfrich-Förster, 1998). The results led to an inference that the (rare) rhythmicity of *disco* individuals usually required at least one PDH-containing s-LN_v cell and its dorsal projection (as described in detail Section VI.A.2).

Monitoring cell-body and neurite stainings after application of the heterologous antibody was performed mainly to exploit PDH immunoreactivity as a *marker* for neurons that also express clock genes in *Drosophila* (if not other insects—read on). This anti-crab-PDH also led to the initial data about projection patterns for a subset of such cells (Figure 12). Subsequent findings, based on *per*- and *tim*-driven axonal markers (Section VI.A.1), confirmed those obtained from PDH staining of s-LN_v's, l-LN_v's, and their axons, although the transgenic approach also revealed that *additional* clock-gene-expressing neurons project to a dorsal-brain region near to that in which s-LN_v neurites terminate.

These findings should be kept in mind, lest the following substory be taken to imply that "all" the circadian-behaviorally relevant *anatomical* and *neurochemical* output pathways (Figures 13 and 17) originate with the small ventrolateral neurons and the PDH neuropeptide contained therein. So, the narrative now moves to a consideration of this substance as more than a marker. For this, it was necessary to isolate the *Drosophila* gene that encodes the peptide. Such cloning was accomplished (Park and Hall, 1998), starting with RNA extraction and proceeding to PCR amplification of DNA copies using primers designed to anneal with the peptide-encoding sequences. The third-chromosomal locus (Figure 18, see color insert) corresponding to the cDNA identified in this manner was called *Pdf* [née *pdf* (see footnote to Table 6), the last letter standing for "factor" instead of the more loaded term "hormone"]. The *Pdf* ORF perforce encodes an 18-mer similar to other invertebrate forms of this peptide; a larger, heretofore unknown oligopeptide is (on paper) translated from the same mRNA from sequences nearer the 5′ end of the transcript (Park and Hall, 1998). This RNA species was found not to cycle appreciably in terms of daily abundance fluctuations; a corollary was that per^{01} did not affect the steady-state level of the transcript, although a *disco* mutation eliminated it from head extracts (Park and Hall, 1998). Curiously, *Pdf* mRNA was more abundant in male than in female extracts (of *D. melanogaster* heads and those of three closely related species), which could have something to do with sex-related differences in adult locomotor rhythmicity (Section IV.C.1) or geotaxis (Section IV.C.6).

The *Pdf* nucleic-acid data and materials promoted identification of a *Drosophila* mutation that eliminates the neuropeptide. Miraculously, this Pdf^{01} mutant had occured in an apparently spontaneous manner and was lurking in a variety of laboratory stocks (Renn *et al.*, 1999). The mutation creates a non-sense codon in the portion of the ORF that specifies the larger peptide (a 63-mer named PAP), which is translated upstream of PDF. This premature stop should eliminate both the 63-mer and PDF, in that the sequence of the overall ORF implies that both peptides are initially translated as one larger oligopeptide, which would be cleaved to release PAP and PDF (Park and Hall, 1998). Applications of antibodies against the separate substances indicated that the Pdf^{01} mutant is null for both peptides (Renn *et al.*, 1999). Importantly, however, the dorsal-brain neurons stained by anti-crab-PDH are still immunoreactive in the mutant. More-over, no dorsal staining was elicited in wild-type brains by an antibody against native *Drosophila* PDF (Park *et al.*, 2000a). These results, along with the absence of dorsal RNA signals in *Pdf in situ* hybridization tests (Park *et al.*, 2000a), indicate that anti-crab-PDH stains cross-reacting material in this brain region. This ostensibly labored point is not made cavalierly: Chronogenetics and molecular biology can help steer investigators away from primrose paths to which they might be guided by immunohistochemical signals raised by antibodies that are designed to stain something of interest and, one hopes, only that. The case of anti-crab-PDH applied to insects of this species and raising spurious "non-PDF" signals in some of the cells stained studied for the presumed tissue distribution of this neuropeptide is but one example of how the staining patterns can be mislead-ing. Another chronobiological false-positive involved application of an antibody raised against a "photoprotein": Might a given structure or cell type relevant to *Drosophila* rhythms contain material known to be light-responsive on other crite-ria (an issue that came up earlier in Section IX.B.3)? Thus, an arrestin protein (cf. Alvarez *et al.*, 1996) was found to be coexpressed with PER in at least some of the famed lateral neurons; this reagent also led to staining of all external photorecep-tors naturally; a null mutation encoded in the *D. melanogaster* gene that encodes this arrestin type wiped out the photoreceptor signals, but left the LN ones just as robust as in the wild-type fly (Van Swinderen and Hall, 1995). Thus, no arrestin is *known* to be functioning in these "deep brain" photoreceptive neurons. For PDF's part, one can properly *ignore the dorsal-brain neurons* in *Drosophila* that are still elicited by the heterologous antibody in the PDF-null mutant in terms of how that particular substance might have functioned chronobiologically in this part of the fly's CNS.

Arguably more interesting applications of the Pdf^{01} mutant involved behavioral tests. Adults homozygous for this mutation and Pdf^{01}/deletion het-erozygotes exhibited poor rhythmicity in locomotor monitorings (Renn *et al.*,

1999). The degrees of arrhythmicity were not as bad as those exhibited by per^{01}, tim^{01}, or *disco* mutants. The latter is not as severely affected as are the former two, which includes the fact that *disco* adults "go arrhythmic" during their first 3 days in DD, whereas per^{01} appears to be immediately aperiodic upon progression from LD into constant conditions (Wheeler *et al.*, 1993). Pdf^{01} flies exhibit similar behavior to that of *disco* in such LD → DD tests. Moreover, a fair fraction of the former mutant type is weakly rhythmic in DD, exhibiting ca. 23-hr periods; a corollary is that Pdf^{01} flies show earlier-than-normal evening peaks of locomotion in LD, in the behavioral context of their solid entrainment in that condition (Renn *et al.*, 1999).

Effects of PDF alterations on biological rhythmicity were also interrogated in misexpression experiments. Most of these drew upon the pan-neural regulatory element of the *elav* gene, which had been fused to *gal4* (Helfrich-Förster *et al.*, 2000; cf. Section V.C.2). When combined with a UAS-$Pdf^{\underline{R}}$ construct in which the peptide encoding sequences came from grasshopper $\underline{Romalea}$, the *elav* driver caused Pdf^R mRNA to be all over the CNS. Incidentally, Pdf^R and its product have not been *studied* in this grasshopper. In contrast, cloning of an analogous gene from the cricket *Gryllus bimaculatus* (Chuman *et al.*, 2002) was connected with assessments of the peptide's expression in optic-lobe neurons (Okamoto *et al.*, 2001) and intracellularly therein (Chuman *et al.*, 2002). Visual-ganglia ablation experiments accompanying the PDF staining studies indicated that cells containing this substance *do not coincide with pacemaker neurons* (Okamoto *et al.*, 2001), which had previously been shown to be located in a relatively distal region of this cricket's optic lobe complex (reviewed by Tomioka *et al.*, 2001). That PDH-like immunoreactivity does not create a universal marker for circadian-pacemaker neurons is reinforced by the finding that such staining is observed in the sillkmoth brain, but not within PER-containing cells (Sauman and Reppert, 1996a).

Back to the combination of *elav-gal4* and UAS-Pdf^R transgenes in *Drosophila*: PDH-like immunoreactivity was much more spatially limited than the normal distibution of *elav* gene products (e.g., Robinow and White, 1991), yet present in many neurons beyond the small number of ventrolateral ones that normally contain this peptide. Eclosion rhythmicity in *elav-gal4*;UAS-Pdf^R cultures was poor, becoming aperiodic after only 2–3 days into constant darkness (Helfrich-Förster *et al.*, 2000). Behavioral monitoring of these doubly transgenic flies revealed a syndrome of anomalies: increased locomotor levels, inappropriate nighttime activity in LD, and period lengthenings, complex rhythmic patterns, or arrhythmicity in DD. Additional *gal4*-containing transgenes were applied in which subsets of the *elav-gal4* pattern are observed. Certain such enhancer-trap;UAS-Pdf^R combinations caused behavioral irregularities and peculiarities like those driven by the pan-neural regulatory sequences (Helfrich-Förster *et al.*, 2000). Comparisons among the effects (or lack thereof) of these *gal4* drivers led to the conclusion that the intrabrain etiology of these behavioral changes involves ectopic expression of *Romalea*'s *Pdf* cDNA in a (bilateral) pair of large dorsal-brain neurons (located rather near the normal terminal of s-LN$_v$ axons) and another large pair of cells in the ventrolateral brain (not located particularly near the clock-gene-expressing LNs).

Perhaps the most readily interpretable effect of these PDF changes is the mediocre but not vanishing rhythmicity associated with Pdf^{01} behavior (Renn *et al.*, 1999). That this PDF-null mutant is not a thoroughgoing arrhythmic could be because neurons containing this neuropeptide supply clock-output functions in addition to those mediated by this particular substance. If this were the case, then ablation of these neurons would have more severe locomotor consequences than those caused by the mutation. Thus, a transgene in which ca. 2.4 kb of *Pdf* 5'-flanking DNA fused to *gal4* was applied in conjunction with UAS-*rpr* or UAS-*hid* cell-killing transgenes. These two kinds of doubly transgenic flies were established as putatively interpretable (in behavioral tests) because *Pdf-gal4* was shown to drive UAS-*lacZ* marking in all the normal PDF cells and only them, including an absence of spurious dorsal-brain signals (Renn *et al.*, 1999; Park *et al.*, 2000a). Driving expression of the cell killers in PDF cells did eliminate immunoreactivity of the peptide, and the behavioral effects of these inferred neuronal ablations were essentially the same as that of the Pdf^{01} mutation (Renn *et al.*, 1999). It follows that anatomical and physiological outputs from portions of the fly's rhythm system *other* than structures and functions associated with PDF neurons are sufficient to mediate weak behavioral periodicity. Indeed, there are many PER and TIM cells left untouched by the effects of Pdf^{01} or the *Pdf-gal4*/cell-killer combinations. For example, *disco* eliminates all *per*-expressing lateral-brain neurons (or clock-gene expression within such cells), that is, LN$_d$ cells as well as the PDF-containing s-LN$_v$'s and l-LN$_v$'s. This could explain why *disco* is more severely arrhythmic than Pdf^{01} in most tests of the former mutant type (Renn *et al.*, 1999; and see Section VI.A.2).

Regarding the meaning of clock outputs emanating from s-LN$_v$ and l-LN$_v$ brain cells, let us suppose that such functions are carried solely by PDF as it is released from the nerve terminals. Thus, the peptide may have an intercellular signaling function with respect to the central-brain targets of s-LN$_v$ axons. The dorsal-brain interneurons would in turn send signals toward, or perhaps directly into, the thoracic nervous system. A conceivable alternative is that PDF release within the dorsal brain creates a humoral signal that ultimately reaches the thoracic ganglia (Figure 17) and modulates the proximate control of locomotion that is exerted by the thoracic neuromuscular system (cf. Handler and Konopka, 1979).

PDF as it putatively emanates from the l-LN$_v$'s may in part to signal pacemaker neurons on the contralateral side of the brain (Figure 17; cf. Figure 12). This could help the two bilaterally symmetrical structures maintain synchronous function—in particular, by delaying the phase of one pacemaker that is drifting too far ahead of its twin in the opposite brain hemisphere. This supposition is made in the context of rhythm-delaying phase shifts that are caused by injection of (crustacean) PDH into one side of the cockroach brain (actually, its optic ganglionic complex) or the other (Petri and Stengl, 1997). These neurochemical manipulations were nicely rationalized by Petri and Stengl (2001) in terms of modeling an "all-delay" PRC like that which stemmed from the peptide injections, although the empirical results and the ensuing model left open the question as to what kind of pacemaker output may mediate *advances* that would contribute to synchronization of the bilateral clock structures. Nevertheless, consider that the Pdf^{01} mutant in *Drosophila* is anomalously phase-advanced in its entrained locomotor rhythm and that it drifts toward arrhythmicity in constant conditions (Renn *et al.*, 1999). These effects of the mutation suggest that at least one of Pdf^{01}'s chronobiological problems involves the lack of an interpacemaker delaying function in LD and a grosser lack of synchronization between such structures in DD. In the latter condition, the pacemaker would still be running well on a cell-by-cell basis within the LNs, and those within one side of the brain may be able to communicate with one another (on behalf of intracluster synchrony) by virtue of their proximity. However, in the absence of PDF, neurons in a behaviorally relevant cluster in the left side of the brain would slip out of synchrony with those in the right side. So the bilateral feature of the system would become slowly desynchronized, and the whole fly would become aperiodic in parallel, in the absence of synchronizing cues from the environment. Consistent with this scenario is that *so* mutant flies (introduced in terms of their eyelessness in Section II.) frequently lack a posterior optic track (Helfrich-Förster and Homberg, 1993) *and* exhibit free-running rhythms with split periodicities (Helfrich, 1986). [This track (abbreviated POT) was introduced diagrammatically in Figure 12.]

It is unknown *which* LN cells (if any) are signaled by PDF release from peptide-containing axons that cross the midline. For example, do the l-LN_v's from which these axons project communicate with s-LN_v's in the other side of the brain, in a story line that clings to apprehension of the latter neuronal type as composing the "major" pacemaker structure? A high-resolution description of the system's anatomy, necessary to proceed further with these suppositions, is lacking. However, there is a good deal of anatomical information available for an *additional* feature of the l-LN_v projection pattern, the aforementioned centrifugal fibers that ramify through the fly's optic lobes.

Most of the findings and interpretations about PDF's influence on rhythms running in the visual system come from analyses of large flies (reviewed by Meinertzhagen and Pyza, 1996, 1999). Even though axon "size and shape" oscillations are also observed in *Drosophila* (Pyza and Meinertzhagen, 1999b)— and certain of its clock mutations have been shown to affect these cyclical morphological changes (Section VI.A.2)—it is unknown whether the "PDH neurite varicosity" rhythm observed in housefly (Pyza and Meinertzhagen, 1997) occurs in its smaller dipteran relative. More generally, the manner by which PDF-containing axons may influence visual-system rhythms in *Drosophila* is obscure, in part because these neurites do not extend as far as the first-order optic lobe (the lamina) in which the anatomical cycles are observed. Perhaps the peptide is released cyclically in *Drosophila*'s medulla optic lobe (Figures 12 and 13) and influences more distally located lamina neurons (Figure 12) by a short-ranging humoral effect. In any case, the only experimental evidence related to PDF action in the visual system involves injections of (the crustacean form of) this substance into large flies: In M. *domestica*, such injections caused axons of the relevant lamina interneurons (Section VI.A.2) to increase in size (Pyza and Meinertzhagen, 1996) and decreased the number of screening-pigment granules in the terminals of photoreceptor axons that synapse with these laminar neurons (Pyza and Meinertzhagen, 1998). During the daytime or subjective day (in unperturbed flies) axonal sizes are maximal and granule numbers are minimal, whereas there are fewer and larger PDH-neurite varicosities at night (Meinertzhagen and Pyza, 1999). In *Calliphora*, an increase in light-on and light-off transient spike in the electroretinogram (ERG) was induced by injection of PDH into the hemolymph. These physiological parameters are on point because modestly higher amplitudes of the spikes were observed during the subjective night compared with those measured in conjunction with ERGs elicited during the subjective day (Chen *et al.*, 1999). PDH-immunoreactive axons project out into the lamina of large flies such as this one (Meinertzhagen and Pyza, 1996), and it is within this first-order optic lobe that the electrical signals defining the ERG transient spikes originate. This discussion is not meant to imply that PDH is the only signaling substance that may influence the circadian oscillations in visual-system morphology and physiology, given the effects on these metrics that are also exerted by other small molecules (Meinertzhagen and Pyza, 1999).

As to whether there is a PDF "release rhythm" occuring at *any* of the nerve terminals containing this substance, the only evidence mildly informative on this point comes from *Drosophila* and involves the centripetal fibers projecting into the dorsal brain from s-LN$_v$ cells (Figure 12). An equally important feature of this study was to establish the *Pdf* gene and its products as regulatees of the fly's clockworks. The mRNA transcribed from *Pdf* has no cyclicity to be affected by clock mutations (Park and Hall, 1998). Moreover, the overall levels of this RNA were unaltered in per^{01} or tim^{01} mutants (Park *et al.*, 2000a). However, there was a distinct posttranslational effect of either clock-null variant on PDF: A high-day/low-night rhythm of nerve-terminal immunoreactivity in the dorsal brain was eliminated in per^{01} or tim^{01}, for which the normally high daytime level was anomalously maintained at opposite cycle times in mutant specimens stained via anti-PDH (Park *et al.*, 2000a). The circadian version of this staining rhythm had its cycle duration shortened by the effect of *per*^{S} in DD. Additionally, in the abnormally behaving transgenic flies with ectopic intrabrain expression of grasshopper PDF (see prior discussion) the nerve-terminal immunoreactivity rhythm was absent (Helfrich-Förster *et al.*, 2000). A reasonable schematic for these results is that PDF is normally dumped in a gated manner during the nighttime from the terminals of the *s*-LN$_v$ axons and that this process is under clock control (although the staining rhythm, which implies a posttranslational phenomenon, may instead involve antigenicity as such or axonal transport of PDF toward the nerve terminals).

Two separate clock factors influenced expression of PDF at a much earlier stage of its production. In larval precursors of the imaginal *s*-LN$_v$ cells, little or no *Pdf* mRNA or peptide was detectable in *Clk*^{Jrk} or *cyc*^{0} mutants (Blau and Young, 1999; Park *et al.*, 2000a). In adults, these mutations knocked out transcript and immunohistochemical signals specifically in the *s*-LN$_v$'s themselves (Park *et al.*, 2000a). The effect of in the imaginal brain *Clk*^{Jrk} was more severe than those of the *cyc*^{0}'s (fewer *Pdf* mRNA- or PDF-containing cells in the former mutant), and neither type of mutation affected the peptide-gene products in *l*-LN$_v$'s. The latter cell type is, however, affected by these arrhythmia-inducing mutations, for the anti-PDF-marked anatomy of *l*-LN$_v$ projections was found to be erratically anomalous in *Clk*^{Jrk} and *cyc*^{0} brains (Park *et al.*, 2000a). How these factors apparently influence pattern formation of certain PDF-containing neurons is obscure. Yet one can surmise that neurochemical differentiation of the *s*-LN$_v$ cells is controlled in part by the CLK and CYC transcription factors interacting (together) with regulatory sequences flanking the *Pdf* ORF. (This assumes that the *s*-LN$_v$'s are *anatomically* normal in *Clk*^{Jrk} or *cyc*^{0}, but only the lack of PDF immunoreactivity is known for those cells in these mutants.) If CLK and CYC directly regulate the *Pdf* gene, this 5'-flanking DNA might contain an E-box (see Section V.C.2), and there is one ca. 1.4 kb upstream of the gene's transcription start (Park *et al.*, 2000a). However, E-box-removing truncations of the 2.4-kb 5'-flanking region within the *Pdf* locus had no effect on marker expression driven by *Pdf-gal4* constructs in brains of the relevant transgenics (Park *et al.*, 2000a). Thus, *Clk* and *cyc* functions may regulate an unknown intermediary, which would go on to control transcription of *Pdf*. The only 5'-flanking truncation of the latter gene that affected any feature of the normal PDF pattern was removal of the distalmost 80% of the 2.4 kb of 5'-flanking material; here, the abdominal-ganglionic expression (of *Pdf*$^+$) was eliminated (Park *et al.*, 2000a). There is no known biological significance for PDF in this posterior CNS region.

Before leaving part of the *Pdf* gene-regulation story, two further sets of facts and findings are worth noting: (i) The *Pdf*(2.4)-*gal4* construct is the transgene that was applied in the various spatial manipulation experiments described in previous sections and earlier in this one, those in which bioassaying the effects of tetanus toxin, cryptochrome, or grasshopper PDF, acting solely within ventrolateral clock-gene-expressing neurons, were summarized (Kaneko *et al.*, 2000b; Emery *et al.*, 2000b; Helfrich-Förster *et al.*, 2000). (ii) Embellishments of the brain–behavioral experiments described in this subsection speak to PDF's regulation of locomotor-activity rhythms and to the inadequacy of viewing the PDF-containing/clock-gene-expressing ventrolateral brain neurons as the sole neural substrate for this phenotype. Thus, a study involving amidation of neuropeptides (Taghert *et al.*, 2001) led to results that imply chronobiological contributions of other neuropeptide-containing cells aside from the PDF ones (in the same spirit as the findings described in Section VI.A.2 and reported by Kaneko *et al.*, 2000b). Oligopeptide amidation, inferentially including such post-translational modification of PDF, was manipulated by transgenically creating mosaics in which an enzyme that carries out these reactions (peptidylglycine α-hydroxylating mono-oxygenase; cf. Kolhekar *et al.*, 1997) was expressed only in subsets of the usual PHM pattern. Various *gal4* drivers were used to restrict PHM to different subsets of peptidergic neurons, and these mosaics displayed aberrant locomotor rhythms to degrees that paralleled complexities of the spatial patterns (Taghert *et al.*, 2001). Certain PHM mosaic types were less rhythmic than is the *Pdf*^{01} mutant (Renn *et al.*, 1999), in that the behavioral severities exhibited by these transgenic flies approached those exhibited by clock-null mutants or flies containing *tim*-driven tetanus toxin (Kaneko *et al.*, 2000b). Adding *Pdf-gal4* (Renn *et al.*, 1999; Park *et al.*, 2000a) to the severely impaired PHM mosaics provided only partial improvement of rhythmicity, whereas addition of *tim-gal4* (Emery *et al.*, 1998; Kaneko *et al.*, 2000b) largely restored the normal locomotor patterns (Taghert *et al.*, 2001).

2. Neurofibromin-1 and Ras/MAPK

The product of a *Drosophila* gene called *Neurofibromin-1* (*Nf1*) is tied to inactivation of the *Ras* oncogene and cAMP-dependent signaling (Lakkis and Tennekoon, 2001). Because various defects in the latter processes cause flies to exhibit rhythm anomalies (Section IV.A.2), it was suggested the *Nf1* mutants might exhibit circadian defects. Williams *et al.* (2001) carried out such locomotor tests and performed other gene manipulations, with the upshot being that *Nf1* seems to function as a circadian output factor. Moreover, elements of its activities are connected with PDF functioning. First, these investigators showed that *Nf1* mutants are behaviorally arrhythmic or infrequently and weakly periodic in DD. Placing *Nf1*$^{+}$ sequences in LN brain cells (by combining *Pdf-gal4* with UAS–*Nf1*) or in a host of putative clock structures (via *tim-gal4*) led to no improvement in locomotor rhythmicity (Williams *et al.*, 2001). This underscores the inadequacy of the LNs for controlling this behavior (if only because all of their neuronal processes terminate within anterior ganglia) and makes one wonder what "non-TIM" structures would be involved (see later discussion). That an output, as opposed to a clock, function is mediated by NF1 was suggested by demonstrations that *per* and *tim* are expressed normally in extracts from and histological preparations of *Nf1* mutants (such as the $Nf1^{P1}$ deletion of this gene). However, these mutations caused clock-controlled CRE-*luc* expression (cf. Section IV.A.2) to be higher than normal and to define anomalously low amplitude cycling, leading to the inference that *Nf1* acts downstream of the clockworks but upstream of dCREB2-mediated transcription (Williams *et al.*, 2001). The latter's regulatory activities could be interposed between NF1's action and revealed rhythmicity, owing to the aberrant (short-period) locomotor rhythm exhibited by a *Creb 2* mutant (Belvin *et al.*, 1999).

Further features of *Nf1*-associated regulation were interrogated by applying mutations that affect the Ras/MAP-kinase signal-transduction pathway: the $Ras1^{e1B}$-null mutant; $rolled^{x162}$, a MAPK-null; and a *son-of-sevenless* mutant (sos^{e2H}), which is devoid of a nucleotide-exchange factor that activates the Ras protein. All three mutations suppressed *Nf1*-induced arrythmicity to one extent or another, with the sos^{e2H} variant being the most effective (Williams *et al.*, 2001). Activation of an inducible PKA-encoding gene—which when chromosomally mutated causes quite a bit of behavioral arrhythmicity in *Drosophila* (see *Pka-C1* in Table 2)—led to weaker restoration of periodic behavior in an *Nf1*-mutant genetic background. However, the remainder of this study turned its attention to signaling factors other than those connected with cAMP. Thus, levels of phosphorylated (hence activated) MAPK were found to be elevated in *Nf1* mutants; in contrast, certain rhythm mutations led to markedly reduced levels of phospho-MAPK. The mutants applied were per^{01}, tim^{01}, and none other than Pdf^{01}; the latter caused the most severe phospho-MAPK reduction (Williams *et al.*, 2001). Remember that PDF at the terminals of dorsal-brain neurites projecting from a subset of the *Pdf*-expressing brain cells, the s-LN$_v$'s (Figure 12), cycles in its immunoreactivity (Section X.B.1, immediately above). A corollary was gleaned from the finding that antibody-mediated staining for phospho-MAPK was strong in the vicinity of PDF-containing axon terminals; moreover, such immunohistochemical signals in the dorsal brain were highest during the second half of the nighttime (Williams *et al.*, 2001), at a cycle time when PDF immunoreactivity in these s-LN$_v$-projecting neurites is relatively low (Park *et al.*, 2000a; Helfrich-Förster *et al.*, 2000). If that PDF trough reflects the time when the neuropeptide is released into this brain region (as hypothesized in the previous subsection), one can surmise that "PDF receptors couple to [the Ras/MAPK] signaling pathway" (Williams *et al.*, 2001). In this scenario, release of the neuropeptide would stimulate G-protein receptors; this cellular event would move its signal into the Ras/MAPK pathway, resulting in phosphorylation of MAPK.

A major regulator of these processes is the *Drosophila* form of neurofibromin-1 (NF1). Where it functions biochemically on the inferred output pathway—which *a priori* would be apprehended to involve signal transduction of some sort—can be inferred to be downstream of central pacemaking and upstream of both dCREB2 and MAPK. How NF1 normally acts with regard to the latter is presumably to downregulate MAP-kinase production, counteracting the PDF effect and contributing to the overall cycling of the inferred processes. Another piece of counteraction can occur experimentally, when accentuated amounts of phospho-MAPK caused by an *Nf1* mutation are brought back down by combining it with a given Ras/MAPK variant. The partial restoration of normality in such double mutants was revealed by locomotor testing. This brings us to the question of where NF1 carries out its actions within the CNS on behalf of locomotor rhythmicity. Presumably the locations are near to but anatomically downstream of the *s*-LN$_v$ cells that project their neurites into the dorsal brain (Figures 12 and 17). This would fit with one of the suppositions discussed and certain of the results just described. Thus (i) PDF receptors at the surface of the relevant "postsynaptic" cells feed into the Ras/MAPK pathway (the quoted neuroanatomical term is used loosely because the neuropeptide signaling could operate in a paracrine manner), and (ii) the effects of rhythm-related mutations and gene-regulatory sequences are nonautonomous in two empirical ways: Putting *Nf1*$^+$ solely into presynaptic *tim*-containing cells, or the *Pdf*-expressing subset of them, did not rescue the arrhythmicity caused by *Nf1* mutations. However, loss of *Pdf*, *tim*, or *per* function attenuated the presumed PDF-stimulated increase in phospho-MAPK. This can be explained as follows: In the first of these mutants, there is of course no PDF to release from the presynaptic *s*-LN$_v$ nerve terminals. In the second two (clock) mutants, PDF immunoreactivity at such terminals is anomalously high during the nighttime and remains at the same apparent level as during the daytime, compared with the staining trough at the former phase observed in the wild-type fly (Park *et al.*, 2000; Helfrich-Förster *et al.*, 2000; Williams *et al.*, 2001). This implies that PDF is minimally released when it should be in *tim*^{01} or *per*^{01} mutants, which would be genocopies of mutationally induced absence of the neuropeptide.

3. *vrille*

The next gene to be discussed in terms of its manipulations and behavioral effects also participates in PDF regulation. This is *vrille* (*vri*), the results of whose analysis seem to place its functions somewhere between the circadian pacemaker and outputs controlled by its actions. However, the case of *vri* is inserted in this section because it was discovered in the spirit of chronomolecular strategies designed to identify additional clock-output factors (as described in the next section; also see Table 6). The *vri* gene encodes an apparent transcription factor of the bZIP type, containing a motif that consists of basic amino acids juxtaposed with a leucine zipper (George and Terracol, 1997). [We may remember (and if not, can root around the fine print accompanying Figure 9) that there is a putative binding site for a bZIP protein in the 5'-flanking region of *tim* (Okada *et al.*, 2001).] *vri* was originally identified by the general role it plays in *Drosophila* development with respect to a signaling pathway involving the action of TGFβ (as encoded by a gene called *decapentaplegic*), and mutations at the *vri* locus (Figure 18) are late-embryonic lethals (George and Terracol, 1997). Therefore, like the cases of *lark*'s and *double-time*'s, *vrille*'s connections to the rhythm system might not have been discoverable in mutant screens that depend on the circadian effects of viable recessive mutations.

The two different-sized transcripts encoded by *vri* cycle in their abundance in LD and DD, peaking around the time that *per* and *tim* mRNAs are maximal at night (Blau and Young, 1999). per^{01} or tim^{01} caused *vri* transcript levels to stay temporally flat at an intermediate level, and Clk^{Jrk} or cyc^{01} caused low constitutive levels of *vri* mRNA. Yet, certain feedback-loop components (Figure 7) were shown to be unnecessary for *vri* mRNA cycling (Yang and Sehgal, 2001) in flies for which cyclical PER and TIM production were wholly dependent on an *elav* GAL4 driver and UAS enhancer sequences (a circumstance in which constitutive production of *per* and *tim* transcript was engineered; cf. Section V.C.2). Nevertheless, the effects of *Clk* and *cyc* mutations can be presumed to reflect on *vri*'s transcriptional regulation in genetically normal circumstances, in particular, that the latter gene would possess an upstream E-box. It does, and transfection of CYC-expressing cultured cells with *vri*-reporter constructs along with supplying CLK function showed that this E-box is necessary for CLK:CYC activation of *vri* (Blau and Young, 1999). In larvae and flies, *vri* is coexpressed with *tim* and *Pdf* in brain cells and (adult) photoreceptor ones. However, *vri* mRNA and β-GAL driven by an enhancer-trapped, *lacZ*-containing transposon inserted at the locus (George and Terracol, 1997) were found in a broader CNS pattern than that defined by the larval LNs (Blau and Young, 1999). This is interpretable in the context of *vrille* being a gene that acts pleiotropically.

Given the lethal effects of *vri* mutations, they are testable for effects on adult-behavioral rhythmicity only as heterozygotes; thus, a recessive-lethal transposon insert and a deletion of the locus each was shown to shorten the circadian period by about ½–1 hr (Blau and Young, 1999). These effects are analogous to those of $per^0/+$, $Clk^-/+$, and $cyc^0/+$ genotypes, although these three heterozygous types exhibit longer-than-normal behavioral cycles (Table 1). The $vri^-/+$ effects are so mild that, had a mutation at this locus been induced in a forward-genetic screen for dominant rhythm-affecting variants (Price *et al.*, 1998; Rothenfluh *et al.*, 2000a), one suspects it would have been discarded.

vri expression was manipulated in the opposite direction by designing a *tim* construct in which 5'-flanking sequences from this clock gene were fused to both UAS ones and those encoding GAL4. The effects of this implicit *tim* → GAL4 → UAS → more-GAL4 positive loop were tested by combining this construct with GAL4-drivable UAS-*vri* transgenes; overall *vri* mRNA levels were found to be pinned at high and largely constitutive levels (Blau and Young, 1999). Behavioral tests of these doubly transgenic flies gave long-period rhythmicity or none, depending on the UAS-*vri* line that was crossed to *tim*(UAS)-*gal4*. In larvae expressing these transgene combinations, *tim* mRNA and TIM levels, as well those of PER protein, were reduced or eliminated in LN cells. If TIM was detectable at low levels, the protein was largely cytoplasmic. Thus, continuously high levels of *vri* function affect the clockworks. The same kind of *vri* manipulation also influences an output factor, PDF: Immunohistochemical signals for the peptide were strongly reduced in larval LNs, although the abundance of *Pdf* mRNA was unaffected (Blau and Young, 1999). Thus, *Clk* and *cyc* activities affect transcript production or accumulation emanating from this peptide-encoding gene (Blau and Young, 1999; Park *et al.*, 2000a), but *vri* has a posttranscriptional effect on PDF accumulation.

4. *takeout*

The *takeout* (*to*) gene was tapped into from a rhythm standpoint analogous to *vri*'s identification (see next section and Table 6). This gene encodes a relatively small, ca. 250-aa TAKEOUT (TO) protein related to certain ligand-binding ones, including one that binds juvenile hormone (Sarov-Blat *et al.*, 2000). TO is also similar to a so-called "0.9 protein" (So *et al.*, 2000), identified long ago as encoded by an *X*-chromosomal transcription unit adjacent to *per* (Bargiello and Young, 1984; Reddy *et al.*, 1984). The 0.9-kb mRNA from which this polypeptide is translated rises shortly before eclosion and decreases within a few hours of that time (Lorenz *et al.*, 1989). The significance of this burst of expression (which is at best under one-time circadian control, contra Reddy *et al.*, 1984) is unknown. The function of the protein is obscure as well; it is an inessential one because the viable "*per*-minus" deletion combination described earlier removes not only that clock gene, but also the 0.9 one (e.g., Reddy *et al.*, 1984). Conceivably the heightened locomotor activity caused by *per⁻* (Hamblen-Coyle *et al.*, 1989), in addition to its elimination of behavioral rhythmicity, has something to do with an absence of the 0.9 protein (although a *third* transcript is also eliminated by the so-called *per⁻* deletion). However, a behavioral role for this small protein is dubious, given that expression of the 0.9-kb mRNA is limited to the epidermis (Lorenz *et al.*, 1989). Another such gene product happened to be identified in a screen of an adult-head cDNA library for transcripts exhibiting daily cycling, but the mRNA in question (encoding a "cuticle protein" called Dacp-1) turned out not to exhibit abundance oscillations (Qiu and Hardin, 1995).

We now refocus attention on the 0.9-protein relative, TO. mRNA encoding the latter material cycles in an entrained and free-running manner, peaking at the end of the night; luciferase-reported cycling (in LD) was similar in a *to*

(5'-flanking)-*luc* transgenic (Sarov-Blat *et al.*, 2000; So *et al.*, 2000). The temporally varying transcriptional regulation of *to* (as just implied) is under clock control, in that per^{01} led to lower-than-normal levels of the mRNA, and tim^{01}, Clk^{Jrk}, and cyc^{01} each caused extreme abundance decrements (So *et al.*, 2000). TO protein, which cycles congruently with the mRNA oscillation, is nearly eliminated by the effects of Clk^{Jrk} or cyc^{01} (Sarov-Blat *et al.*, 2000). There is an E-box within the 5'-flanking region of *to*, and it (along with 15 surrounding nucleotides) was found to be activated (in terms of reporter expression) in yeast cells engineered to produce CLK and CYC proteins (So *et al.*, 2000). This result was shown to be intriguingly misleading in terms of *takeout* regulation in the actual fly: A larger E-box-containing DNA fragment from this gene was found to be insufficient for mRNA cycling because an intrafly transgene in which this 5'-flanking DNA was fused to *luc* drove very weak luciferase expression and cycling (So *et al.*, 2000). Moreover, the same stretch of upstream DNA (*to*'s E-box and surrounding nucleotides) was not activated by the CLK:CYC combination in yeast, and constructs from which those putative regulatory sequences were deleted did not diminish expression of the *luc* reporter in transfected *Drosophila* cells (supplemented with CLK) or robustly cyclical expression in transformed flies (So *et al.*, 2000).

The biological significance of *to* was inferred from the behavioral effects of a mutation at the locus, which astonishingly was found to be present in a mundane marker strain (cf. the serendipitous identification of Pdf^{01}). The *to* mutant is a partial intralocus deletion that removes 3' sequences and leads to lower-than-normal mRNA levels as well as depleted, noncycling TO protein (Sarov-Blat *et al.*, 2000). The effects of this mutation were assessed in the context of *to*'s spatial expression. mRNA encoded by the gene is found within the brain, the compound eye, and the antenna of adults (So *et al.*, 2000; Sarov-Blat *et al.*, 2000). In their bodies, *in situ* hybridization signals were observed in the cardia and crop (Sarov-Blat *et al.*, 2000), portions of the thoracic alimentary system in which there are daily cycles of PER and TIM immunoreactivities (Giebultowicz *et al.*, 2001). The latter feature of the expression pattern seems related to the fact that subjecting flies to several hours of starvation induces higher-than-normal and spatially broader levels of *to* mRNA and TO protein (these effects were observed both in the head and the thorax); refeeding reversed this starvation effect, which was most prominent when flies were deprived of food around the cycle time that *to*-gene-product levels are maximal (Sarov-Blat *et al.*, 2000). The clock connection so implied was supported by the noninducibility of *to* expression in starvation tests of per^{01}, tim^{01}, or cyc^{01} mutants. The *to* one ground to a rapid, death-anticipating behavioral halt in food-deprived locomotor monitorings; in contrast, wild-type flies tended to be rhythmically active over the course of 3 days of LD cycles in starvation conditions (Sarov-Blat *et al.*, 2000). Transgenic flies in which a *tim-gal4* drove *takeout* coding sequences (UAS-*to*) in a mutant genetic background were rescued in terms of the rapid diminishing of LD rhythmicity and viability caused by the *to* mutation alone in a foodless situation.

C. Genes whose products oscillate and are regulated by clock components

Well before cyclings of clock-gene products were discovered, various substances were long known to go up and down in their concentration over the course of daily cycles. Even in *Drosophila*, a certain protein was inferred to oscillate in terms of an enzyme activity (associated with the *ebony* mutant, as noted in Table 3). This was more than 10 years in advance of the first demonstration that PER-protein abundance appears to fluctuate circadianly (Siwicki *et al.*, 1988). One of the best-known molecular cyclings in various species is tht of melatonin, which has been studied since chronobiological antiquity. Enzymes involved in melatonin synthesis—such as N-acetyltransferase (NAT) or hydroxyindole-O-methyltransferase (HIOMT)—seemed as if they could be encoded by cycling mRNAs. Inevitably, this came true (for the NAT-encoding gene) in studies of vertebrates (reviewed by Foulkes *et al.*, 1997; Li *et al.*, 1998; Hall, 1998b).

It would not stun the student of this investigatory area if an analogous gene in *Drosophila* (Hintermann *et al.*, 1996) turns out to be similarly regulated. Why? Well, melatonin is also present in *Drosophila* (Finocchiaro *et al.*, 1988) among other insects. In one study (involving fly extracts taken at two cycle times spaced 12 hr apart), melatonin concentration was found to increase as nighttime approaches in conjunction with a rise in NAT activity (Callebert *et al.*, 1991). Intriguingly, no melatonin and hardly any HIOMT activity were detectable in extracts of two separate *per*-null mutants, although NAT levels were unaffected (Callebert *et al.*, 1991).

NAT was studied subsequently with more chronobiological intensity, along with handling of the second-chromosomal *Dat* gene that encodes this enzyme. Its more complete name is arylalkylamine-NAT or AANAT, and we may dimly remember the mutation at the *Dat* locus as it was applied in sleep-related experiments (Section IV.C.2). Prior to these phenogenetic studies (Table 3), biochemical ones had shown that neither AANAT activity, protein, nor *Dat* mRNA was found to exhibit any systematic abundance fluctuations (Hintermann *et al.*, 1996; Brodbeck *et al.*, 1998). Concomitantly (almost), there was minimal evidence for melatonin cycling (Hintermann *et al.*, 1996), in that the concentration of this substance rose some 60% above the plateau level at only one time of day (in a nine-timepoint LD experiment)—literally during the daytime (contra Callebert *et al.*, 1991).

These two sets of investigators generated data that speak to melatonin-related functions, if not chronobiological ones (however, absorb the implications of Greenspan *et al.*, 2001). Thus, injections of this substance into adult females caused increased latency to mating initiation and lower-than-normal rate of oviposition (Finocchiaro et al., 1988). Dat RNA was found to be expressed in the embryonic CNS and gut (Hintermann et al., 1996) by application of an in situ hybridization probe that recognized both of the mRNAs deriving from this gene. It also generates two forms of the AANAT protein that differ with respect to N-terminal residues. The more abundant gene product (AANAT 1b-encoding mRNA) was shown to be first expressed during late embryogenesis in the brain, the VNC, and the midgut; AANAT immunoreactivity (via application of an antibody that would recognize both the 1b and 1a forms) was observed in the adult brain and the midgut (Brodbeck et al., 1998). The transcript type that encodes the less abundant AANAT1a form of the enzyme does not appear until a late pupal stage; it persists in adults and is located predominantly in the brain (Brodbeck et al., 1998).

The mutation at this locus was brought to bear on the matter of mRNA isoforms emanating from the gene. Thus, the Dat^{lo} mutant (whose ca. 85% drop in overall enzyme activity was exploited in the sleep-rebound experiments alluded to above) was found to exhibit lowered abundance of only the AANAT1b isoform; this could be rationalized in terms of where two transposons are inserted within the locus in this mutant (Brodbeck et al., 1998). One feature of the brain expression—in a region near, you guessed it, the clock-gene-expressing LN cells—was inferred to be AANAT1a, owing to the lack of effect of Dat^{lo} on staining in this anterolateral location (Brodbeck et al., 1998). Certain additional features of the immunohistochemical pattern (in the wild-type fly) were inferred conceivably to correspond to brain cells that express serotonin or octopamine. Recall the odd connection of the latter's biosynthesis to the fly's clock system, that is, a per^{01}-induced decrement (Livingstone and Tempel, 1983), although also bear in mind that this clock mutation was shown not to affect the grossly determined level of NAT activity (Callebert et al., 1991).

These melatonin- and NAT-related results, inchoate as they may seem, provide a platform for consideration of clock-controlled processes. The idea is that production of certain gene products and of associated small molecules could be regulated downstream of the clockworks. The macromolecules in question would therefore cycle in their abundance, yet they would not seem to possess pacemaker regulating functions, if only because at least some of them are plain old enzymes. For example, a catalytic function involved in melatonin synthesis (viz. HIOMT, if not NAT) can be inferred to be a regulatee of period gene function. Now, per as well as its clock-gene cohorts regulate their own cyclical expressions. So one takes a short mental step to imagine that these central-pacemaker factors also control the oscillating activities of genes functioning in pathways proceeding outward from the clock. Some such output factors (Table 6) are known variously as clock-regulated genes (Crg's) or Drosophila rhythmically expressed genes (Dreg's). To identify them, searches were undertaken to detect gene products that exhibit daily oscillations in their concentration. Moreover, the molecular screens could readily include value-free strategies in which more than candidate factors (those already known to be contained within the animal) would be identified.

Some words of caution are warranted before the fruits of the molecular searches are described: If a mere enzyme gene were identified, it would not necessarily be an output factor, given that the DBT, SGG, and CK2 kinases seem to function as part of the clockworks (Table 1 and Figures 3 and 7). In addition, if a circadianly oscillating mRNA were found that encodes a heretofore unknown protein whose cellular function cannot be surmised, it might be eventually found to play a pacemaker role instead of being an output factor alone or at all. The issue just raised is squarely on point in the case of *vrille*. Its chronobiological identification, independent of finding developmentally lethal mutations at the locus, involved a search for normal genes whose mRNAs cycle. PCR-based differential display was applied (cf. Section IV.C.2); primers that could find "any" amplified cDNA, no matter what it might encode, were used to look for copies of RNAs that were down during the early night and up during the late night, or vice versa (Blau and Young, 1999). In parallel, RNA was extracted from *per*^{01} flies. Thus, recognition of an intranighttime cycling candidate in wild-type flies could be pitted against hypothetical amplification of equivalent amounts of cDNA originating from this clock mutant at the separate cycle times. The result in this case was a cDNA fragment that is normally downregulated at ZT20 versus ZT14, whereas the inferred mRNA abundances were equivalent and ostensibly intermediate in extracts of *per*^{01} taken at these two timepoints (Blau and Young, 1999). Sequencing this cDNA fragment launched the chronobiological side of the *vri* story. As it was told in the previous section, the conclusion so far is that this factor is a *Crg*, but it can also be perceived as a pacemaker constituent. In a way, the semantics is moot. More important is the manner by which clock genes regulate expression of a gene like *vri*, and whether manipulation of its sequences not only causes alterations of biological rhythmicity, but also modifies the levels of clock-gene products. It was additionally informative that the engineered *vri*-expression changes affects the level of *Pdf*-encoded neuropeptide, which seems almost certainly to function along a pacemaker-output pathway (Figure 17). This issue was initially raised a light-year's distance earlier in the current monograph, when the *Drosophila* gene that encodes a fragile-X-related protein was considered as to whether it participates in the control of circadian behavior and eclosion (definitely "yes" for the former phenotype), and whether, in turn, the relevant *dfmr1* mutations (Table 1) cause anomalies of PER and TIM cycling (a contentious situation, as discussed in Section IV.A.2). That the DFMR1 (putative) RNA-binding protein does not itself exhibit a daily cycle in its apparent abundance (Inoue *et al.*, 2002) perhaps pushes this factor more toward the output category (Table 6).

Come to think of it, might the similarly noncycling *Pdf* gene somehow influence clock functions as well as causing poor behavioral rhythmicity when mutated? Pdf^{01} has not been tested for effects on expression of *per*, *tim*, or other clock genes. Perhaps this is because it seems difficult to imagine how a neuropeptide would fit into the clockworks, with their "top-of-the-hierarchy" regulatory functions (Figure 7). In contrast, a transcriptional factor such as VRI is, in immediate retrospect, a candidate for influencing *per* and *tim* expression, which the *vri* variants do. These issues remain in force in the case of *takeout*. Its molecular discovery led to identification of the *to* mutant described earlier. However, this hypomorphic mutation has been tested so far only for effects on chronobiological outputs, perhaps because the low levels of a small, putatively ligand-binding (and humorally related?) protein in the mutant are not easily apprehended as damaging the functions of high-level pacemaking regulators. At all events, *to*'s isolation strategy is valuable to register. For this, a PCR-based cDNA subtraction procedure was effected in which mRNAs from the cyc^{01} mutant were in effect subtracted from wild-type RNA. This led to 108 subtracted clones, which were candidates for corresponding to genes whose mRNAs are at anomalously low abundance in cyc^{01}; three of these clones were so affected by this mutation (So *et al.*, 2000). Recall in this regard the severely subnormal levels of *Pdf* mRNA and peptide observed in cyc^{01} and cyc^{02} brains. All three of the subtraction-isolated sequences were found to encode novel proteins, one of which is TAKEOUT (So *et al.*, 2000).

A different molecular subtraction approach had been applied previous to execution of the *takeout* isolation strategy and its reliance on cyc^0 effects. The earlier screen used RNA extracted from wild-type flies at two timepoints. Reverse transcription and subtractive hybridization were performed using ZT15 RNA as the source of tracer cDNA and ZT3 RNA as driver material. After performing three rounds of subtraction, 40 clones were identified that gave DNA-annealing signals with probes prepared from the subtracted cDNA but no signals with nonsubstracted probes. One such cDNA clone was found to hybridize at a low level with RNA extracted at ZT3 compared with a higher intensity signal at ZT15 (Rouyer *et al.*, 1997). The corresponding gene, dubbed *Crg-1*, produces three differentially spliced mRNAs that encode protein relatives of known transcription factors. All three mRNA species cycle in their abundance in concert with each other and with that of *per* transcript. per^{01} or tim^{01} mutations eliminated *Crg-1* cycling in LD; the abundance fluctuation of the newly identified gene product persists in DD at diminished amplitude, but the peak versus trough differences were appreciable enough to reveal that per^S caused a 20-hr free-running periodicity for one of the *Crg-1* mRNA forms (Rouyer *et al.*, 1997). Even though PER and TIM can be regarded to function at least in part as transcription factors, the abundance fluctuations of *Crg-1* mRNAs appear not to be underpinned by oscillating transcription of this *X*-chromosomal gene (Figure 18): Temporally controlled nuclear run-ons indicated little or no systematic fluctuation in its transcription rate (So *et al.*, 1997), not enough to explain the 3- to 4-fold higher amplitude of steady-state *Crg-1* mRNA levels during the early night compared with the early day. Assessing *Crg-1* mRNA in tissue sections of adult heads (Rouyer *et al.*, 1997) indicated coexpression with *per* in compound-eye photoreceptors (Rachidi *et al.*, 1997). Comparison of *in situ* hybridization signals for both genes in the optic ganglia and the central brain suggested similar expression patterns (Rouyer *et al.*, 1997), although it could not be definitively determined whether *per* neurons and glia also contain *Crg-1* transcripts on a cell-by-cell basis.

However, suppose that this gene is coexpressed in *per* LN cells, several of which also produce the PDF neuropeptide. Might it be that *Pdf* gene expression is regulated in part by *Crg-1*? This speculation is put forward, given that *Pdf* regulation by the CLK and CYC transcription factors seems to be indirect (Park *et al.*, 2000a). Conceivably, therefore, a CLK:CYC → *Crg-1* → *Pdf* pathway exists. The intermediate step just proposed could explain why cycling CLK levels do not lead to oscillations of *Pdf* mRNA (Park and Hall, 1998; Park *et al.*, 2000a). This supposition might have to include: (i) noncycling of CRG-1, which could occur if that protein does not turn over rapidly, its implied cyclical synthesis notwithstanding; and (ii) an additional intermediary, functioning between CLK's transcriptional regulation of a hypothetical entity that goes on to exert *post*transcriptional control of *Crg-1* mRNA cycling. These thoughts, idle as they may be, are laid out just so we keep in mind that it is difficult or useless to label a gene whose products cycle, at least at one level of gene expression, as a clock component versus a *crg*. *Crg-1* is a clock-regulated gene by definition, but the transcription factor it encodes leaves open the possibility that these proteins play a pacemaker-regulatory role as well.

Moving from the temporally based molecular subtraction that led to *Crg-1* one more historical step backward takes us to the first case of RNA screening performed in a search for fly genes that are subject to clock control. Here, brute-force, temporally controlled assessments of transcript abundances were made, using a collection of cDNA clones previously identified as complementary to transcripts present in the adult head but not in *Drosophila* embryos. Of 261 cDNAs applied as probes in four-timepoint Northern blottings, 20 detected RNA species that exhibit abundance fluctuations in LD (Van Gelder *et al.*, 1995). The score of implied *Dreg* genes (some of which are listed in Table 6) were categorized into 17 whose encoded transcripts were high early in the nighttime and 3 that were maximally expressed during the late night or early daytime. Most of the *Dreg*'s produced mRNAs detectable only in fly heads, although 10% of the cDNA probes hybridized to body-only transcripts and another 10% to (separately extracted) head plus body RNA (Van Gelder *et al.*, 1995). Seventy percent of the *Dreg*'s were expressed in embryos, belying the tightness of the molecular subtractions (implied earlier) that led to this clone collection. Partial (and in a few cases, complete) sequencing of these cDNA revealed only one known protein, alcohol dehydrogenase (Table 6).

Two of the "morning" *Dreg*'s cycled in DD; the third one did not. For all three of these cDNAs, alteration of fly-feeding time shifted the mRNA rhythms in LD, whereas *per*'s transcript cycling was unaffected (Van Gelder *et al.*, 1995). This possibly bizarre feature of *Dreg* regulation nevertheless sidles up to an element of the *takeout* story. One of the morning *Dreg*'s, whose mRNA cycling free-runs, had its abundance oscillation eliminated by per^{01} in a light:dark cycle as well as in constant darkness, but LD oscillations of the other two gene products in this category persisted in the face of this mutation (Van Gelder *et al.*, 1995).

An evening *Dreg* was chosen for relatively intensive chronobiological analysis, probably because this factor makes an mRNA that cycles in DD, failed to do so in per^{01} (in LD or DD), and did not have its transcript peak shifted by a change in feeding time (Van Gelder *et al.*, 1995). This is *Dreg-5* (Table 6), whose mRNA rhythm tracks that of *per*; the former cycling is eliminated by per^{01}, such that *Dreg-5*'s transcript abundance cruises along at the wild-type peak level (Van Gelder and Krasnow, 1996). The mRNA is found in extracts of embryos and of adult heads, including those of an *eyes-absent* mutant, but was undetectable in fly bodies (Van Gelder *et al.*, 1995; Van Gelder and Krasnow, 1996). The open-reading frame contained within this second-chromosomal gene (Figure 18) should encode a 298-aa protein. It seems to be posttranslationally modified because an anti-DREG-5 antibody detected two protein species in Western blottings. The 58- and 64-kD forms were presumed to come from this gene because both of them oscillated in abundance with essentially the same phase as that of the *Dreg-5* mRNA rhythm (Van Gelder and Krasnow, 1996). What these proteins are doing biologically, potentially to influence some feature of *Drosophila* rhythmicity, is unknown.

To take another tack toward identifying rhythmically expressed genes, *luciferase*-containing transposons were constructed and transformed into flies. The inherent features of these constructs were designed for constitutive expression of the *luc* sequences, but they could report putatively oscillating gene activity in live flies within strains resulting from mobilization of the transgenes (Table 6). By screening ca. 1,200 such derivative lines for bioluminescence fluctuations, rhythmic luciferase activity was detected in 6% of them (Stempfl *et al.*, 2002). Rhythmically fluctuating reporter levels were shown to be altered by various clock mutations, but oscillating luminescence in certain lines was affected by only a subset of such pacemaker variants (e.g., minimal or no effects of *per* mutations in one case, but near elimination of molecular rhythmicity by a *tim*-null mutation). By isolating genes near certain of the transposon insertions and determining their temporal mRNA expression, it was found that some of loci adjacent to the trapped enhancers are indeed rhythmically expressed Sequences of the open-reading frames for certain of these genes encode proteins heretofore unidentified chronobiologically or otherwise (Table 6). However, one transposon was shown to be inserted near and caused mutant expression of an already-known developmental gene called *numb* (Jan and Jan, 2000), which was revealed to be rhythmically expressed in adult flies as a newly uncovered feature of its molecular biology. The novel *numb*^{nuts} allele as well as previously described mutations at this locus were shown cause modest increases in the cycle durations of locomotor-activity rhythms (Stempfl *et al.*, 2002). In addition to its known role in cell-fate determination (Jan and Jan, 2000), this gene and the phosphotyrosine-binding membrane protein it encodes were inferred to function in the circadian system.

An additional gene of this kind, an apparent transcription factor encoded by a gene dubbed *regular* (*rgr*), was identified in *luc*-based, enhancer-trapped screen carried out in parallel (Scully *et al.*, 2002). Luciferase activity in the *rgr* enhancer-trapped strain exhibits reporter cycling in LD and DD, as does the native mRNA (peaking in the nighttime at a phase similar to that of *per*). LUC cycling was abolished by *per*^{01} in the *rgr* strain, and mRNA from this gene "stayed high" throughout the nighttime in the *Clk*^{Jrk} mutant (Scully *et al.*, 2002). A mutational effect of the transposon insert was inferred to be severe hypomorphy, but *tim-luc* cycling was normal in *rgr/rgr* flies, as if the transposon trapped an output factor as opposed to a clock one (Table 6). Tissue marking capabilities were included in the *luc* transposon designed by Scully *et al.* (2002). The *rgr* gene was thus inferred to be expressed in several CNS neurons of larvae and adults, although this marking did not include LN$_v$ cells (Figure 11). Intriguingly, however, the axons of certain "*regular* neurons" were closely juxtaposed to the PDF-containing neurites that project into the dorsal brain from *s*-LN$_v$ cells (Figure 12).

Molecular screening for rhythm-related factors can nowadays draw on genomics as yet another way to search for novel genes putatively involved in these processes. For this, one applies DNA microarrays. To these "genome chips"—on which are arrayed fragments of many to most of the genes or cDNA sequences known for the organism in question—one applies labeled copies (cRNAs) of transcripts harvested from whatever biological perspective is at hand. For the case in point, RNA was extracted at six equally spaced timepoints from *Drosophila* heads that had been removed from flies progressing through a 24-hr period in constant darkness (McDonald and Rosbash, 2001). Any of these transcript copies that might prove temporally interesting are instantly identifiable as to the genes that gave rise to them. This is in contrast to screens based on RNA subtractions (Rouyer *et al.*, 1997; So *et al.*, 2000) or differential display (Blau and Young, 1999; Shaw *et al.*, 2000), wherein one needs to clone and sequence material to which attention was drawn by transcript abundances that vary according to the biological standpoint that prompted the molecular search.

This particular time-based microarry screening identified 134 genes whose mRNAs exhibit free-running oscillations in their concentration (McDonald and Rosbash, 2001). Among these, satisfyingly, were *period, timeless, Clock, cryptochrome, vrille*, and *takeout*, listed here in historical order with respect to their molecular identifications and demonstrations (by standard methods) of gene-product cyclings, or by identifying them using more old-fashioned methods for detection of oscillating RNA species (in the cases of *vri* and *to*). Approximately

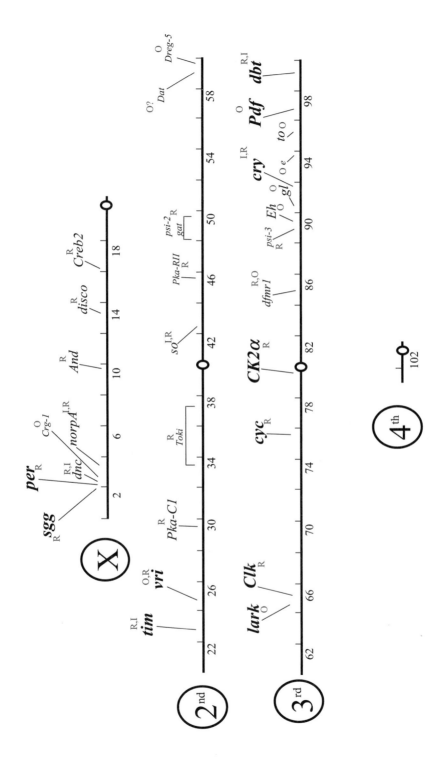

Figure 18. Mutations and genes involved in the biological rhythms of *D. melanogaster*. The many different kinds of mutants (Tables 1 and 2) that exhibit abnormalities of various rhythmic phenomena (Tables 1–3) are indicated by showing the gene symbols and mutated loci among the chromosomes of this species. In three cases, the genes have not been mutated, but were cloned and molecularly characterized with fair degrees of intensity (*Cgr-1*, *Dreg-5*) or manipulated (*Eh*) to cause eclosion anomalies (Table 6). The chronobiological impacts of the mutants and genes are indicated by red upper-case letters (one or more, depending on the locus): I for environmental input(s), R for rhythmicity (possibly including pacemaker functions), O for clock output(s). These different types of rhythm-related genes are distributed throughout three chromosomes only, in that dedicated screens for rhythm variants on the fourth one have not been performed (although a dominant such mutation might have turned out to be located there); and no extant mutants on chromosome 4 were tested for possible rhythm abnormalities (cf. Table 2), nor have any genes within this small proportion of the genome been manipulated (cf. Griffith *et al.*, 1993) to ask whether such defects would occur. The euchromatic portion of this insect's genome is divided into the 102 segments shown, based on cytological analyses of the banding patterns observed in the giant salivary-gland chromosomes of third-instar larvae. This numbering scheme (details of which are available at http://flybase.bio.indiana.edu/) is such that it is about half as long as either chromosome 2 (divided into regions 21–60) or chromosome 3 (regions 61–100); circles associated with the linear depiction of a given chromosome designate centromeres. Most of the mutations depicted were mapped by meiotic recombination to an at-least roughly positioned locus, followed (again, in most cases) by application of deletions (*Df*'s) and/or duplications that achieved higher resolution positioning of the mutated sites; the three "cloned-only" genes were localized by *in situ* annealing of nucleic acid probes to salivary-gland chromosome squashes. That procedure pinned down the sequence encoding *CK2α* (see Figure 3) to region 80A, very near the genetic position determined for the *Timekeeper* (*Tik*) mutation (Table 1), which can be regarded as synonymous with the *CK2α* gene shown here on chromosome 3. By analogy, the third chromosomal *dfmr* gene (encoding a so-called fragile-X-like protein) has also been dubbed *dfxr* (see Table 1). Exceptions to high-resolution gene localizations are: (a) *Toki* on chromosome 2, which was mapped to (somewhere within) the bracketed interval shown by recombination only; (b) *psi-2* and *gat*, both of which were recombination-mapped roughly and to the same second-chromosomal region; the deletion-based positioning of *psi-2* to the cytogenetic interval 49E7–50A2 is catalogued in "flybase" (each of the major 102 chromosomal regions numbered is subdivided into lettered ones, within which are essentially band-designating numbers); but this archival source (see above) is in error on this point because the results of locomotor testing *Df/psi-2* flies were ambiguous (Jackson, 1982); thus, the best that can be said for *psi-2*'s location is that it is within the bracketed region shown (to the left of map position 72, which corresponds approximately to cytogenetic region 51); the position of *gat* is indicated in this diagram at the same locus based on its noncomplementation with *psi-2*; for this, the latter mutation was shown to cause a lengthening of courtship-song cycle duration in *psi-2/psi-2* males (Table 3, cf. Figure 6B); homozygous *gat* males were aperiodic for this character, and *psi-2/gat* males exhibited longer-than-normal son-rhythm periods (Jackson, 1982); (c) *psi-3* on chromosome 3, like the two mutations just discussed, was mapped by recombination only. The rhythm mutations distributed among the three chromosomes shown are categorized into: (a) major variants (largest lettering) that arguably exhibit significant abnormalities of circadian-clock function (Table 1 and Figures 7 and 8), inputs to the pacemaker (Figure 16), its outputs (Table 6 and Figure 17), or some combination of these phenomena; (b) secondary variants (medium-sized lettering) defined by (originally) nonrhythm mutants applied in chronobiological tests, and mutations resulting from rhythm screening (Table 1) or retrospective testing (Table 2), which might cause mildly or erratically abnormal rhythmicity (at least for certain phenotypes), but have been applied in tests of more than one chronobiological attribute; and (c) minor mutants (smallest lettering), which exhibit unimpressive rhythm defects, possibly in conjunction with minimal analysis of the phenotype(s) in question; most of the genes in this tertiary category are defined by only one mutant allele, leaving open the possibility that further mutations induced or encountered at these loci may cause more substantial defects (compare the effects, for example, of *per*^Clk vs. *per*^T at the *period* locus, or *tim*^S2 vs. *tim*^UL at the *timeless* locus, as noted in Table 1).

one-fourth of the remaining 128 genes had been preidentified by fruitfly genomics, but they encode putative proteins of no known function (e.g., amino-acid sequences that fail to connect with the informational content of genes in this or other organisms for which biological or biochemical roles have been established or deduced). This (on-paper) subtraction of the "unknowns" left about 100 cyclically expressed genes. Intriguingly, most of them fell into known categories, which were discussible in terms of *Drosophila's* chronobiology (McDonald and Rosbash, 2001). The entries in Table 6 include a bowdlerized summary of these categories, based in part on the assumption that the various types of genes are clock-controlled. Taking this seminal study as exemplary (see citations to three further chrono-chip ones below), here are several of the gene types identified by McDonald and Rosbash (2001):

(i) Transcription factors, which may help relay the results of gene-regulation activities functioning within the clockworks into output phenomena [cf. Rouyer *et al.* (1997) and Blau and Young (1999) and the previously discussed inference that regulation of the *Pdf* neuropeptide gene is only indirectly under *Clock* and *cycle* control].

(ii) Neuropeptides and enzymes involved in processing such factors (cf. Taghert *et al.*, 2001).

(iii) Genes known or proposed to possess olfactory functions (in vertebrates and in *Drosophila*), which brings to mind the fly's circadian rhythm of odor sensitivity (Krishnan *et al.*, 1999, 2001).

(iv) Ligand-binding proteins, including members of the "TAKEOUT family" (Sarov-Blat *et al.*, 2000; So *et al.*, 2000).

(v) Proteins that contribute to cuticle formation, known to involve cyclical processes in other insects.

(vi) A protein called SLOB, which interacts with a certain type of potassium-channel polypeptide, which is encoded by the *slowpoke* gene (e.g., Schopperle *et al.*, 1998), making one wonder whether oscillating potassium conductances in certain *Drosophila* neurons [cf. Nitabach *et al.* (2002) plus discussion thereof by van den Pol and Obrietan (2002)] may underline daily fluctuations of membrane potential and firing frequencies that are a hallmark of neurophysiologically monitored circadian rhythmicity in other species (e.g., Whitmore and Block, 1996).

(vii) A "signal transduction" kinase (called s6kII), which functions downstream of other kinases including MAPK (Merienne *et al.*, 2001; Williams *et al.*, 2001).

(viii) Molecular relatives of mammalian detoxification enzymes, which conceptually connect with many gene products that exhibit daily cycling in rodent liver (e.g., Kornmann *et al.*, 2001).

(ix) Proteins that play roles in the fly's immunity and host defense, heretofore unsuspected to be under circadian control (but why not?).

This collection of presumed rhythm-related genes and their encoded products is a potential goldmine for a host of further chronobiological and molecular-chronogenetic investigations. In this regard, additional time-based applications of microarray technology have identified further genic candidates (Claridge-Chang *et al.*, 2001; Ueda *et al.*, 2002; Lin *et al.*, 2002b). The concordance among certain of the lists of genes that produce cycling RNAs in DD may seem displeasingly low [reviewed by Jackson and Schroeder (2001) and Etter and Ramaswami (2002)], although the numbers of candidates happen to be similar for two of the screens (McDonald and Rosbash, 2001; Claridge-Chang *et al.*, 2001). Subsequently, Ueda *et al.* (2002) turned up a whopping 712 genes that show daily cycles of their encoded mRNAs in LD; this list got boiled down to 115 free-running cyclers, similar in terms of raw numbers to the conclusions of McDonald and Rosbash (2001) and Claridge-Chang *et al.* (2001). In contrast, the empiricsm of, and analytical approach taken by, Lin *et al.* (2002b) resulted in only 72 *Drosophila* genes that cycle in LD, a mere 22 of which exhibited that expression property in DD; most but not all of the latter number had their cycling severely impinged on by the effects of per^{01}. The global overlap *among the four sets* of genes detected by putting a chronobiological twist to the microarray approach is again on the face of it unimpressive (and depends in part on the various analytical criteria applied). However, it is almost certain that the congruences between a given pair of these studies involves more than coincidence.

This quartet of research groups proceeded beyond the "chip screens" alone to tie various elements of their gene identifications into *Drosophila*'s rhythm system. Thus, cRNAs were prepared from clock-mutant head extracts. The results of McDonald and Rosbash (2001), the first such published involving *Drosophila*, will again be treated as exemplifying these kinds of mutational effects. With regard to RNA taken 4 hr and 16 hr into the subjective night, all of the 134 gene products that showed abundance differences in wild-type flies were equivalent in their inferred concentrations at these two timepoints (McDonald and Rosbash, 2001). This result was backed up by subjecting temporally collected RNA samples to the aforementioned differential-display analysis (albeit based on embellishments of these tactics that had been conceived and executed by Kornmann *et al.*, 2001). The results of this molecular screen identified 171 RNA bands (amplicons resulting from application of many PCR primer pairs) that cycled circadianly. None showed oscillations in mRNA extracted from Clk^{Jrk} flies. Both microarray probing and differential display revealed another feature of *Clock*-gene control (McDonald and Rosbash, 2001): a few hundred noncycling RNA species with higher- or lower-than-normal abundances in the mutant extracts (the bulk of these findings stemmed from executing the differential-display procedures). By analogy, a *per*-null mutation was found by Lin *et al.* (2002b) to influence the expression levels of more than 600 nonoscillating transcripts. The genes that produce these mRNAs could also function with the organism's rhythm system, inasmuch as cyclical fluctuation of a given gene's products is not the quintessence of a chronobiological connection. As noted in various sections of this review, genes such a *cycle*, *double-time*, and *Pdf* generate transcripts that are temporally flat in their concentrations. However, if a gene is regulated downstream of *Clock*'s actions, it may play a rhythm role. Given that severe mutations in this transcription-factor-encoding gene (Table 1) do not cause overt and wildly pleiotropic biological defects—unlike the case of *vrille*, for example—any regulatee of *Clk* can be gingerly proposed to carry out "special" functions that could be primarily devoted to rhythm control. Alternatively or in addition, several components of *Clock*'s significance may fall outside the scope of *Drosophila*'s chronobiology. Recall that *Clk*^{Jrk} flies exhibit certain neuroanatomical (hence, developmental) defects that do not blatantly connect with the control of ongoing rhythmicity (Park *et al.*, 2000a), and this mutant is abnormal for drug sensitization in a manner that has no known connection with a daily cycle (Andretic *et al.*, 1999).

XI. CONCLUSION

Molecular screening for factors that may be chronobiologically significant had the sanguine outcome of identifying, in several instances, novel genes that might have remained throughgoing puzzles had they been identified by fruitfly genomics alone. In other words, who would necessarily bother to investigate *Drosophila* genes encoding such featureless products if they had not been detected by more than their nucleotide sequences?—in the current cases, also by virtue of being rhythmically expressed. That such expression patterns are "clock-controlled" is suggested by the effects of circadian pacemaker mutations on the gene-product oscillations, although for some of these genes not all the clock mutations tested affect the molecular rhythmicity in question, or only some of the applicable mutations have been assessed for effects on the mRNA fluctuations.

In any case, these experimental connections between mutants and and molecules take us back near the beginning of the insect chronogenetic story as a whole. One of its earliest chapters involved induction of the seminal *period* mutations. They identified a gene analogous to several of the factors identified as cycling nucleic acids, in that *per* began its molecular life as a satisfyingly novel entity. Yet it is arguable that the genetic story summarized in Figure 18 has moved well beyond the mysteries that initially revolved around this gene, the first clock one known. How *per* acts (Figure 7), how it interacts with its subsequently identified pacemaker mates (Figures 7–9), and how all these factors interface with chronobiologically relevant features of the environment (Figures 5 and 16) as well as with proximate regulators of revealed organismal rhythms (Figure 17) tell a tale that one hopes was worth recounting here in what is likely the densest review ever prepared on behalf of this genetic subject.

Acknowledgments

I am extremely grateful for the corrections of and many other comments on embryonic versions of this article that stemmed from manuscript readings (including fine-print scrutiny) by Charalambos P. Kyriacou and Ralf Stanewsky. I appreciate several pieces of information and advice that came from discussions with Ravi Allada, Harold B. Dowse, Isaac Edery, Charlotte Helfrich-Förster, Michael J. McDonald, F. Rob Jackson, Maki Kaneko, Charalambos P. Kyriacou, Sebastian Martinek, Michael Rosbash, Marla B. Sokolowski, Ralf Stanewsky, Vipin Suri, and Michael W. Young. Acknowledgments are due Paul E. Hardin; Isaac Edery and Paul E. Hardin; Charlotte Helfrich-Förster; Alexandre A. Peixoto; and Kenji Tomioka for granting permissions to adapt figures published by these authors in their respective five papers. I thank Edward Dougherty and Joshua Robbins for digitizing those images and others that were generated from scratch for the purposes of this piece.

References

Allada, R., White, N. E., So, W. V., Hall, J. C., and Rosbash, M. (1998). A mutant *Drosophila* homolog of mammalian *Clock* disrupts circadian rhythms and transcription of *period* and *timeless*. *Cell* **93**, 791–804.

Alt, S., Ringo, J., Talyn, B., Bray, W., and Dowse, H. (1998). The *period* gene controls courtship song cycles in *Drosophila melanogaster*. *Anim. Behav.* **56**, 87–97.

Alvarez, C. E., Robison, K., and Gilbert, W. (1996). Novel $G_q\alpha$ isoform is a candidate transducer of rhodopsin signaling in a *Drosophila* testes-autonomous pacemaker. *Proc. Natl. Acad. Sci. USA* **93**, 12278–12282.

Andretic, R., and Hirsh, J. (2000). Circadian modulation of dopamine receptor responsiveness in *Drosophila melanogaster*. *Proc. Natl. Acad. Sci. USA* **97**, 1873–1878.

Andretic, R., Chaney, S., and Hirsh, J. (1999). Requirement of circadian genes for cocaine sensitization in *Drosophila*. *Science* **285**, 1066–1068.

Antoch, M. P., Song, E.-J., Chang, A.-M., Vitaterna, M. H., Zhao, Y., Wilsbacher, L. D., Sangoram, A. M., King, D. P., Pinto, L. H., and Takahashi, J. S. (1997). Functional identification of the mouse circadian *Clock* gene by transgenic BAC rescue. *Cell* **89**, 655–667.

Atkinson, P. W., and James, A. A. (2002). Germline transformants spreading out to many insect species. *Adv. Genet.* **47**, 49–86.

Bae, K., Lee, C., Sidote, D., Chang, K.-Y., and Edery, I. (1998). Circadian regulation of a *Drosophila* homolog of the mammalian *Clock* gene: PER and TIM function as positive regulators. *Mol. Cell. Biol.* **18**, 6142–6151.

Bae, K., Lee, C., Hardin, P. E., and Edery, I. (2000). dCLOCK is present in limiting amounts and likely mediates daily interactions between the dCLOCK–CYC transcription factor and the PER–TIM complex. *J. Neurosci.* **20**, 1746–1753.

Baehrecke, E. H. (1996). Ecdysone signaling cascade and regulation of *Drosophila* metamorphosis. *Arch. Insect Biochem. Physiol.* **33**, 231–244.

Baker, J. D., McNabb, S. L., and Truman, J. W. (1999). The hormonal coordination of behavior and physiology at adult ecdysis in *Drosophila melanogaster*. *J. Exp. Biol.* **202**, 3037–3048.

Bangs, P., Franc, N., and White, K. (2000). Molecular mechanisms of cell death and phagocytosis in *Drosophila*. *Cell Death Differ.* **7**, 1027–1034.

Bao, S., Rihel, J., Bjes, E., Fan, J.-Y., and Price, J. L. (2001). The *Drosophila double-time* mutation delays the nuclear accumulation of *period* protein and affects the feedback regulation of *period* mRNA. *J. Neurosci.* **21**, 7117–7126.

Bargiello, T. A., and Young, M. W. (1984). Molecular genetics of a biological clock in *Drosophila*. *Proc. Natl. Acad. Sci. USA* **81**, 2142–2146.

Bargiello, T. A., Jackson, F. R., and Young, M. W. (1984). Restoration of circadian behavioural rhythms by gene transfer in *Drosophila*. *Nature* **312**, 752–754.

Bargiello, T. A., Saez, L., Baylies, M. K., Gasic, G., Young, M. W., and Spray, D. C. (1987). The *Drosophila* clock gene *per* affects intercellular junctional communication. *Nature* **328**, 686–691.

Barrett, R. K., and Page, T. L. (1989). Effects of light on circadian pacemaker development. I. The freerunning period. *J. Comp. Physiol. A* **165**, 41–49.

Baylies, M. K., Bargiello, T. A., and Young, M. W. (1987). Changes in abundance and structure of the *per* gene product can alter periodicity of the *Drosophila* clock. *Nature* **326**, 390–392.

Baylies, M. K., Vosshall, L. B., Sehgal, A., and Young, M. W. (1992). New short *period* mutations of the Drosophila clock gene *per*. *Neuron* **9**, 575–581.

Baylies, M. K., Weiner, L., Vosshall, L. B., Saez, L., and Young, M. W. (1993). Genetics, molecular and cellular studies of the *per* locus and its products in *Drosophila melanogaster*. *In* "Molecular Genetics of Biological Rhythms" (M. W. Young, ed.), pp. 123–153. Marcel Dekker, New York.

Beaver, L. M., Gvakharia, B. O., Vollintine, T. S., Hege, D. M., Stanewsky, R., and Giebultowicz, J. M. (2002). Loss of circadian clock function decreases reproductive fitness in males of *Drosophila melanogaster*. *Proc. Natl. Acad. Sci. USA.* **99**, 2134–2139.

Bebas, P., and Cymborowski, B. (1999). Effects of constant light on male sterility in *Spodptera littoralis* (Lepidoptera: Noctuidae). *Physiol. Entomol.* **24**, 165–170.

Bebas, P., Cymborowski, B., and Giebultowicz, J. M. (2001). Circadian rhythm of sperm release in males of the cotton leafworm, *Spodptera littoralis*: In vivo and in vitro study. *J. Insect Physiol.* **47**, 859–866.

Bebas, P., Cymborowski, B., and Giebultowicz, J. M. (2002). Circadian rhythm of acidification in insect vas deferens regulated by expression of vacuolar H-ATPase. *J. Exp. Biol.* **205**, 37–44.

Belcher, K., and Brett, W. J. (1973). Relationship between a metabolic rhythm and emergence rhythm in *Drosophila melanogaster*. *J. Insect Physiol.* **19**, 277–286.

Belvin, M. P., Zhou, H., and Yin, J. C. P. (1999). The *Drosophila* dCREB2 gene affects the circadian clock. *Neuron* **22**, 777–787.

Benna, C., Scannapieco, P., Piccin, A., Sandrelli, F., Zordan, M., Rosato, E., Kyriacou, C. P., Valle, G., and Costa, R. (2000). A second *timeless* gene in *Drosophila* shares greater sequence similarity with mammalian *tim*. *Curr. Biol.* **10**, R512–R513.

Blair, S. S. (1994). A role for the segment polarity gene *shaggy-zeste white 3* in the specification of regional identity in the developing wing of *Drosophila*. *Dev. Biol.* **162**, 229–244.

Blanchardon, E., Grima, B., Klarsfeld, A., Chelot, E., Hardin, P. E., Préat, T., and Rouyer, F. (2001). Defining the role of *Drosophila* lateral neurons in the control of activity and eclosion rhythms by targeted genetic ablation and PERIOD overexpression. *Eur. J. Neurosci.* **13**, 871–888.

Blau, J., and Young, M. W. (1999). Cycling *vrille* expression is required for a functional *Drosophila* clock. *Cell* **99**, 661–671.

Block, G., Toma, D. P., and Robinson, G. E. (2001). Behavioral rhythmicity, age, division of labor and *period* expression in the honey bee brain. *J. Biol. Rhythms* **16**, 444–456.

Boutros, M., and Mlodzik, M. (1999). Dishevelled: At the crossroads of divergent intracellular signaling pathways. *Mech. Dev.* **83**, 27–37.

Brand, A. H., and Perrimon, N. (1993). Targeting gene expression as a means of altering cell fates and generating dominant phenotypes. *Development* **118**, 401–415.

Brandes, C., Plautz, J. D., Stanewsky, R., Jamison, C. F., Straume, M., Wood, K. V., Kay, S. A., and Hall, J. C. (1996). Novel features of Drosophila period transcription revealed by real-time luciferase reporting. *Neuron* **16**, 687–692.

Brett, W. J. (1955). Persistent diurnal rhythmicity in Drosophila emergence. *Ann. Entomol. Soc. Am.* **48**, 119–131.

Brodbeck, D., Amherd, R., Callaerts, P., Hintermann, E., Meyer, U. A., and Affolter, M. (1998).

Molecular and biochemical characterization of the *aaNAT1* (*Dat*) locus in *Drosophila melanogaster*: differential expression of two gene products. *DNA Cell Biol.* **17**, 621–633.

Busto, M., Iyengar, B., and Campos, A. R. (1999). Genetic dissection of behavior: modulation of locomotion by light in the *Drosophila melanogaster* larva requires genetically distinct visual system functions. *J. Neurosci.* **19**, 3337–3344.

Callebert, J., Launay, J.-M., and Jallon, J.-M. (1991). Melatonin pathway in wild type and mutant (*period*) strains of *Drosophila*. *In* "Advances in Pineal Research, Vol. 5" (J. Arendt and P. Pévet, eds.), pp. 81–84. John Libbey, London.

Castiglione-Morelli, M. A., Guantieri, V., Villani, V., Kyriacou, C. P., Costa, R., and Tamburro, A. M. (1995). Conformational study of the Thr–Gly repeat in the *Drosophila* clock protein, PERIOD. *Proc. R. Soc. Lond. B* **260**, 155–163.

Ceriani, M. F., Darlington, T. K., Staknis, D., Más, P., Petti, A. A., Weitz, C. J., and Kay, S. A. (1999). Light-dependent sequestration of TIMELESS by CRYPTOCHROME. *Science* **285**, 553–556.

Chen, B., Meinertzhagen, I. A., and Shaw, S. R. (1999). Circadian rhythms in light-evoked responses of the fly's compound eye, and the effects of neuromodulators 5-HT and the peptide PDF. *J. Comp. Physiol. A* **185**, 393–404.

Chen, D.-M., Christianson, J. S., Sapp, R. J., and Stark, W. S. (1992). Visual receptor cycle in normal and *period* mutant *Drosophila*: microspectrophotometry, electrophysiology, and ultrastructural morphometry. *Vis. Neurosci.* **9**, 125–135.

Chen, Y. F., Hunter-Ensor, M., Schotland, P., and Sehgal, A. (1998). Alterations of *per* RNA in noncoding regions affect periodicity of circadian behavioral rhythms. *J. Biol. Rhythms* **13**, 364–379.

Cheng, Y., and Hardin, P. E. (1998). *Drosophila* photoreceptors contain an autonomous circadian oscillator that can function without *period* mRNA cycling. *J. Neurosci.* **18**, 741–750.

Cheng, Y., Gvakharia, B., and Hardin, P. E. (1998). Two alternatively spliced transcripts from the *Drosophila period* gene rescue rhythms having different molecular and behavioral characteristics. *Mol. Cell. Biol.* **18**, 6505–6514.

Chuman, Y., Matsushima, A., Sato, S., Tomioka, K., Tominaga, Y., Meinertzhagen, I. A., Shimohigashi, Y., and Shimohigashi, M. (2002). cDNA cloning and nuclear localization of circadian neuropeptide designated as pigment-dispersing factor PDF in the cricket *Gryllus bimaculatus*. *J. Biochem.* **131**, 895–903.

Citri, Y., Colot, H. V., Jacquier, A. C., Yu, Q., Hall, J. C., Baltimore, D., and Rosbash, M. (1987). A family of unusually spliced biologically active transcripts encoded by a *Drosophila* clock gene. *Nature* **326**, 42–47.

Claridge-Chang, A., Wijnen, H., Naef, F., Boothroyd, C., Nikolaus Rajewsky, N., and Young, M. W. (2001). Circadian regulation of gene expression systems in the *Drosophila* head. *Neuron* **32**, 657–671.

Clayton, D. L., and Paietta, J. V. (1972). Selection for circadian eclosion time in *Drosophila melanogaster*. *Science* **178**, 994–995.

Cobb, M., Connolly, K., and Burnet, B. (1987). The relationship between locomotor activity and courtship in the *melanogaster* species sub-group of *Drosophila*. *Anim. Behav.* **35**, 705–713.

Colot, H. V., Hall, J. C., and Rosbash, M. (1988). Interspecific comparison of the *period* gene of *Drosophila* reveals large blocks of non-conserved coding DNA. *EMBO J.* **7**, 3929–3937.

Cooper, M. K., Hamblen-Coyle, M. J., Liu, X., Rutila, J. E., and Hall, J. C. (1994). Dosage compensation of the *period* gene in *Drosophila melanogaster*. *Genetics* **138**, 721–732.

Costa, R., and Kyriacou, C. P. (1998). Functional and evolutionary implications of natural variation in clock genes. *Curr. Opin. Neurobiol.* **8**, 659–664.

Costa, R., Peixoto, A. A., Thackeray, J. R., Dalgleish, R., and Kyriacou, C. P. (1991). Length polymorphism in the threonine–glycine-encoding repeat region of the *period* gene in *Drosophila*. *J. Mol. Evol.* **32**, 238–246.

Costa, R., Peixoto, A. A., Barbujani, G., and Kyriacou, C. P. (1992). A latitudinal cline in a *Drosophila* clock gene. *Proc. R. Soc. Lond. B* **250,** 43–49.

Coté, G. G., and Brody, S. (1986). Circadian rhythms in *Drosophila melanogaster:* Analysis of period as a function of gene dosage at the *per (period)* locus. *J. Theor. Biol.* **121,** 487–503.

Coto-Montes, A., and Hardeland, R. (1999). Diurnal rhythm of protein carbonyl as an indicator of oxidative damage in *Drosophila melanogaster:* Influence of clock gene alleles and deficiencies in the formation of free-radical scavengers. *Biol. Rhythm Res.* **30,** 383–391.

Crews, S. T., Thomas, J. B., and Goodman, C. S. (1988). The Drosophila *single-minded* gene encodes a nuclear protein with sequence similarity to the *per* gene product. *Cell* **52,** 143–151.

Crossley, S. A. (1988). Failure to confirm rhythms in *Drosophila* courtship song. *Anim. Behav.* **36,** 1098–1109.

Curtin, K. D., Huang, Z. J., and Rosbash, M. (1995). Temporally regulated nuclear entry of the *Drosophila period* protein contributes to the circadian clock. *Neuron* **14,** 363–372.

Curtis, N. J., Ringo, J. M., and Dowse, H. B. (1999). Morphology of the pupal heart, adult heart, and associated tissues in the fruit fly, *Drosophila melanogaster. J. Morphol.* **240,** 225–235.

Darlington, T., Wager-Smith, K., Ceriani, M. F., Staknis, D., Gekakis, N., Steeves, T., Weitz, C., Takahashi, J. S., and Kay, S. A. (1998). Closing the circadian loop: CLOCK-induced transcription of its own inhibitors, *per* and *tim. Science* **280,** 1599–1603.

Darlington, T. K., Lyons, L. C., Hardin, P. E., and Kay, S. A. (2000). The *period* E-box is sufficient to drive circadian transcription *in vivo. J. Biol. Rhythms* **15,** 462–471.

Dembinska, M. E., Stanewsky, R., Hall, J. C., and Rosbash, M. (1997). Circadian cycling of a *period–lacZ* fusion protein in *Drosophila:* evidence for an instability cycling element in PER. *J. Biol. Rhythms* **12,** 157–172.

Demetriades, M. C., Thackeray, J. R., and Kyriacou, C. P. (1999). Courtship song rhythms in *Drosophila yakuba. Anim. Behav.* **57,** 379–386.

DiBartolomeis, S. M., Akten, B., Genova, G., Roberts, M. A., and Jackson, F. R. (2002). Molecular analysis of the *Drosophila miniature-dusky (m-dy)* gene complex: *m-dy* mRNAs encode transmembrane proteins with similarity to C. *elegans* cuticulin. *Mol. Genet. Genom.* **267,** 564–576.

Dockendorff, T. C., Su, H. S., McBride, S. M., Yang, Z., Choi, C. H, Siwicki, K. K., Sehgal, A., and Jongens, T. A. (2002). *Drosophila* lacking *dfmr1* activity show defects in circadian output and fail to maintain courtship interest. *Neuron* **34,** 973–984.

Dowse, H. B., Kass, L., and Ringo, J. M. (1988). Studies on a congenital heart defect in *Drosophila. Behav. Genet.* **18,** 714–715.

Dowse, H. B., and Ringo, J. M. (1987). Further evidence that the circadian clock in *Drosophila* is a population of ultradian oscillators. *J. Biol. Rhythms* **2,** 65–76.

Dowse, H. B., and Ringo, J. M. (1989). Rearing *Drosophila* in constant darkness produces phenocopies of *period* circadian clock mutants. *Physiol. Zool.* **62,** 785–803.

Dowse, H. B., and Ringo, J. M. (1993). Is the circadian clock a "metaoscillator?" Evidence from studies of circadian rhythms in *Drosophila. In* "Molecular Genetics of Biological Rhythms" (M. W. Young, ed.), pp. 195–220. Marcel Dekker, New York.

Dowse, H. B., Hall, J. C., and Ringo, J. M. (1987). Circadian and ultradian rhythms in *period* mutants of *Drosophila melanogaster.* **17,** 19–35.

Dowse, H. B., Dushay, M. S., Hall, J. C., and Ringo, J. M. (1989). High-resolution analysis of locomotor activity rhythms in *disconnected,* a visual-system mutant of *Drosophila melanogaster. Behav. Genet.* **19,** 529–542.

Dowse, H., Ringo, J., Power, J., Johnson, E., Kinney, K., and White, L. (1995). A congenital heart defect in *Drosophila* caused by an action-potential mutation. *J. Neurogenet.* **10,** 153–168.

Draizen, T. A., Ewer, J., and Robinow, S. (1999). Genetic and hormonal regulation of the death of peptidergic neurons in the *Drosophila* central nervous system. *J. Neurobiol.* **38,** 455–465.

Dubnau, J., and Tully, T. (1998). Gene discovery in *Drosophila:* new insights for learning and memory. *Annu. Rev. Neurosci.* **21,** 407–444.

Dunlap, J. C. (1999). Molecular bases for circadian clocks. *Cell* **96**, 271–290.

Dushay, M. S., Rosbash, M., and Hall, J. C. (1989). The *disconnected* visual system mutations in *Drosophila* drastically disrupt circadian rhythms. *J. Biol. Rhythms* **4**, 1–27.

Dushay, M. S., Konopka, R. J., Orr, D., Greenacre, M. L., Kyriacou, C. P., Rosbash, M., and Hall, J. C. (1990). Phenotypic and genetic analysis of *Clock*, a new circadian rhythm mutant in *Drosophila melanogaster*. *Genetics* **125**, 557–578.

Dushay, M. S., Rosbash, M., and Hall, J. C. (1992). Mapping the *Clock* mutation rhythm mutation to the *period* locus of *Drosophila melanogaster* by germline transformation. *J. Neurogenet.* **8**, 173–179.

Edery, I., Rutila, J. E., and Rosbash, M. (1994a). Phase shifting of the circadian clock by induction of the *Drosophila period* protein. *Science* **263**, 237–240.

Edery, I., Zwiebel, L. J., Dembinska, M. E., and Rosbash, M. (1994b). Temporal phosphorylation of the *Drosophila period* protein. *Proc. Natl. Acad. Sci. USA* **91**, 2260–2264.

Egan, E. S., Franklin, T. M., Hildebrand-Chae, M. J., McNeil, G. P., Roberts, M. A., Schroeder, A. J., Zhang, X., and Jackson, F. R. (1999). An extraretinally expressed insect cryptochrome with similarity to the blue light photoreceptors of mammals and plants. *J. Neurosci.* **19**, 3665–3673.

Ellis, H. M., Spann, D. R., and Posakony, J. W. (1990). *extramacrochaetae*, a negative regulator of sensory organ development in *Drosophila*, defines a new class of helix–loop–helix proteins. *Cell* **61**, 27–38.

Emery, I. F., Noveral, J. M., Jamison, C. F., and Siwicki, K. K. (1997). Rhythms of *Drosophila period* gene expression in culture. *Proc. Natl. Acad. Sci. USA* **94**, 4092–4096.

Emery, P., So, W. V., Kaneko, M., Hall, J. C., and Rosbash, M. (1998). CRY, a *Drosophila* clock and light-regulated cryptochrome, is a major contributor to circadian rhythm resetting and photosensitivity. *Cell* **95**, 669–679.

Emery, P., Stanewsky, R., Hall, J. C., and Rosbash, M. (2000a). *Drosophila* cryptochrome—a unique circadian-rhythm photoreceptor. *Nature* **404**, 456–457.

Emery, P., Stanewsky, R., Helfrich-Forster, C., Emery-Le, M., Hall, J. C., and Rosbash, M. (2000b). *Drosophila* CRY is a deep brain circadian photoreceptor. *Neuron* **26**, 493–504.

Engelmann, W., and Honegger, H. W. (1966). Tagesperiodische Schlüpfrhythmik einer augenlosen *Drosophila melanogaster*-Mutante [Daily eclosion rhythms in an eyeless *Drosophila melanogaster* mutant]. *Naturwissenschaften* **22B**, 588.

Etter, P. D., and Ramaswami, M. (2002). The ups and downs of daily life: profiling circadian gene expression in *Drosophila*. *BioEssays* **24**, 494–498.

Ewer, J., and Truman, J. W. (1996). Increases in cyclic 3′,5′-guanosine monophosphate (cGMP) occur at ecdysis in an evolutionary conserved crustacean cardioactive peptide-immunoreactive insect neuronal network. *J. Comp. Neurol.* **370**, 330–341.

Ewer, J., Rosbash, M., and Hall, J. C. (1988). An inducible promoter fused to the *period* gene in *Drosophila* conditionally rescues adult *per*-mutant arrhythmicity. *Nature* **333**, 82–84.

Ewer, J., Hamblen-Coyle, M. J., Rosbash, M., and Hall, J. C. (1990). Requirement for *period* gene expression in the adult and not during development for locomotor activity rhythms of imaginal *Drosophila melanogaster*. *J. Neurogenet.* **7**, 31–73.

Ewer, J., Frisch, B., Hamblen-Coyle, M. J., Rosbash, M., and Hall, J. C. (1992). Expression of the *period* clock gene within different cells types in the brain of *Drosophila* adults and mosaic analysis of these cells' influence on circadian behavioral rhythms. *J. Neurosci.* **12**, 3321–3349.

Ewer, J., Oliva, M., and Park, J. H. (2001). Neuroendocrine control of larval ecdysis. *J. Neurogenet.* **15**, 19.

Ewing, A. W. (1988). Cycles in the courtship song of male *Drosophila melanogaster* have not been detected. *Anim. Behav.* **36**, 1091–1097.

Field, M. D., Maywood, E. S., O'Brien, J. A., Weaver, D. R., Reppert, S. M., and Hastings, M. H. (2000). Analysis of clock proteins in mouse SCN demonstrates phylogenetic divergence of the circadian clockwork and resetting mechanisms. *Neuron* **25**, 437–447.

Finocchiaro, L., Callebert, J., Launay, J. M., and Jallon, J.-M. (1988). Melatonin biosynthesis in *Drosophila*: its nature and its effects. *J. Neurochem.* **50**, 382–387.

Fleissner, G., Loessel, R., Fleissner, G., Waterkamp, M., Kleiner, O., Batschauer, A., and Homberg, U. (2001). Candidates for extraocular photoreceptors in the cockroach suggest homology to the lamina and lobula organs in beetles. *J. Comp. Neurol.* **433**, 401–414.

Flint, K. K., Rosbash, M., and Hall, J. C. (1993). Transfer of dye among salivary gland cells is not affected by genetic variations of the *period* clock gene in *Drosophila melanogaster*. *J. Membr. Biol.* **136**, 333–342.

Fluegel, W. (1981). Oviposition rhythm in *Drosophila melanogaster* is influenced by acetic acid. *J. Insect Physiol.* **27**, 705–710.

Ford, M. J., Yoon, C. K., and Aquadro, C. F. (1994). Molecular evolution of the *period* gene in *Drosophila athabasca*. *Mol. Biol. Evol.* **11**, 169–182.

Foulkes, N. S., Borjigin, J., Synder, S. H., and Sassone-Corsi, P. (1997). Rhythmic transcription: The molecular basis of circadian melatonin synthesis. *Trends Neurosci.* **20**, 487–492.

Frank, K. D., and Zimmerman, W. F. (1969). Action spectra for phase shifts of a circadian rhythm in *Drosophila*. *Science* **163**, 688–689.

Friedman, T. B., and Johnson, D. H. (1977). Temporal control of urate oxidase activity in *Drosophila*: Evidence of an autonomous timer in Malpighian tubules. *Science* **197**, 477–479.

Friedman, T. B., and Johnson, D. H. (1978). Autonomous timer in Malpighian tubules. *Science* **200**, 1185–1186.

Frisch, B., Hardin, P. E., Hamblen-Coyle, M. J., Rosbash, M., and Hall, J. C. (1994). A promoterless *period* gene mediates behavioral rhythmicity and cyclical *per* expression in a restricted subset of the Drosophila nervous system. *Neuron* **12**, 555–570.

Frisch, B., Fleissner, G., Fleissner, G., and Hall, J. C. (1996). PER-like immunoreactivity in the brain of the beetle *Pachymorpha sexguttata*: histological characterization of cells with possible importance for the circadian clock system. *Cell Tiss. Res.* **286**, 411–429.

Froy, O., Chang, D. C., and Reppert, S. M. (2002). Redox potential: differential roles in dCRY and mCRY1 functions. *Curr. Biol.* **12**, 147–152.

Gäde, G., Hoffmann, K. H., and Spring, J. H. (1997). Hormonal regulation in insects: facts, gaps, and future directions. *Physiol. Rev.* **77**, 963–1032.

Gailey, D. A., Villella, A., and Tully, T. (1991). Reassessment of the effects of biological rhythm mutations on learning in *Drosophila melanogaster*. *J. Comp. Physiol. A* **169**, 685–697.

García-Fernández, J. M., Siwicki, K. K., and Foster, R. G. (1994). *Per*-like immunoreactive neurons within the brain of *Xenopus* innervate both the retina and pineal organ. *In* "Advances in Pineal Research, Vol. 8". (M. Møller and P. Pévet, eds.), pp. 19–24. Libbey, London.

Gekakis, N., Saez, L., Delahaye-Brown, A.-M., Myers, M. P., Sehgal, A., Young, M. W., and Weitz, C. J. (1995). Isolation of *timeless* by PER protein interaction: defective interaction between TIMELESS protein and long-period mutant PER^L. *Science* **270**, 811–815.

George, H., and Terracol, R. (1997). The *vrille* gene of Drosophila is a maternal enhancer of *decapentaplegic* and encodes a new member of the bZIP family of transcription factors. *Genetics* **146**, 1345–1363.

Giebultowicz, J. M. (1999). Insect circadian rhythms: Is it all in their heads? *J. Insect Physiol.* **45**, 791–800.

Giebultowicz, J. M. (2000). Molecular mechanism and cellular distribution of insect circadian clocks. *Annu. Rev. Entomol.* **45**, 767–791.

Giebultowicz, J. M. (2001). Peripheral clocks and circadian timing: Insights from insects. *Phil. Trans. R. Soc. Lond. B* **356**, 1791–1799.

Giebultowicz, J. M., and Hege, D. M. (1997). Circadian clock in Malpighian tubules. *Nature* **386**, 664.

Giebultowicz, J. M., Stanewsky, R., Hall, J. C., and Hege, D. M. (2000). Transplanted *Drosophila* excretory tubules maintain circadian clock cycling out of phase with the host. *Curr. Biol.* **10,** 107–110.

Giebultowicz, J. M., Ivanchenko, M., and Vollintine, T. (2001). Organization of the insect circadian system: spatial and developmental expression of clock genes in peripheral tissues of *Drosophila melanogaster*. *In* "Insect Timing: Circadian Rhythmicity and Seasonality" (D. L. Denlinger, J. Giebultowicz, and D. S. Saunders, eds.), pp. 31–42. Elsevier, Amsterdam.

Gillette, M. U. (1996). Regulation of entrainment pathways by the suprachisamatic circadian clock: sensitivities to second messengers. *In* "Progress in Brain Research, Vol. 111". (R. Buijs, H. Romijn, C. Penartz, and M. Mirmiran, eds.), pp. 119–130. Elsevier, Amsterdam.

Gleason, J. M., and Powell, J. R. (1997). Interspecific and intraspecific comparisons of the *period* locus in the *Drosophila willistoni* sibling species. *Mol. Biol. Evol.* **14,** 741–753.

Glossop, N. R. J., Lyons, L. C., and Hardin, P. E. (1999). Interlocked feedback loops within the *Drosophila* circadian oscillator. *Science* **286,** 766–768.

Gonzalez, M. Z. (1923). Experimental studies on the duration of life of certain mutant genes of *Drosophila melanogaster*. *Am. Nat.* **57,** 289–325.

Gotter, A. L., Levine, J. D., and Reppert, S. M. (1999). Sex-linked *period* genes in the silkmoth *Antheraea pernyi*: Implications for circadian clock regulation and the evolution of sex chromosomes. *Neuron* **24,** 953–965.

Gotter, A. L., Manganaro, T., Weaver, D. R., Kolakowski, L. F., Jr., Possidente, B., Sriram, S., MacLaughlin, D. T., and Reppert, S. M. (2000). A time-less function for mouse *timeless*. *Nat. Neurosci.* **8,** 755–756.

Greenacre, M. L., Ritchie, M. G., Byrne, B. C., and Kyriacou, C. P. (1993). Female song preference and the *period* gene in *Drosophila*. *Behav. Genet.* **23,** 85–90.

Greenspan, R. J. (1990). The emergence of neurogenetics. *Semin. Neurosci.* **2,** 145–157.

Greenspan, R. J., Tononi, G., Cirelli, C., and Shaw, P. J. (2001). Sleep and the fruit fly. *Trends Neurosci.* **24,** 142–145.

Griffith, L. C., Verselis, L. M., Aitken, K. M., Kyriacou, C. P., Danho, W., and Greenspan, R. J. (1993). Inhibition of calcium/calmodulin-dependent protein kinase in *Drosophila* disrupts behavioral plasticity. *Neuron* **10,** 501–509.

Gu, Y. Z., Hogenesch, J. B., and Bradfield, C. A. (2000). The PAS superfamily: Sensors of environmental and developmental signals. *Annu. Rev. Pharmacol. Toxicol.* **40,** 519–561.

Guantieri, V., Pepe, A., Zordan, M., Kyriacou, C. P., Costa, R., and Tamburro, A. M. (1999). Different *period* gene repeats take 'turns' at fine-tuning the circadian clock. *Proc. R. Soc. Lond. B* **266,** 2283–2288.

Gvakharia, B. O., Kilgore, J. A., Bebas, P., and Giebultowicz, J. M. (2000). Temporal and spatial expression of the *period* gene in the reproductive system of the codling moth. *J. Biol. Rhythms* **15,** 4–12.

Hall, J. C. (1995). Tripping along the trail to the molecular mechanisms of biological clocks. *Trends Neurosci.* **18,** 230–240.

Hall, J. C. (1997). Circadian pacemakers blowing hot & cold—but they're clocks, not thermometers. *Cell* **90,** 9–12.

Hall, J. C. (1998a). Genetics of biological rhythms in *Drosophila*. *Adv. Genet.* **38,** 135–184.

Hall, J. C. (1998b). Molecular neurogenetics of biological rhythms. *J. Neurogenet.* **12,** 115–181.

Hall, J. C. (2000). Cryptochrome: Sensory reception, transduction and clock functions subserving circadian systems. *Curr. Opin. Neurobiol.* **10,** 456–466.

Hall, J. C., and Kyriacou, C. P. (1990). Genetics of biological rhythms in *Drosophila*. *Adv. Insect Physiol.* **22,** 221–298.

Hamblen, M., Zehring, A. A., Kyriacou, C. P., Reddy, P., Yu, Q., Wheeler, D. A., Zwiebel, L. J., Konopka, R. J., Rosbash, M., and Hall, J. C. (1986). Germ-line transformation involving DNA from the *period* locus in *Drosophila melanogaster*: Overlapping genomic fragments that restore circadian and ultradian rhythmicity to *per[0]* and *per[−]* mutants. *J. Neurogenet.* **3**, 249–291.

Hamblen, M. J., White, N. E., Emery, P. T. J., Kaiser, K., and Hall, J. C. (1998). Molecular and behavioral analysis of four *period* mutants in *Drosophila melanogaster* encompassing extreme short, novel long, and unorthodox arrhythmic types. *Genetics* **149**, 165–178.

Hamblen-Coyle, M., Konopka, R. J., Zwiebel, L. J., Colot, H. V., Dowse, H. B., Rosbash, M., and Hall, J. C. (1989). A new mutation at the *period* locus of *Drosophila melanogaster* with some novel effects on circadian rhythms. *J. Neurogenet.* **5**, 229–256.

Hamblen-Coyle, M. J., Wheeler, D. A., Rutila, J. E., Rosbash, M., and Hall, J. C. (1992). Behavior of period-altered circadian rhythm mutants of *Drosophila* in light:dark cycles. *J. Insect Behav.* **5**, 417–446.

Handler, A. M., and Konopka, R. J. (1979). Transplantation of a circadian pacemaker in *Drosophila*. *Nature* **279**, 236–238.

Hao, H. P., Allen, D. L., and Hardin, P. E. (1997). A circadian enhancer mediates PER-dependent mRNA cycling in *Drosophila melanogaster*. *Mol. Cell. Biol.* **17**, 3687–3693.

Hao, H., Glossop, N. R. J., Lyons, L., Qui, J., Morrish, B., Cheng, Y., Helfrich-Förster, C., and Hardin, P. E. (1999). A 69 bp circadian regulatory sequence (CRS) mediates *per*-like developmental, spatial, and circadian expression and behavioral rescue in *Drosophila*. *J. Neurosci.* **19**, 987–994.

Hardeland, R. (1972). Species differences in the diurnal rhythmicity of courtship behaviour within the *melanogaster* group of the genus *Drosophila*. *Anim. Behav.* **20**, 170–174.

Hardin, P. E. (1994). Analysis of *period* mRNA cycling in *Drosophila* head and body tissues indicates that body oscillators behave differently from head oscillators. *Mol. Cell. Biol.* **4**, 7211–7218.

Hardin, P. E., Hall, J. C., and Rosbash, M. (1990). Feedback of the *Drosophila period* gene product on circadian cycling of its messenger RNA levels. *Nature* **343**, 536–540.

Hardin, P. E., Hall, J. C., and Rosbash, M. (1992a). Behavioral and molecular analyses suggest that circadian output is disrupted by *disconnected* mutants in *D. melanogaster*. *EMBO J.* **11**, 1–6.

Hardin, P. E., Hall, J. C., and Rosbash, M. (1992b). Circadian oscillations in *period* gene mRNA levels are transcriptionally regulated. *Proc. Natl. Acad. Sci. USA* **89**, 11711–11715.

Hassan, J., Busto, M., Iyengar, B., and Campos, A. R. (2000). Behavioral characterization and genetic analysis of the *Drosophila melanogaster* larval response to light as revealed by a novel individual assay. *Behav. Genet.* **30**, 59–69.

Hediger, M., Niessen, M., Wimmer, E. A., Dubendorfer, A., and Bopp, D. (2001). Genetic transformation of the housefly *Musca domestica* with the lepidopteran derived transposon *piggyBac*. *Insect Mol. Biol.* **10**, 113–119.

Hege, D. M., Stanewsky, R., Hall, J. C., and Giebultowicz, J. M. (1997). Rhythmic expression of a PER-reporter in the Malpighian tubules of decapitated *Drosophila*: evidence for a brain-independent circadian clock. *J. Biol. Rhythms* **12**, 300–308.

Heilig, J. S., Freeman, M., Laverty, T., Lee, K. J., Campos, A. R., Rubin, G. M., and Steller, H. (1991). Isolation and characterization of the *disconnected* gene of *Drosophila melanogaster*. *EMBO J.* **10**, 809–815.

Heisenberg, M., Borst, A., Wagner, S., and Byers, D. (1985). *Drosophila* mushroom body mutants are deficient in olfactory learning. *J. Neurogenet.* **2**, 1–30.

Helfrich, C. (1986). Role of the optic lobes in the regulation of the locomotor activity rhythms in *Drosophila melanogaster*: behavioral analysis of neural mutants. *J. Neurogenet.* **3**, 321–343.

Helfrich, C., and Engelmann, W. (1983). Circadian rhythm of the locomotor activity in *Drosophila melanogaster* and its mutants "sine oculis" and "small optic lobes". *Physiol. Entomol.* **8**, 257–272.

Helfrich, C., and Engelmann, W. (1987). Evidences for circadian rhythmicity in the *per[0]* mutant of *Drosophila melanogaster*. *Z. Naturforsch.* **42c**, 1335–1338.

Helfrich-Förster, C. (1995). The period clock gene is expressed in CNS neurons which also produce a neuropeptide that reveals the projections of circadian pacemaker cells within the brain of *Drosophila melanogaster. Proc. Natl. Acad. Sci. USA* **92,** 612–616.

Helfrich-Förster, C. (1996). *Drosophila* rhythms: from brain to behavior. *Semin. Cell Dev. Biol.* **7,** 791–802.

Helfrich-Förster, C. (1997a). Photic entrainment of *Drosophila*'s activity rhythm occurs via retinal and extraretinal pathways. *Biol. Rhythms Res.* **28**(Suppl.), 119.

Helfrich-Förster, C. (1997b). Development of pigment-dispersing hormone-immunoreactive neurons in the nervous system of *Drosophila melanogaster. J. Comp. Neurol.* **380,** 335–354.

Helfrich-Förster, C. (1998). Robust circadian rhythmicity of *Drosophila melanogaster* requires the presence of lateral neurons: A brain–behavioral study of *disconnected* mutants. *J. Comp. Physiol. A* **182,** 435–453.

Helfrich-Förster, C. (2000). Differential control of morning and evening components in the activity rhythm of *Drosophila melanogaster*—sex-specific differences suggest a different quality of activity. *J. Biol. Rhythms* **15,** 135–154.

Helfrich-Förster, C. (2001). The activity rhythm of *Drosophila melanogaster* is controlled by a dual oscillator system. *J. Insect Physiol.* **47,** 877–887.

Helfrich-Förster, C., and Homberg, U. (1993). Pigment-dispersing hormone-immunoreactive neurons in the nervous system of wild-type *Drosophila melanogaster* and of several mutants with altered circadian rhythmicity. *J. Comp. Neurol.* **337,** 177–190.

Helfrich-Förster, C., Stengl, M., and Homberg, U. (1998). Organization of the circadian system in insects. *Chronobiol. Int.* **15,** 567–594.

Helfrich-Förster, C., Täuber, M., Park, J. H., Mühlig-Versen, M., Schneuwly, S., and Hofbauer, A. (2000). Ectopic expression of the neuropeptide pigment-dispersing factor alters behavioral rhythms in *Drosophila melanogaster. J. Neurosci.* **20,** 3339–3353.

Helfrich-Förster, C., Winter, C., Hofbauer, A., Hall, J. C., and Stanewsky, R. (2001). The circadian clock of fruit flies is blind after elimination of all known photoreceptors. *Neuron* **30,** 249–261.

Helfrich-Förster, C., Wülf, J., and De Belle, J. S. (2002). Mushroom body influence on locomotor activity and circadian rhythms in *Drosophila melanogaster. J. Neurogenet.* **16,** 73–109.

Hendricks, J. C., Finn, S. M., Panckeri, K. A., Chavkin, J., Williams, J. A., Sehgal, A., and Pack, A. I. (2000). Rest in *Drosophila* is a sleep-like state. *Neuron* **25,** 129–138.

Hesterlee, S., and Morton, D. B. (1996). Insect physiology: the emerging story of ecdysis. *Curr. Biol.* **6,** 648–650.

Hilton, H., and Hey, J. (1996). DNA sequence variation at the *period* locus reveals the history of species and speciation events in the *Drosophila virilis* group. *Genetics* **144,** 1015–1025.

Hintermann, E., Grieder, N. C., Amherd, R., Brodbeck, D., and Meyer, U. A. (1996). Cloning of an arylalkylamine *N*-acetyltransferase (aaNAT1) from *Drosophila melanogaster* expressed in the nervous system and the gut. *Proc. Natl. Acad. Sci. USA* **93,** 12315–12320.

Hinton, B. T., and Palladino, M. A. (1995). Epididymal epithelium: Its contribution to the formation of a luminal fluid microenvironment. *Microsc. Res. Tech.* **30,** 67–81.

Hodgetts, R. B., and Konopka, R. J. (1973). Tyrosine and catecholamine metabolism in wild-type *Drosophila melanogaster* and a mutant, *ebony. J. Insect Physiol.* **19,** 1211–1220.

Hofbauer, A., and Buchner, E. (1989). Does *Drosophila* have seven eyes? *Naturwissenschaften* **76,** 335–336.

Hoffman, E. C., Reyes, H., Chu, F.-F., Sander, F., Conley, L. H., Brooks, B. A., and Hankinson, O. (1991). Cloning of a factor required for the activity of the Ah (dioxin) receptor. *Science* **252,** 954–958.

Horodyski, F. M. (1996). Neuroendocrine control of insect ecdysis by eclosion hormone. *J. Insect Physiol.* **42,** 917–924.

Horodyski, F. M., Riddiford, L. M., and Truman, J. W. (1989). Isolation and expression of the eclosion hormone gene from the tobacco hornworm, *Manduca sexta. Proc. Natl. Acad. Sci. USA* **86,** 8123–8127.

Horodyski, F. M., Ewer, J., Riddiford, L. M., and Truman, J. W. (1993). Isolation, characterization and expression of the eclosion hormone gene of *Drosophila melanogaster. Eur. J. Biochem.* **215,** 221–228.

Hotta, Y., and Benzer, S. (1969). Abnormal electroretinograms in visual mutants of *Drosophila. Nature* **222,** 354–356.

Hotta, Y., and Keng, Z. C. (1984). Genetic dissection of larval photoreceptors in *Drosophila. In* "Animal behavior: Neurophysiological and Ethological Approaches" (K. Aoki, ed.), pp. 49–60. Springer-Verlag, Berlin.

Huang, Z. J., Edery, I., and Rosbash, M. (1993). PAS is a dimerization domain common to *Drosophila* Period and several transcription factors. *Nature* **364,** 259–261.

Huang, Z. J., Curtin, K. D., and Rosbash, M. (1995). PER protein interactions and temperature compensation of a circadian clock in *Drosophila. Science* **267,** 1169–1172.

Hunter-Ensor, M., Ousley, A., and Sehgal, A. (1996). Regulation of the Drosophila protein timeless suggests a mechanism for resetting the circadian clock by light. *Cell* **84,** 677–685.

Inoue, B., Shimoda, M., Nishinokubi, I., Siomi, M. C., Okamura, M., Nakamura, A., Kobayashi, S., Ishida, N., and Siomi, H. (2002). A role for the *Drosophila* fragile X-related gene in circadian output. *Curr. Biol.* **12,** 1331–1335.

Ishikawa, T., Matsumoto, A., Kato, T. Jr., Togashi, S., Ryo, H., Ikenaga, M., Todo, T., Ueda, R., and Tanimura, T. (1999). DCRY is a *Drosophila* photoreceptor protein implicated in light entrainment of circadian rhythm. *Genes Cells* **4,** 57–65.

Ivanchenko, M., Stanewsky, R., and Giebultowicz, J. M. (2001). Circadian photoreception in *Drosophila:* Functions of cryptochrome in peripheral and central clocks. *J. Biol. Rhythms* **16,** 205–215.

Jackson, F. R. (1978). Autonomous timer in Malpighian tubules. *Science* **200,** 1186.

Jackson, F. R. (1982). A genetic analysis of temporally-programmed behavior in *Drosophila melanogaster:* circadian rhythms, ultradian rhythms, and the role of biological clocks in learning. Ph.d. thesis, University of California at Los Angeles, Los Angeles, CA.

Jackson, F. R. (1983). The isolation of biological rhythm mutations on the autosomes of *Drosophila melanogaster. J. Neurogenet.* **1,** 3–15.

Jackson, F. R., and Newby, L. M. (1993). Products of the *Drosophila miniature-dusky* gene complex function in circadian rhythmicity and wing development. *Comp. Biochem. Physiol.* **104A,** 749–756.

Jackson, F. R., and Schroeder, A. J. (2001). A timely expression profile. *Dev. Cell* **1,** 730–731.

Jackson, F. R., Gailey, D. A., and Siegel, R. W. (1983). Biological rhythm mutations affect an experience-dependent modification of male courtship behavior in *Drosophila melanogaster. J. Comp. Physiol. A* **151,** 545–552.

Jackson, F. R., Bargiello, T. A., Yun, S.-H., and Young, M. W. (1986). Product of *per* locus of *Drosophila* shares homology with proteoglycans. *Nature* **320,** 185–188.

James, A. A., Ewer, J., Reddy, P., Hall, J. C., and Rosbash, M. (1986). Embryonic expression of the *period* clock gene in the central nervous system of *Drosophila melanogaster. EMBO J.* **5,** 2313–2320.

Jan, Y.-N., and Jan, L. Y. (2000). Polarity in cell division: What frames thy fearful asymmetry? *Cell* **100,** 599–602.

Jauch, E., Melzig, J., Brkulj, M., and Raabe, T. (2002). *In vivo* functional analysis of *Drosophila* protein kinase CK2 β subunit. *Genes* **298,** 29–39.

Johnson, C. H., and Hastings, J. W. (1986). The elusive mechanism of the circadian clock. *Am. Sci.* **74,** 29–36.

Joshi, D. S. (1999). Latitudinal variation in locomotor activity rhythm in adult *Drosophila ananassae. Can. J. Zool.* **77,** 865–870.

Joshi, S., Hodgar, R., Kanojia, M., Chatale, B., Parihar, V., and Joshi, D. S. (2002). Mutations for activity level in *Drosophila jambulina* perturbed its pacemaker that controls circadian eclosion rhythm. *Naturwissenschaften* **89**, 67–70.

Kamito, T., Tanaka, H., Sato, B., Nagasaw, H., and Suzuki, A. (1992). Nucleotide sequence of cDNA for the eclosion hormone of the silkworm, *Bombyx mori*, and the expression in a brain. *Biochem. Biophys. Res. Commun.* **182**, 514–519.

Kaneko, M. (1998). Neural substrates of *Drosophila* rhythms revealed by mutants and molecular manipulations. *Curr. Opin. Neurobiol.* **8**, 652–658.

Kaneko, M. (1999). Neural substrates of circadian rhythms in developing and adult *Drosophila*, Ph.d. thesis. Brandeis University, Waltham, MA.

Kaneko, M., and Hall, J. C. (2000). Neuroanatomy of cells expressing clock genes in *Drosophila*: Transgenic manipulation of the *period* and *timeless* genes to mark the perikarya of circadian pacemaker neurons and their projections. *J. Comp. Neurol.* **422**, 66–94.

Kaneko, M., Helfrich-Förster, C., and Hall, J. C. (1997). Spatial and temporal expression of the *period* and *timeless* genes in the developing nervous system of *Drosophila*: newly identified pacemaker candidates and novel features of clock gene product cycling. *J. Neurosci.* **17**, 6745–6760.

Kaneko, M., Hamblen, M. J., and Hall, J. C. (2000a). Involvement of the *period* gene in developmental time-memory: Effect of the *per^{Short}* mutation on phase shifts induced by light pulses delivered to *Drosophila* larvae. *J. Biol. Rhythms* **15**, 13–30.

Kaneko, M., Park, J. H., Cheng, Y., Hardin, P. E., and Hall, J. C. (2000b). Disruption of synaptic transmission or clock-gene-product oscillations in circadian pacemaker cells of *Drosophila* cause abnormal behavioral rhythms. *J. Neurobiol.* **43**, 207–233.

Kawakami, A., Kataoka, H., Oka, T., Mizoguchi, A., Kimura-Kawakami, M., Adachi, T., Iwami, M., Nagasawa, H., Suzuki, A., and Ishizaki, H. (1990). Molecular cloning of the *Bombyx mori* prothoracicotropic hormone. *Science* **247**, 1333–1335.

Kilduff, T. S. (2000). What rest in flies can tell us about sleep in mammals. *Neuron* **26**, 295–298.

Kim, E. Y., Bae, K., Ng, F. S., Glossop, N. R. J., Hardin, P. E., and Edery, I. (2002). *Drosophila* CLOCK protein is under posttranscriptional control and influences light-induced activity. *Neuron* **34**, 69–81.

Kim, A. J., Cha, G. H., Kim, K., Gilbert, L. I., and Lee, C. C. (1997). Purification and characterization of the prothoracicotropic hormone of *Drosophila melanogaster*. *Proc. Natl. Acad. Sci. USA* **94**, 1130–1135.

Kimura, M. T., and Yoshida, T. (1995). A genetic analysis of photoperiodic reproductive diapause in *Drosophila triauraria*. *Physiol. Entomol.* **20**, 253–256.

King, D. P., Zhao, Y., Sangoram, A. M., Wilsbacher, L. D., Taaka, M., Antoch, M. P., Steeves, T. D. L., Vitaterna, M. H., Kornhauser, J. M., Lowrey, P. L., Turek, F. W., and Takahashi, J. S. (1997). Positional cloning of the mouse circadian *Clock* gene. *Cell* **89**, 641–653.

Klarsfeld, A., and Rouyer, F. (1998). Effects of circadian mutations and LD periodicity on the life span of *Drosophila melanogaster*. *J. Biol. Rhythms* **13**, 471–478.

Klemm, E., and Ninnemann, H. (1976). Detailed action spectra for the delay phase shift in pupal emergence of *Drosophila pseudoobscura*. *Photochem. Photobiol.* **24**, 364–371.

Kliman, R. M., and Hey, J. (1993). DNA sequence variation at the *period* locus within and among species of the *Drosophila melanogaster* complex *Genetics* **133**, 375–387.

Kloss, B., Price, J. L., Saez, L., Blau, J., Rothenflluh, A., and Young, M. W. (1998). The *Drosophila* clock gene *double-time* encodes a protein closely related to human casein kinase Iε. *Cell* **94**, 97–107.

Kloss, B., Rothenfluh, A., Young, M. W., and Saez, L. (2001). Phosphorylation of PERIOD is influenced by cycling physical associations of DOUBLE-TIME, PERIOD, and TIMELESS in the *Drosophila* clock. *Neuro* **30**, 699–706.

Kolhekar, A. S., Roberts, M. S., Jiang, N., Johnson, R. C., Mains, R. E., Eipper, B. A., and Taghert, P. H. (1997). Neuropeptide amidation in *Drosophila*: separate genes encode the two enzymes catalyzing amidation. *J. Neurosci.* **17**, 1363–1376.

Konopka, R. J. (1979). Genetic dissection of the *Drosophila* circadian system. *Fed. Proc.* **38,** 2602–2605.

Konopka, R. J. (1987). Neurogenetics of *Drosophila* circadian rhythms. In "Evolutionary Genetics of Invertebrate Behavior" (M. D. Huettel, ed.), pp. 215–221. Plenum, New York.

Konopka, R. J. (1988). A variegating long-period clock mutant of *Drosophila melanogaster. Life Sci. Adv. (Genet.)* **7,** 39–41.

Konopka, R. J., and Benzer, S. (1971). Clock mutants of *Drosophila melanogaster. Proc. Natl. Acad. Sci. USA* **68,** 2112–2116.

Konopka, R. J., and Orr, D. (1980). Effects of a clock mutation on the subjective day—Implications for a membrane model of the *Drosophila* circadian clock. In "Development and Neurobiology of *Drosophila*" (O. Siddiqi, P. Babu, L. M. Hall, and J. C. Hall, eds.), pp. 409–415. Plenum, New York.

Konopka, R. J., and Wells, S. (1980). *Drosophila* clock mutations affect the morphology of a brain neurosecretory cell group. *J. Neurobiol.* **11,** 411–415.

Konopka, R., Wells, S., and Lee, T. (1983). Mosaic analysis of a *Drosophila* clock mutant. *Mol. Gen. Genet.* **190,** 284–288.

Konopka, R. J., Pittendrigh, C., and Orr, D. (1989). Reciprocal behaviour associated with altered homeostasis and photosensitivity of *Drosophila* clock mutants. *J. Neurogenet.* **6,** 1–10.

Konopka, R. J., Smith, R. F., and Orr, D. (1991). Characterization of *Andante*, a new *Drosophila* clock mutant, and its interactions with other clock mutants. *J. Neurogenet.* **7,** 103–114.

Konopka, R. J., Hamblen-Coyle, M. J., Jamison, C. F., and Hall, J. C. (1994). An ultrashort clock mutation at the *period* locus of *Drosophila melanogaster* that reveals some new features of the fly's circadian system. *J. Biol. Rhythms.* **9,** 189–216.

Konopka, R. J., Kyriacou, C. P., and Hall, J. C. (1996). Mosaic analysis in the *Drosophila* CNS of circadian and courtship-song rhythms affected by a *period* clock mutation. *J. Neurogenet.* **11,** 117–139.

Kornmann, B., Preitner, N., Rifat, D., Fleury-Olela, F., and Schibler, U. (2001). Analysis of circadian liver expression by ADDER, a highly sensitive method for display of differentially expressed mRNAs. *Nucleic Acid Res.* **29,** E51–E61.

Kostál, V., and Shimada, K. (2001). Malfunction of circadian clock in the *non-periodic-diapause* mutants of the drosophilid fly, *Chymomyza costata. J. Insect Physiol.* **47,** 1269–1274.

Kostál, V., Noguchi, H., Shimada, K., and Hayakawa, Y. (2000). Circadian component influences the photoperiodic induction of diapause in a drosophilid fly, *Cymomyza costata. J. Insect Physiol.* **46,** 887–896.

Krishnan, B., Dryer, S. E., and Hardin, P. E. (1999). Circadian rhythms in olfactory responses of *Drosophila melanogaster. Nature* **400,** 375–378.

Krishnan, B., Levine, J. D., Lynch, K. S., Dowse, H. B., Funes, P., Hall, J. C., Hardin, P. E., and Dryer, S. E. (2001). A novel role for cryptochrome in a *Drosophila* circadian oscillator. *Nature* **411,** 313–317.

Kyriacou, C. P., and Hall, J. C. (1980). Circadian rhythm mutations in *Drosophila melanogaster* affect short-term fluctuations in the male's courtship song. *Proc. Natl. Acad. Sci. USA* **77,** 6929–6933.

Kyriacou, C. P., and Hall, J. C. (1982). The function of courtship song rhythms in *Drosophila. Anim. Behav.* **30,** 794–801.

Kyriacou, C. P., and Hall, J. C. (1984). Learning and memory mutations impair acoustic priming of mating behaviour in *Drosophila. Nature* **308,** 62–65.

Kyriacou, C. P., and Hall, J. C. (1985). Action potential mutations stop a biological clock in *Drosophila. Nature* **314,** 171–173.

Kyriacou, C. P., and Hall, J. C. (1986). Interspecific genetic control of courtship song production and reception in *Drosophila. Science* **232,** 494–497.

Kyriacou, C. P., and Hall, J. C. (1989). Spectral analysis of *Drosophila* courtship song rhythms. *Anim. Behav.* **37,** 850–859.

Kyriacou, C. P., and Hastings, M. (2001). Keystone clocks (molecular clocks: regulation of circadian behavioral rhythms). *Trends Neurosci.* **24,** 434–435.

Kyriacou, C. P., and Rosato, E. (2000). Squaring up the E-box. *J. Biol. Rhythms* **15,** 483–490.

Kyriacou, C. P., Burnet, B., and Connolly, K. (1978). The behavioural basis of over-dominance in competitive mating success at the *ebony* locus of *Drosophila melanogaster. Anim. Behav.* **26,** 1194–1206.

Kyriacou, C. P., Oldroyd, M., Wood, J., Sharp, M., and Hill, M. (1990a). Clock mutations alter developmental timing in *Drosophila. Heredity* **64,** 395–401.

Kyriacou, C. P., van den Berg, M. J., and Hall, J. C. (1990b). *Drosophila* courtship song cycles in normal and *period* mutant males revisited. *Behav. Genet.* **20,** 617–644.

Kyriacou, C. P., Greenacre, M. L., Thackeray, J. R., and Hall, J. C. (1993). Genetic and molecular analysis of song rhythms in *Drosophila. In* "Molecular Genetics of Biological Rhythms" (M. W. Young, ed.), pp. 171–193. Marcel Dekker, New York.

Lachaise, D., Harry, M., Solignac, M., Lemeunier, F., Benassi, V., and Cariou, M. L. (2000). Evolutionary novelties in islands: *Drosophila santomea,* a new *melanogaster* sister species from Sao Tome. *Proc. R. Soc. Lond. B* **267,** 1487–1495.

Lakkis, M. M., and Tennekoon, G. I. (2001). Neurofibromatosis type 1: II. Answers from animal models. *J. Neurosci. Res.* **65,** 191–194.

Lankinen, P. (1993a). North–south differences in circadian eclosion rhythm in European populations of *Drosophila subobscura. Heredity* **71,** 210–218.

Lankinen, P. (1993b). Characterization of *linne,* a new autosomal eclosion rhythm mutant in *Drosophila melanogaster. Behav. Genet.* **23,** 359–367.

Lankinen, P., and Riihimaa, A. J. (1992). Weak circadian eclosion rhythmicity in *Chymomyza costata* (Diptera: Drosophilidae), and its independence of diapause type. *J. Insect Physiol.* **38,** 803–811.

Lankinen, P., and Riihimaa, A. J. (1997). Effects of temperature on weak eclosion rhythmicity in *Chymomyza costata* (Diptera: Drosophilidae). *J. Insect Physiol.* **43,** 251–260.

Lee, C. G., Parikh, V., Itsukaichi, T., Bae, K., and Edery, I. (1996). Resetting the *Drosophila* clock by photic regulation of PER and a PER–TIM complex. *Science* **271,** 1740–1744.

Lee, C., Bae, K., and Edery, I. (1998). The *Drosophila* CLOCK protein undergoes daily rhythms in abundance, phosphorylation and interactions with the PER–TIM complex. *Neuron* **21,** 857–867.

Lee, C., Bae, K., and Edery, I. (1999). PER and TIM inhibit the DNA binding activity of a *Drosophila* CLOCK–CYC/dBMAL1 heterodimer without disrupting formation of the heterodimer: a basis for circadian transcription. *Mol. Cell. Biol.* **19,** 5316–5325.

Leloup, J. C., and Goldbeter, A. (2000). Modeling the molecular regulatory mechanism of circadian rhythms in *Drosophila. BioEssays* **22,** 84–93.

Levine, J. D., Casey, C. I., Kalderon, D. D., and Jackson, F. R. (1994). Altered circadian pacemaker functions and cyclic AMP rhythms in the Drosophila learning mutant *dunce. Neuron* **13,** 967–974.

Levine, J. D., Sauman, I., Imbalzano, M., Reppert, S. M., and Jackson, F. R. (1995). Period protein from the giant silkmoth Antheraea pernyi functions as a circadian clock element in *Drosophila melanogaster. Neuron* **15,** 147–157.

Li, X., Borjigin, J., and Snyder, S. H. (1998). Molecular rhythms in the pineal gland. *Curr. Opin. Neurobiol.* **8,** 648–651.

Lin, F.-J., Song, W., Meyer-Bernstein, E., Naidoo, N., and Sehgal, A. (2001). Photic signaling by cryptochrome in the *Drosophila* circadian system. *Mol. Cell. Biol.* **21,** 7287–7294.

Lin, J.-M., Kilman, V., Keegan, K., Paddock, B., Emery-Le, M., Rosbash, M., and Allada, R. (2002a). A role for casein kinase 2α in the *Drosophila* circadian clock. *Nature.* In press.

Lin, Y., Han, M., Shimada, B., Wang, L., Gibler, T. M., Amarakone, A., Awad, T. A., Stormo, G. D., Van Gelder, R. N., and Taghert, P. H. (2002b). Influence of the *period*-dependent circadian clock on diurnal, circadian, and aperiodic gene expression in *Drosophila melanogaster. Proc. Natl. Acad. Sci USA* **99,** 9562–9567.

Liu, X., Lorenz, L., Yu, Q., Hall, J. C., and Rosbash, M. (1988). Spatial and temporal expression of the *period* gene in *Drosophila melanogaster*. *Genes Dev.* **2**, 228–238.

Liu, X., Yu, Q., Huang, Z., Zwiebel, L. J., Hall, J. C., and Rosbash, M. (1991). The strength and periodicity of D. melanogaster circadian rhythms are differentially affected by alterations in *period* gene expression. *Neuron* **6**, 753–766.

Liu, X., Zwiebel, L. J., Hinton, D., Benzer, S., Hall, J. C., and Rosbash, M. (1992). The *period* gene encodes a predominantly nuclear protein in adult *Drosophila*. *J. Neurosci.* **12**, 2735–2744.

Livingstone, M. S. (1981). Two mutations in *Drosophila* differentially affect the synthesis of octopamine, dopamine and serotonin by altering the activities of two different amino acid decarboxylases. *Soc. Neurosci. Abstr.* **7**, 351.

Livingstone, M. S., and Tempel, B. L. (1983). Genetic dissection of monoamine neurotransmitter synthesis in *Drosophila*. *Nature* **303**, 67–70.

Lorenz, L. J., Hall, J. C., and Rosbash, M. (1989). Expression of a *Drosophila* mRNA is under circadian clock control during pupation. *Development* **107**, 869–880.

Lumme, J. (1981). Localization of the genetic unit controlling the photoperiodic adult diapause in *Drosophila littoralis*. *Hereditas* **94**, 241–244.

Lumme, J., and Keränen, L. (1978). Photoperiodic diapause in *Drosophila lummei* Hackman is controlled by an X-chromosomal factor. *Hereditas* **89**, 261–262.

Lumme, J., and Oikarinen, A. (1977). The genetic basis of geographically variable photoperiodic diapause in *Drosophila littoralis*. *Hereditas* **86**, 129–142.

Lumme, J., and Pohjola, L. (1980). Selection against photoperiodic diapause started from monohybrid crosses in *Drosophila littoralis*. *Hereditas* **92**, 377–378.

Lyons, L. C., Darlington, T. K., Kay, S. A., and Hardin, P. E. (2000). Specific sequences outside of the E-box are required for proper *per* expression and behavioral rescue. *J. Biol. Rhythms* **15**, 472–482.

Mack, J., and Engelmann, W. (1981). Circadian control of the locomotor activity in eye mutants of *Drosophila melanogaster*. *J. Interdisciplin. Cycle Res.* **12**, 313–323.

Mahaffey, J. W., Griswold, C. M., and Cao, Q. M. (2001). The *Drosophila* genes *disconnected* and *disco-related* are redundant with respect to larval head development and accumulation of mRNAs from deformed target genes. *Genetics* **157**, 225–236.

Majercak, J., Kalderon, D., and Edery, I. (1997). *Drosophila melanogaster* deficient in protein kinase a manifests behavior-specific arrhythmia but normal clock function. *Mol. Cell. Biol.* **17**, 5915–5922.

Majercak, J., Sidote, D., Hardin, P. E., and Edery, I. (1999). How a circadian clock adapts to seasonal decreases in temperature and day length. *Neuron* **24**, 219–230.

Malpel, S., Klarsfeld, A., and Rouyer, F. (2002). Larval optic nerve and adult extra-retinal photoreceptors sequentially associate with clock neurons during *Drosophila* brain development. *Development* **129**, 1443–1453.

Marrus, S. B., Zeng, H., and Rosbash, M. (1996). Effect of constant light and circadian entrainment of per^S flies: Evidence for light-mediated delay of the negative feedback loop in *Drosophila*. *EMBO J.* **15**, 6877–6886.

Martinek, S., and Young, M. W. (2000). Specific genetic interference with behavioral rhythms in *Drosophila* by expression of inverted-repeats. *Genetics* **156**, 1717–1725.

Martinek, S., Inonog, S., Manoukian, A. S., and Young, M. W. (2001). A role for the segment polarity gene *shaggy*/GSK-3 in the *Drosophila* circadian clock. *Cell* **105**, 769–779.

Matsumoto, A., Motoshige, T., Murata, T., Tomioka, K., Tanimura, T., and Chiba, Y. (1994). Chronobiological analysis of a new clock mutant, *Toki*, in *Drosophila melanogaster*. *J. Neurogenet.* **9**, 141–155.

Matsumoto, A., Tomioka, K., Chiba, Y., and Tanimura, T. (1999). tim^{rit} lengthens circadian period in a temperature-dependent manner through suppression of PERIOD protein cycling and nuclear localization. *Mol. Cell. Biol.* **19**, 4343–4354.

McCabe, C., and Birley, A. (1998). Oviposition in the *period* genotype of *Drosophila melanogaster*. *Chronobiol. Int.* **15,** 119–133.

McClung, C., and Hirsh, J. (1999). The trace amine tyramine is essential for sensitization to cocaine in *Drosophila*. *Curr. Biol.* **9,** 853–860.

McDonald, M. J., and Rosbash, M. (2001). Microarray analysis and organization of circadian gene expression in *Drosophila*. *Cell* **107,** 567–578.

McDonald, M. J., Rosbash, M., and Emery, P. (2001). Wild-type circadian rhythmicity is dependent on closely spaced E boxes in the *Drosophila timeless* promoter. *Mol. Cell. Biol.* **21,** 1207–1217.

McNabb, S. L., Baker, J. D., Agapite, J., Steller, H., Riddiford, L. M., and Truman, J. W. (1997). Disruption of behavioral sequence by targeted death of pepidergic neurons in *Drosophila*. *Neuron* **19,** 813–823.

McNeil, G. P., Zhang, X., Genova, G., and Jackson, F. R. (1998). A molecular rhythm mediating circadian clock output in *Drosophila*. *Neuron* **20,** 297–303.

McNeil, G. P., Zhang, X., Roberts, M., and Jackson, F. R. (1999). Maternal function of a retroviral-type zinc finger protein is essential for *Drosophila* development. *Dev. Genet.* **25,** 387–396.

McNeil, G. P., Schroeder, A. J., Mary, A., Roberts, M. A., and Jackson, F. R. (2001). Genetic analysis of functional domains within the *Drosophila* LARK RNA-binding protein. *Genetics* **159,** 229–240.

Megighian, A., Zordan, M., and Costa, R. (2001). Giant neuron pathway neurophysiological activity in *per*^{0} mutants of *Drosophila melanogaster*. *J. Neurogenet.* **15,** 221–231.

Meinertzhagen, I. A., and Pyza, E. (1996). Daily rhythms in cells of the fly's optic lobe: taking time out from the circadian clock. *Trends Neurosci.* **19,** 285–291.

Meinertzhagen, I. A., and Pyza, E. (1999). Neurotransmitter regulation of circadian structural changes in the fly's visual system. *Microsc. Res. Tech.* **45,** 96–105.

Merienne, K., Pannetier, S., Harel-Bellan, A., and Sassone-Corsi, P. (2001). Mitogen-regulated RSK2–CBP interaction controls their kinase and acetylase activities. *Mol. Cell. Biol.* **21,** 7089–7096.

Montell, C. (1999). Visual transduction in *Drosophila*. *Annu. Rev. Cell Dev. Biol.* **15,** 231–268.

Morales, J., Hiesinger, P. R., Schroeder, A. J., Kume, K., Verstreken, P., Jackson, F. R., Nelson, D. L., and Hassan, B. A. (2002). *Drosophila* fragile X protein, DFXR, regulates neuronal morphology and function in the brain. *Neuron* **34,** 961–972.

Murata, T., Matsumoto, A., Tomioka, K., and Chiba, Y. (1995). *Ritsu*: A rhythm mutant from a natural population of *Drosophila melanogaster*. *J. Neurogenet.* **9,** 239–249.

Myers, M. P., Wagner-Smith, K., Wesley, C. S., Young, M. W., and Sehgal, A. (1995). Positional cloning and sequence analysis of the *Drosophila* clock gene, *timeless*. *Science* **270,** 805–808.

Myers, M. P., Wager-Smith, K., Rothenfluh-Hilfiker, A., and Young, M. W. (1996). Light-induced degradation of TIMELESS and entrainment of the *Drosophila* circadian clock. *Science* **271,** 1736–1740.

Myers, M. P., Rothenfluh, A., Chang, M., and Young, M. W. (1997). Comparison of chromosomal DNA composing *timeless* in *D. melanogaster* and *D. virilis* suggests a new, conserved structure for the TIMELESS protein. *Nucleic Acids Res.* **25,** 4710–4714.

Naidoo, N., Song, W., Hunter-Ensor, M., and Sehgal, A. (1999). A role for the proteosome in the light response of the timeless clock protein. *Science* **285,** 1737–1741.

Nambu, L. R., Lewis, J. O., Wharton, K. A., Jr., and Crews, S. T. (1991). The Drosophila *single-minded* gene encodes a helix–loop–helix protein that acts as a master regulator of CNS midline development. *Cell* **67,** 1157–1167.

Nanda, K. K., and Hamner, K. C. (1958). Studies on the nature of the endogenous rhythm affecting photoperiodic response of Biloxi soy bean. *Bot. Gaz.* **120,** 14–25.

Nässel, D. R. (2000). Functional roles of neuropeptides in the insect central nervous system. *Natur-wissenschaften* **87,** 439–449.

Nässel, D. R., Shiga, S., Mohrherr, C. J., and Rao, K. R. (1993). Pigment-dispersing hormone-like peptide in the nervous system of the flies *Phormia* and *Drosophila*: immunocytochemistry and partial characterization. *J. Comp. Neurol.* **331,** 183–198.

Newby, L. M., and Jackson, F. R. (1991). *Drosophila ebony* mutants have altered circadian activity rhythms but normal eclosion rhythms. *J. Neurogenet.* **7,** 85–101.

Newby, L. M., and Jackson, F. R. (1993). A new biological rhythm mutant of *Drosophila melanogaster* that identifies a gene with an essential embryonic function. *Genetics* **135,** 1077–1090.

Newby, L. M., and Jackson, F. R. (1995). Developmental and genetic mosaic analysis of *Drosophila m-dy* mutants: tissue foci for behavioral and morphogenetic defects. *Dev. Genet.* **16,** 85–93.

Newby, L. M., and Jackson, F. R. (1996). Regulation of a specific circadian clock output pathway by Lark, a putative RNA-binding protein with repressor activity. *J. Neurobiol.* **31,** 117–128.

Newby, L. M., White, L., DiBartolomeis, S. M., Walker, B. J., Dowse, H. B., Ringo, J. M., Khuda, N., and Jackson, F. R. (1991). Mutational analysis of the Drosophila *miniature-dusky* (*m-dy*) locus: effects on cell size and circadian rhythms. *Genetics* **128,** 571–582.

Nielsen, J., Peixoto, A. A., Piccin, A., Costa, R., Kyriacou, C. P., and Chalmers, D. (1994). Big flies, small repeats: the "Thr-Gly" region of the *period* gene in Diptera. *Mol. Biol. Evol.* **11,** 839–853.

Nitabach, M. N., Blau, J., and Holmes, T. C. (2002). Electrical silencing of *Drosophila* pacemaker neurons stops the free-running circadian clock. *Cell* **109,** 485–495.

Noguti, T., Adachi-Yamada, T., Katagiri, T., Kawakami, A., Iwami, M., Ishibashi, J., Kataoka, H., Suzuki, A., Go, M., and Ishizaki, H. (1995). Insect prothoracicotropic hormone: A new member of the vertebrate growth factor superfamily. *FEBS Lett.* **376,** 251–256.

Ohata, K., Nishiyama, H., and Tsukahara, Y. (1998). Action spectrum of the circadian clock photoreceptor in *Drosophila melanogaster*. In "Biological Clocks, Mechanisms and Applications" (Y. Touitou, ed.), pp. 167–170. Elsevier, Amsterdam.

Okada, T., Sakai, T., Murata, T., Kako, K., Sakamoto, K., Ohtomi, M., Katsura, T., and Ishida, N. (2001). Promoter analysis for daily expression of *Drosophila timeless* gene. *Biochem. Biophys. Res. Commun.* **283,** 577–582.

Okamoto, A., Mori, H., and Tomioka, K. (2001). The role of the optic lobe in circadian loco-motor rhythm generation in the cricket, *Gryllus bimaculatus*, with special reference to PDH-immunoreactive neurons. *J. Insect Physiol.* **47,** 889–895.

Okano, S., Kanno, S., Takao, M., Eker, A. P. M., Isono, K., Tsukahara, Y., and Yasui, A. (1999). A putative blue-light receptor from *Drosophila melanogaster*. *Photochem. Photobiol.* **69,** 108–113.

Orr, D. P.-Y. (1982). Genetic analysis of the circadian clock system of *Drosophila melanogaster*, Ph.d. thesis. California Institute of Technology, Pasadena, CA.

Ousley, A., Zafarulluh, K., Chen, Y., Emerson, M., Hickman, L., and Sehgal, A. (1998). Conserved regions of the *timeless* (*tim*) clock gene in Drosophila analyzed through phylogenetic and functional studies. *Genetics* **148,** 815–825.

Page, T. L., Mans, C., and Griffeth, G. (2001). History dependence of circadian pacemaker period in the cockroach. *J. Insect Physiol.* **47,** 1085–1093.

Palmer, C. M., Kendrick, T. E., and Hotchkiss, S. K. (1985). The phototactic response of clock mutant *Drosophila melanogaster*. In "First Symposium in the Biological Sciences" (W. L. Scott and F. L. Strand, eds.), pp. 323–324. New York Academy of Sciences, New York.

Park, J. H., and Hall, J. C. (1998). Isolation and chronobiological analysis of a neuropeptide pigment-dispersing factor gene in *Drosophila melanogaster*. *J. Biol. Rhythms* **13,** 219–228.

Park, J. H., Helfrich-Förster, C., Lee, G., Liu, L., Rosbash, M., and Hall, J. C. (2000a). Differential regulation of circadian pacemaker output by separate clock genes in *Drosophila*. *Proc. Natl. Acad. Sci. USA* **97,** 3608–3613.

Park, S. K., Sedore, S. A., Cronmiller, C., and Hirsh, J. (2000b). Type II cAMP-dependent protein kinase-deficient *Drosophila* are viable but show developmental, cirdadian, and drug response phenotypes. *J. Biol. Chem.* **275,** 20588–20596.

Park, Y., Zitnan, D., Gill, S. S., and Adams, M. E. (1999). Molecular cloning and biological activity of ecdysis-triggering hormones in *Drosophila melanogaster*. *FEBS Lett.* **463,** 133–138.

Peifer, M., Pai, L.-M., and Casey, M. (1994). Phosphorylation of the *Drosophila* adherens junction protein ARMADILLO: Roles for WINGLESS and ZESTE-WHITE 3 kinase. *Dev. Biol.* **166,** 543–556.

Peixoto, A. A. (2002). Evolutionary behavioral genetics in *Drosophila*. *Adv. Genet.* **47,** 117–150.

Peixoto, A. A., Costa, R., Wheeler, D. A., Hall, J. C., and Kyriacou, C. P. (1992). Evolution of the threonine–glycine repeat region of the *period* gene of the *melanogaster* subgroup. *J. Mol. Evol.* **35,** 411–419.

Peixoto, A. A., Campesan, S., Costa, R., and Kyriacou, C. P. (1993). Molecular evolution of a repetitive region within the *per* gene of *Drosophila*. *Mol. Biol. Evol.* **10,** 127–139.

Peixoto, A. A., Hennessy, J. M., Townson, I., Hasan, G., Rosbash, M., Costa, R., and Kyriacou, C. P. (1998). Molecular coevolution within a *Drosophila* clock gene. *Proc. Natl. Acad. Sci. USA* **95,** 4475–4480.

Perrimon, N., and Smouse, D. (1989). Multiple functions of a *Drosophila* homeotic gene, *zeste-white 3*, during segmentation and neurogenesis. *Dev. Biol.* **135,** 287–305.

Petersen, G., Hall, J. C., and Rosbash, M. (1988). The *period* gene of *Drosophila* carries species-specific behavioral instructions. *EMBO J.* **7,** 3939–3947.

Petri, B., and Stengl, M. (1997). Pigment dispersing hormone shifts the phase of the circadian pacemaker of the cockroach *Leucophaea maderae*. *J. Neurosci.* **17,** 4087–4093.

Petri, B., and Stengl, M. (2001). Phase response curves of a molecular model oscillator: implications for mutual coupling of paired oscillators. *J. Biol. Rhythms* **16,** 125–141.

Pflugfelder, G. O., and Heisenberg, M. (1995). *optomotor-blind* of *Drosophila melanogaster*: a neurogenetic approach to optic lobe development and optomotor behavior. *Comp. Biochem. Physiol.* **110A,** 185–202.

Piccin, A., Couchman, M., Clayton, J. D., Chalmers, D., Costa, R., and Kyriacou, C. P. (2000). The clock gene *period* of the housefly, *Musca domestica*, rescues behavioral rhythmicity in *Drosophila melanogaster*; evidence for intermolecular coevolution. *Genetics* **154,** 747–758.

Pittendrigh, C. S. (1974). Circadian oscillations in cells and the circadian organization of multicellular systems. *In* "The Neurosciences Third Study Program" (F. O. Schmitt and F. G. Worden, eds.), pp. 437–458. MIT Press, Cambridge, MA.

Pittendrigh, C. S. (1981). Circadian organization and the photoperiodoc phenomena. *In* "Biological Blocks in Seasonal Reproductive Cycles" (B. K. Follett and D. E. Follett, eds.), pp. 1–35. Wright, Bristol, UK.

Pittendrigh, C. S., and Minis, D. H. (1972). Circadian systems: longevity as a function of circadian resonance in *Drosophila melanogaster*. *Proc. Natl. Acad. Sci. USA* **69,** 1537–1539.

Plautz, J. D., Kaneko, M., Hall, J. C., and Kay, S. A. (1997a). Independent photoreceptive circadian clocks throughout *Drosophila*. *Science* **278,** 1632–1635.

Plautz, J. D., Straume, M., Stanewsky, R., Jamison, C. F., Brandes, C., Dowse, H. B., Hall, J. C., and Kay, S. A. (1997b). Quantitative analysis of *Drosophila period* gene transcription in living animals. *J. Biol. Rhythms* **12,** 204–217.

Power, J., Ringo, J., and Dowse, H. (1995a). The role of light in the initiation of circadian activity rhythms of adult *Drosophila melanogaster*. *J. Neurogenet.* **9,** 227–238.

Power, J. M., Ringo, J. M., and Dowse, H. B. (1995b). The effects of *period* mutations and light on the activity rhythms of *Drosophila melanogaster*. *J. Biol. Rhythms* **10,** 267–280.

Price, J. L., Dembinska, M. E., Young, M. W., and Rosbash, M. (1995). Suppression of PERIOD protein abundance and circadian cycling by the *Drosophila* clock mutation *timeless*. *EMBO J.* **14,** 4044–4049.

Price, J. L., Blau, J., Rothenfluh, A., Abodeely, M., Kloss, B., and Young, M. W. (1998). *double-time* is a new *Drosophila* clock gene that regulates PERIOD protein accumulation. *Cell* **94,** 83–95.

Pyza, E., and Meinertzhagen, I. A. (1995). Day/night size changes in lamina cells are influenced by the *period* gene in *Drosophila*. *Soc. Neurosci. Abstr.* **21,** 408.

Pyza, E., and Meinertzhagen, I. A. (1996). Neurotransmitters regulate rhythmic size changes amongst cells in the fly's optic lobes. *J. Comp. Physiol. A* **178,** 33–45.

Pyza, E., and Meinertzhagen, I. A. (1997). Neurites of *period*-expressing PDH cells in the optic lobe of the housefly exhibit circadian oscillations in morphology. *Eur. J. Neurosci.* **9,** 1784–1788.

Pyza, E., and Meinertzhagen, I. A. (1998). Neurotransmitters alter the numbers of synapses and organelles in the photoreceptor terminals of the housefly, *Musca domestica*. *J. Comp. Physiol. A* **183,** 719–727.

Pyza, E., and Meinertzhagen, I. A. (1999a). The role of clock genes and glial cells in expressing circadian rhythms in the fly's lamina. *In* "Neurobiology of *Drosophila* Abstracts" U. Heberlein and H. Keshishian, eds.), p. 145. Cold Spring Harbor Laboratories, Cold Spring Harbor, NY.

Pyza, E., and Meinertzhagen, I. A. (1999b). Daily rhythmic changes of cell size and shape in the first optic neuropil in *Drosophila melanogaster*. *J. Neurobiol.* **40,** 77–88.

Qiu, J., and Hardin, P. E. (1995). Temporal and spatial expression of an adult cuticle protein gene from *Drosophila* suggests that its protein product may impart some specialized cuticle function. *Dev. Biol.* **167,** 416–425.

Qiu, J., and Hardin, P. E. (1996a). *per* mRNA cycling is locked to lights-off under photoperiodic conditions that support circadian feedback loop function. *Mol. Cell. Biol.* **16,** 4182–4188.

Qiu, J., and Hardin, P. E. (1996b). Developmental state and the circadian clock interact to influence the timing of eclosion in *Drosophila melanogaster*. *J. Biol. Rhythms* **11,** 75–86.

Rachidi, M., Lopes, C., Benichou, J.-C., and Rouyer, F. (1997). Analysis of *period* circadian expression in the *Drosophila* head by *in situ* hybridization. *J. Neurogenet.* **11,** 255–263.

Rajashekhar, K. P., and Singh, R. N. (1994). Neuroarchitecture of the tritocerebrum of *Drosophila melanogaster*. *J. Comp. Neurol.* **349,** 633–645.

Reddy, P., Zehring, W. A., Wheeler, D. A., Pirrotta, V., Hadfield, C., Hall, J. C., and Rosbash, M. (1984). Molecular analysis of the *period* locus in Drosophila melanogaster and identification of a transcript involved in biological rhythms. *Cell* **38,** 701–710.

Reddy, P., Jacquier, A. C., Abovich, N., Petersen, G., and Rosbash, M. (1986). The *period* clock locus of D. melanogaster codes for a proteoglycan. *Cell* **46,** 53–61.

Regier, J. C., Fang, Q. Q., Mitter, C., Peigler, R. S., Friedlander, T. P., and Solis, M. A. (1998). Evolution and phylogenetic utility of the *period* gene in Lepidoptera. *Mol. Biol. Evol.* **15,** 1172–1182.

Reifegerste, R., and Moses, K. (1999). Genetics of epithelial polarity and pattern in the *Drosophila* retina. *BioEssays* **21,** 275–285.

Renn, S. C. P., Park, J. H., Rosbash, M., Hall, J. C., and Taghert, P. H. (1999). A *pdf* neuropeptide gene mutation and ablation of PDF neurons each cause severe abnormalities of behavioral circadian rhythms in *Drosophila*. *Cell* **99,** 791–802.

Rensing, L. (1964). Daily rhythmicity of corpus allatum and neurosecretory cells in *Drosophila melanogaster* (Meig). *Science* **144,** 1586–1587.

Rensing, L., and Hardeland, R. (1967). Zur Wirkung der circadianen Rhythmik auf der Entwicklung von *Drosophila* [About the effects of circadian rhythmicity on the development of *Drosophila*]. *J. Insect Physiol.* **13,** 1547–1568.

Rensing, L., Brucnken, W., and Hardeland, R. (1968). On the genetics of a circadian rhythm in *Drosophila*. *Experientia* **24,** 509–510.

Reppert, S. M., and Weaver, D. R. (2001). Molecular analysis of mammalian circadian rhythms. *Annu. Rev. Physiol.* **63,** 647–676.

Reppert, S. M., Tasi, T., Roca, A. L., and Sauman, I. (1994). Cloning of a structural and functional homolog of the circadian clock gene period from the giant silkmoth Antheraea pernyi. *Neuron* **13,** 1167–1176.

Riego-Escovar, J., Raha, D., and Carlson, J. R. (1995). Requirement for a phospholipase C in odor response: overlap between olfaction and vision in *Drosophila*. *Proc. Natl. Acad. Sci. USA* **92**, 2864–2868.

Riihimaa, A. J., and Kimura, M. T. (1988). A mutant strain of *Chymomyza costata* (Diptera: Drosophilidae) insensitive to diapause-inducing action of photoperiod. *Physiol. Entomol.* **13**, 441–445.

Riihimaa, A. J., and Kimura, M. T. (1989). Genetics of the photoperiodic larval diapause in *Chymomyza costata* (Diptera: Drosophilidae). *Hereditas* **110**, 193–200.

Ritchie, M. G., and Kyriacou, C. P. (1994). Reproductive isolation and the *period* gene of *Drosophila*. *Mol. Ecol.* **3**, 595–599.

Ritchie, M. G., Halsey, E. J., and Gleason, J. M. (1999). *Drosophila* song as a species-specific mating signal and the behavioural importance of Kyriacou & Hall cycles in *D. melanogaster* song. *Anim. Behav.* **58**, 649–657.

Robinow, S., and White, K. (1991). Characterization and spatial distribution of the ELAV protein during *Drosophila melanogaster* development. *J. Neurobiol.* **22**, 443–461.

Roche, J. P., Talyn, B. C. P., and Dowse, H. B. (1998). Courtship bout duration in *per* circadian period mutants in *Drosophila melanogaster*. *Behav. Genet.* **28**, 391–394.

Rong, Y. S., and Golic, K. G. (2000). Gene targeting by homologous recombination in *Drosophila*. *Science* **288**, 2013–2018.

Rong, Y. S., and Golic, K. G. (2001). A targeted gene knockout in Drosophila. *Genetics* **157**, 1307–1312.

Rong, Y. S., Titen, S. W., Xie, H. B., Golic, M. M., Bastiani, M., Bandyopadhyay, P., Olivera, B. M., Brodsky, M., Rubin, G. M., and Golic, K. G. (2002). Targeted mutagenesis by homologous recombination in *D. melanogaster*. *Genes Dev.* **16**, 1568–1581.

Rørth, P. (1996). A modular misexpression screen in *Drosophila* detecting tissue-specific phenotypes. *Proc. Natl. Acad. Sci. USA* **93**, 12418–12422.

Rørth, P., Szabo, K., Bailey, A., Laverty, T., Rehm, J., Rubin, G., Weigmann, K., Milan, M., Benes, V., Ansorge, W., and Cohen, S. (1998). Systematic gain-of-function genetics in *Drosophila*. *Development* **125**, 1049–1057.

Rosato, E., Peixoto, A. A., Barbujani, G., Costa, R., and Kyriacou, C. P. (1994). Molecular polymorphism in the *period* gene of *Drosophila simulans*. *Genetics* **138**, 693–707.

Rosato, E., Peixoto, A. A., Gallippi, A., Kyriacou, C. P., and Costa, R. (1996). Mutational mechanisms, phylogeny, and evolution of a repetitive region within a clock gene of *Drosophila melanogaster*. *J. Mol. Evol.* **42**, 392–408.

Rosato, E., Peixoto, A. A., Costa, R., and Kyriacou, C. P. (1997a). Linkage disequilibrium, mutational analysis and natural selection in the repetitive region of the clock gene, *period,* in *Drosophila melanogaster*. *Genet. Res.* **69**, 89–99.

Rosato, E., Trevisan, A., Sandrelli, F., Zordan, M., Kyriacou, C. P., and Costa, R. (1997b). Conceptual translation of *timeless* reveals alternative initiating methionines in *Drosophila*. *Nucleic Acids Res.* **25**, 455–457.

Rosato, E., Codd, V., Mazzotta, G., Piccin, A., Zordan, M., Costa, R., and Kyriacou, C. P. (2001). Light-dependent interactions between *Drosophila* CRY and the clock protein PER mediated by the carboxy terminus of CRY. *Curr. Biol.* **11**, 909–917.

Rosewell, K. L., Siwicki, K. K., and Wise, P. M. (1994). A *period* (*per*)-like protein exhibits daily rhythmicity in the suprachiasmatic nucleus of the rat. *Brain Res.* **659**, 231–236.

Rothenfluh, A., Abodeely, M., Price, J. L., and Young, M. W. (2000a). Isolation and analysis of six *timeless* alleles that cause short- or long-period circadian rhythms in Drosophila. *Genetics* **156**, 665–675.

Rothenfluh, A., Abodeely, M., and Young, M. W. (2000b). Short-period mutations of *per* affect a *double-time*-dependent step in the *Drosophila* circadian clock. *Curr. Biol.* **10**, 1399–1402.

Rothenfluh, A., Young, M. W., and Saez, L. (2000c). A TIMELESS-independent function for PERIOD proteins in the *Drosophila* clock. *Neuron* **26**, 505–514.

Rouyer, F., Rachidi, M., Pikielny, C., and Rosbash, M. (1997). A new gene encoding a putative transcription factor regulated by the *Drosophila* circadian clock. *EMBO J.* **16**, 3944–3954.

Rutila, J. E., Edery, I., Hall, J. C., and Rosbash, M. (1992). The analysis of new short-period circadian rhythm mutants suggests features of *Drosophila melanogaster period* gene function. *J. Neurogenet.* **8**, 101–113.

Rutila, J. E., Zeng, H., Le, M., Curtin, K. D., Hall, J. C., and Rosbash, M. (1996). The *tim*^{SL} mutant of the Drosophila rhythm gene *timeless* manifests allele-specific interactions with *period* gene mutants. *Neuron* **17**, 921–929.

Rutila, J. E., Maltseva, O., and Rosbash, M. (1998a). The *tim*^{SL} mutant affects a restricted portion of the *Drosophila melanogaster* circadian cycle. *J. Biol. Rhythms* **13**, 380–392.

Rutila, J. E., Suri, V., Le, M., So, W. V., Rosbash, M., and Hall, J. C. (1998b). CYCLE is a second bHLH-PAS clock protein essential for circadian rhythmicity and transcription of *Drosophila period* and *timeless*. *Cell* **93**, 805–814.

Saez, L., and Young, M. W. (1988). *In situ* localization of the *per* clock protein during development of *Drosophila melanogaster*. *Mol. Cell. Biol.* **8**, 5378–5385.

Saez, L., and Young, M. W. (1996). Regulated nuclear localization of the *Drosophila* clock proteins PERIOD and TIMELESS. *Neuron* **17**, 911–920.

Saez, L., Young, M. W., Baylies, M. K., Gasic, G., Bargiello, T. A., and Spray, D. C. (1992). Per—No link to gap junctions. *Nature* **360**, 542.

Sakai, T., and Ishida, N. (2001). Circadian rhythms of female mating activity governed by clock genes in *Drosophila*. *Proc. Natl. Acad. Sci. USA* **98**, 9221–9225.

Sarov-Blat, L., So, W. V., Liu, L., and Rosbash, M. (2000). The *Drosophila takeout* gene is a novel link between circadian rhythms and feeding behavior. *Cell* **101**, 647–656.

Sauman, I., and Reppert, S. M. (1996a). Circadian clock neurons in the silkmoth *Antheraea pernyi*: Novel mechanism of Period protein regulation. *Neuron* **17**, 889–900.

Sauman, I., and Reppert, S. M. (1996b). Molecular characterization of prothoracicotropic hormone (PTTH) from the giant silkmoth *Antheraea pernyi*: developmental appearance of PTTH-expressing cells and relation to circadian clock cells in central brain. *Dev. Biol.* **178**, 418–429.

Sauman, I., and Reppert, S. M. (1998). Brain control of embryonic circadian rhythms in the silkmoth *Antheraea pernyi*. *Neuron* **20**, 741–748.

Sauman, I., Tasi, T., Roca, A. L., and Reppert, S. M. (1996). Period protein is necessary for circadian control of egg hatching behavior in the silkmoth Antheraea pernyi. *Neuron* **20**, 901–909.

Sauman, I., Tsai, T., Roca, A. L., and Reppert, S. M. (2000). Erratum. *Neuron* **27** (1), following p. 189.

Saunders, D. S. (1982). "Insect Clocks," 2nd ed., Pergamon Press, Oxford.

Saunders, D. S. (1990). The circadian basis of ovarian diapause regulation in *Drosophila melanogaster*: Is the *period* gene causally involved in photoperiodic time measurement. *J. Biol. Rhythms* **5**, 315–331.

Saunders, D. S., and Gilbert, L. I. (1990). Regulation of ovarian diapause in *Drosophila melanogaster* by photoperiod and moderately low temperature. *J. Insect Physiol.* **36**, 195–200.

Saunders, D. S., Henrich, V. C., and Gilbert, L. I. (1989). Induction of diapause in *Drosophila melanogaster*: Photoperiodic regulation and the impact of arrhythmic clock mutations on time measurement. *Proc. Natl. Acad. Sci. USA* **86**, 3748–3752.

Saunders, D. S., Richard, D. S., Applebaum, S. W., Ma, M., and Gilbert, L. I. (1990). Photoperiodic diapause in *Drosophila melanogaster* involves a block to the juvenile hormone regulation of ovarian maturation. *Gen. Comp. Endocrinol.* **79**, 174–184.

Saunders, D. S., Gillanders, S. W., and Lewis, R. D. (1994). Light-pulse phase response curves for the locomotor activity rhythm in *period* mutants of *Drosophila melanogaster*. *J. Insect Physiol.* **40**, 957–968.

Sawin, E. P., Dowse, H. B., Hamblen-Coyle, M. J., Hall, J. C., and Sokolowski, M. B. (1994). A search for locomotor activity rhythms in *Drosophila melanogaster* larvae. *J. Insect Behav.* **7**, 249–262.

Sawin-McCormack, E. P., Sokolowski, M. B., and Campos, A. R. (1995). characterization and genetic analysis of *Drosophila melanogaster* photobehavior during larval development. *J. Neurogenet.* **10**, 119–135.

Sawyer, L. A., Hennessey, M. J., Peixoto, A. A., Rosato, E., Parkinson, H., Costa, R., and Kyriacou, C. P. (1997). Natural variation in *Drosophila* clock gene and temperature compensation. *Science* **278**, 2117–2120.

Saxena, A., Padmanabha, R., and Glover, C. V. (1987). Isolation and sequencing of cDNA clones encoding alpha and beta subunits of *Drosophila melanogaster* casein kinase II. *Mol. Cell. Biol.* **7**, 3409–3417.

Schopperle, W. M., Holmqvist, M. H., Zhou, Y., Wang, J., Griffith, L. C., Keselman, I., Kusinitz, F., Dagan, D., and Levitan, I. B. (1998). Slob, a novel protein that interacts with the Slowpoke calcium-dependent potassium channel. *Neuron* **20**, 565–573.

Schotland, P., Hunter-Ensor, M., Lawrence, T., and Sehgal, A. (2000). Altered entrainment and feedback loop function effected by a mutant *period* protein. *J. Neurosci.* **20**, 958–968.

Scully, A. L., Zelhof, A. C., and Kay, S. A. (2002). A P element with novel fusion of reporters identifies *regular*, a C_2H_2 zinc-finger gene downstream of the circadian clock. *Mol. Cell. Neurosci.* **19**, 501–514.

Sehgal, A., Man, B., Price, J. L., Vosshall, L. B., and Young, M. W. (1991). New clock mutations in *Drosophila*. *Ann. N. Y. Acad. Sci.* **618**, 1–10.

Sehgal, A., Price, J., and Young, M. W. (1992). Ontogeny of a biological clock in *Drosophila*. *Proc. Natl. Acad. Sci. USA* **89**, 1423–1427.

Sehgal, A., Price, J. L., Man, B., and Young, M. W. (1994). Loss of circadian behavioral rhythms and *per* RNA oscillations in the *Drosophila* mutant *timeless*. *Science* **263**, 1603–1606.

Sehgal, A., Rothenfluh-Hilfiker, A., Hunter-Ensor, M., Chen, Y., Myers, M. P., and Young, M. W. (1995). Rhythmic expression of *timeless*: A basis for promoting circadian cycles in *period* gene autoregulation. *Science* **270**, 808–810.

Selby, C. P., and Sancar, A. (1999). A third member of the photolyase/blue-light photoreceptor family in *Drosophila*: A putative circadian photoreceptor. *Photochem. Photobiol.* **69**, 105–107.

Serikaku, M. A., and O'Tousa, J. E. (1994). *sine oculis* is a homeobox gene required for Drosophila visual system development. *Genetics* **138**, 1137–1150.

Shafer, O. T., Rosbash, M., and Truman, J. W. (2002). Sequential nuclear accumulation of the clock proteins Period and Timeless in the pacemaker neurons of *Drosophila melanogaster*. *J. Neurosci.* **22**, 5946–5954.

Shaw, P. J., Cirelli, C., Greenspan, R. J., and Tononi, G. (2000). Correlates of sleep and waking in *Drosophila melanogaster*. *Science* **287**, 1834–1837.

Shaw, P. J., Tononi, G., Greenspan, R. J., and Robinson, D. F. (2002). Stress response genes protect against lethal effects of sleep deprivation in *Drosophila*. *Nature* **417**, 287–291.

Sheeba, V., Sharma, V. K., Chandrashekaren, M. K., and Joshi, A. (1999a). Effect of different light regimes on pre-adult fitness in *Drosophila melanogaster* populations reared in constant light for over six hundred generations. *Biol. Rhythm Res.* **30**, 424–433.

Sheeba, V., Sharma, V. K., Chandrashekaren, M. K., and Joshi, A. (1999b). Persistence of eclosion rhythm after 600 generations in an aperiodic environment. *Naturwissenschaften* **86**, 448–449.

Shigeyoshi, Y., Meyer-Bernstein, E., Yagita, K., Fu, W., Chen, Y., Takumi, T., Schotland, P., Sehgal, A., and Okamura, H. (2002). Restoration of circadian behavioural rhythms in a period null *Drosophila* mutant (per^{01}) by mammalian period homologues *mPer1* and *mPer2*. *Genes Cells* **7**, 163–171.

Shimada, K. (1999). Genetic linkage analysis of photoperiodic clock genes in *Chymomyza costata* (Diptera: Drosophilidae). *Entomol. Sci.* **183**, 970–972.

Shimizu, T., Miyatake, T., Watari, Y., and Arai, T. (1997). A gene pleiotropically controlling developmental and circadian period in the melon fly, *Bactrocera cucurbitae* (Diptera: Tephritidae). *Heredity* **79**, 600–605.

Shin, H.-S., Bargiello, T. A., Clark, B. R., Jackson, F. R., and Young, M. W. (1985). An unusual coding sequence from a *Drosophila* clock gene is conserved in vertebrates. *Nature* **317**, 445–448.

Sidote, D., and Edery, I. (1999). Heat-induced degradation of PER and TIM in *Drosophila* bearing a conditional allele of the heat shock transcription factor gene. *Chronobiol. Int.* **16**, 519–525.

Sidote, D., Majercak, J., Parikh, V., and Edery, I. (1998). Differential effects of light and heat on the *Drosophila* circadian clock proteins PER and TIM. *Mol. Cell. Biol.* **18**, 2004–2013.

Siegfried, E., Chou, T.-B., and Perrimon, N. (1992). *wingless* signaling acts through *zeste-white 3*, the Drosophila homolog of *glycogen-synthase kinase-3*, to regulate *engrailed* and establish cell fate. *Cell* **71**, 1167–1179.

Simpson, P., and Carteret, C. (1989). A study of *shaggy* reveals spatial domains of expression of *achaete-scute* alleles on the thorax of *Drosophila*. *Development* **106**, 57–66.

Siwicki, K. K., Eastman, C., Petersen, G., Rosbash, M., and Hall, J. C. (1988). Antibodies to the *period* gene product of Drosophila reveal diverse tissue distribution and rhythm changes in the visual system. *Neuron* **1**, 141–150.

Siwicki, K. K., Strack, S., Rosbash, M., Hall, J. C., and Jacklet, J. W. (1989). An antibody to the Drosophila *period* protein recognizes circadian pacemaker neurons in *Aplysia* and *Bulla*. *Neuron* **3**, 51–58.

Siwicki, K. K., Flint, K. K., Hall, J. C., Rosbash, M., and Spray, D. C. (1992a). The *Drosophila period* gene and dye coupling in larval salivary glands: a reevaluation. *Biol. Bull.* **183**, 340–341.

Siwicki, K. K., Schwartz, W. J., and Hall, J. C. (1992b). An antibody to the *Drosophila* period protein labels antigens in the suprachiasmatic nucleus of the rat. *J. Neurogenet.* **8**, 33–42.

Smith, P. H. (1987). Naturally occurring arrhythmicity in eclosion and activity in *Lucilia cuprina*: Its genetic basis. *Physiol. Entomol.* **12**, 99–107.

Smith, R. F., and Konopka, R. J. (1981). Circadian clock phenotypes of chromosome aberrations with a breakpoint at the *per* locus. *Mol. Gen. Genet.* **183**, 243–251.

Smith, R. F., and Konopka, R. J. (1982). Effects of dosage alterations at the *per* locus on the circadian clock of *Drosophila*. *Mol. Gen. Genet.* **185**, 30–36.

So, W. V., and Rosbash, M. (1997). Post-transcriptional regulation contributes to *Drosophila* clock gene mRNA cycling. *EMBO J.* **16**, 7146–7155.

So, W. V., Sarov-Blat, L., Kotarski, C. K., McDonald, M. J., Allada, R., and Rosbash, M. (2000). *takeout*, a novel *Drosophila* gene under circadian clock transcriptional regulation. *Mol. Cell. Biol.* **20**, 6935–6944.

Sokolove, P. G., and Bushell, W. N. (1978). The chi square periodogram: its utility for analysis of circadian rhythms. *J. Theoret. Biol.* **72**, 131–160.

Stanewsky, R., Frisch, B., Brandes, C., Hamblen-Coyle, M. J., Rosbash, M., and Hall, J. C. (1997a). Temporal and spatial expression patterns of transgenes containing increasing amounts of the *Drosophila* clock gene *period* and a *lacZ* reporter: mapping elements of the PER protein involved in circadian cycling. *J. Neurosci.* **17**, 676–696.

Stanewsky, R., Jamison, C. F., Plautz, J. D., Kay, S. A., and Hall, J. C. (1997b). Multiple circadian-regulated elements contribute to cycling *period* gene expression in *Drosophila*. *EMBO J.* **16**, 5006–5018.

Stanewsky, R., Kaneko, M., Emery, P., Beretta, B., Wager-Smith, K., Kay, S. A., Rosbash, M., and Hall, J. C. (1998). The *cry^b* mutation identifies cryptochrome as a circadian photoreceptor in *Drosophila*. *Cell* **95**, 681–692.

Stanewsky, R., Lynch, K. S., Brandes, C., and Hall, J. C. (2002). Mapping of elements involved in regulating normal *period* and *timeless* RNA expression patterns in *Drosophila melanogaster*. *J. Biol. Rhythms* **17**, 293–306.

Stark, W. S., Sapp, R., and Schilly, D. (1998). Rhabdomere turnover and rhodopsin cycle: Maintenance of retinula cells in *Drosophila melanogaster*. *J. Neurocytol.* **17**, 499–509.

Steller, H., Fischbach, K.-F., and Rubin, G. M. (1987). *disconnected*: A locus required for neuronal pathway formation in the visual system of Drosophila. *Cell* **50**, 1139–1153.

Stempfl, T., Vogel, M., Szabo, G., Wülbeck, C., Liu, J., Hall, J. C., and Stanewsky, R. (2002). Identification of circadian-clock regulated enhancers and genes of _Drosophila melanogaster_ by transposon mobilization and luciferase reporting of cyclical gene expression. _Genetics_ **160**, 571–593.

Störtkuhl, K. F., Hovemann, B. T., and Carlson, J. R. (1999). Olfactory adaptation depends on the Trp Ca^{2+} channel in _Drosophila_. _J. Neurosci._ **19**, 4839–4846.

Strack, S., and Jacklet, J. W. (1993). Antiserum to an eye-specific protein identifies photoreceptor and circadian pacemaker neuron projections in _Aplysia_. _J. Neurobiol._ **24**, 552–570.

Suri, V. (2000). Function of TIMELESS and DOUBLETIME in the _Drosophila_ pacemaker mechanism, Ph.d. thesis. Brandeis University, Waltham, MA.

Suri, V., Qian, Z., Hall, J. C., and Rosbash, M. (1998). Evidence that the TIM light response is relevant to light-induced phase shifts in _Drosophila melanogaster_. _Neuron_ **21**, 225–234.

Suri, V., Lanjuin, A., and Rosbash, M. (1999). TIMELESS-dependent positive and negative autoregulation in the _Drosophila_ circadian clock. _EMBO J._ **18**, 675–686.

Suri, V., Hall, J. C., and Rosbash, M. (2000). Two novel _doubletime_ mutants alter circadian properties and eliminate the delay between RNA and protein in _Drosophila_. _J. Neurosci._ **20**, 7547–7555.

Taghert, P. H., Hewes, R. S., Park, J. H., O'Brien, M. A., Han, M., and Peck, M. E. (2001). Multiple amidate neuropeptides are required for normal circadian rhythmicity in _Drosophila_. _J. Neurosci._ **21**, 6673–6686.

Tauber, E., and Kyriacou, C. P. (2001). Insect photopriodism and circadian clocks: models and mechanisms. _J. Biol. Rhythms_ **16**, 381–390.

Tauber, H., and Hardeland, R. (1977). Circadian rhythmicity of tyrosine aminotransferase activity in _Drosophila melanogaster_. _Insect Biochem._ **7**, 503–505.

Taylor, B. L., and Zhulin, I. B. (1999). PAS domains: Internal sensors of oxygen, redox potential, and light. _Microbiol. Mol. Biol. Rev._ **63**, 479–506.

Thackeray, J. R., and Kyriacou, C. P. (1990). Molecular evolution in the _Drosophila yakuba period_ locus. _J. Mol. Evol._ **31**, 389–401.

Toma, D. P., Block, G., Moore, D., and Robinson, G. E. (2000). Changes in _period_ mRNA levels in the brain and division of labor in honey bee colonies. _Proc. Natl. Acad. Sci. USA_ **97**, 6914–6919.

Toma, D. P., White, K. P., Hirsch, J., and Greenspan, R. J. (2002). Identification of genes involved in _Drosophila melanogaster_ geotaxis, a complex behavioral trait. _Nat. Genet._ **31**, 349–353.

Tomioka, K., Uwozumi, K., and Matsumoto, N. (1997). Light cycles given during development affect freerunning _period_ of circadian locomotor rhythm of _period_ mutants in _Drosophila melanogaster_. _J. Insect Physiol._ **43**, 297–305.

Tomioka, K., Sakamoto, M., Harui, Y., Matsumoto, N., and Matsumoto, A. (1998). Light and temperature cooperate to regulate the circadian locomotor rhythm of wild type and _period_ mutants of _Drosophila melanogaster_. _J. Insect Physiol._ **44**, 587–596.

Tomioka, K., Saifullah, A. S. M., and Koga, M. (2001). The circadian clock system of hemimetabolous insects. _In_ "Insect Timing: Circadian Rhythmicity to Seasonality" (D. L. Denlinger, J. Giebultowicz, and D. S. Saunders, eds.), pp. 43–54. Elsevier, Amsterdam.

Truman, J. W. (1992a). The eclosion hormone system of insects. _Prog. Brain Res._ **92**, 361–374.

Truman, J. W. (1992b). Developmental neuroethology of insect metamorphosis. _J. Neurobiol._ **23**, 1404–1422.

Truman, J. W., and Morton, D. B. (1990). The eclosion hormone system: An example of coordination of endocrine activity during the molting cycle of insects. _Prog. Clin. Biol. Res._ **342**, 300–308.

Ueda, H. R., Matsumoto, A., Kawamura, M., Iino, M., Tanimura, T., and Hashimoto, S. (2002). Genome-wide transcriptional orchestration of circadian rhythms in _Drosophila_. _J. Biol. Chem._ **277**, 14048–14052.

Vafopoulou, X., and Steel, C. G. H. (1991). Circadian regulation of synthesis of ecdysteroids by prothoracic glands of the insect _Rhodnius prolixus_: evidence of a dual oscillator system. _Gen. Comp. Endocrinol._ **83**, 27–34.

van den Pol, A. N., and Obrietan, K. (2002). Short circuiting the circadian clock. *Nat. Neurosci.* **5**, 616–618.

Van Doren, M., Ellis, H. M., and Posakony, J. W. (1991). The *Drosophila extramacrochaetae* protein antagonizes sequence-specific DNA binding by *daughterless/achaete-scute* protein complexes. *Development* **113**, 245–255.

Van Doren, M., Powell, P. A., Pasternak, D., Singson, A., and Posakony, J. W. (1992). Spatial regulation of proneural gene activity: auto- and cross-activation of *achaete* is antagonized by *extramacrochaetae*. *Genes Dev.* **6**, 2592–2605.

Van Gelder, R. N., and Krasnow, M. A. (1996). A novel circadianly expressed *Drosophila melanogaster* gene dependent on the *period* gene for its rhythmic expression. *EMBO J.* **15**, 1625–1631.

Van Gelder, R. N., Bae, H., Palazzolo, M. J., and Krasnow, M. A. (1995). Extent and character of circadian gene expression in *Drosophila melanogaster*: identification of twenty oscillating mRNAs in the fly head. *Curr. Biol.* **5**, 1424–1436.

van Swinderen, B., and Hall, J. C. (1995). Analysis of conditioned courtship in *dusky-Andante* rhythm mutants of *Drosophila*. *Learn. Mem.* **2**, 49–61.

Veenstra, J. A. (1994). Isolation and structure of the *Drosophila* corazonin gene. *Biochem. Biophys. Res. Commun.* **204**, 292–296.

Vitaterna, M. H., King, D. P., Chang, A.-M., Kornhauser, J. M., Lowrey, P. L., McDonald, J. D., Dove, W. F., Pinto, L. H., Turek, F. W., and Takahashi, J. S. (1994). Mutagenesis and mapping of a mouse gene essential for circadian behavior. *Science* **264**, 719–725.

Vosshall, L. B., and Young, M. W. (1995). Circadian rhythms in Drosophila can be driven by *period* expression in a restricted group of central brain cells. *Neuron* **15**, 345–360.

Vosshall, L. B., Price, J. L., Sehgal, A., Saez, L., and Young, M. W. (1994). Block in nuclear localization of *period* protein by a second clock mutation, *timeless*. *Science* **263**, 1606–1609.

Wan, L., Dockendorff, T. C., Jongens, T. A., and Dreyfuss, G. (2000). Characterization of dFMR1, a *Drosophila melanogaster* homolog of the fragile X mental retardation protein. *Mol. Cell. Biol.* **20**, 8536–8547.

Wang, G. K., Ousley, A., Darlington, T. K., Chen, D., Chen, Y., Fu, W., Hickman, L. J., Kay, S. A., and Sehgal, A. (2001). Regulation of the cycling of *timeless* (*tim*) RNA. *J. Neurobiol.* **47**, 161–175.

Wang, R. L., and Hey, J. (1996). The speciation history of *Drosophila pseudoobscura* and close relatives: Inferences from DNA sequence variation at the *period* locus. *Genetics* **144**, 1113–1126.

Warman, G. R., Newcomb, R. D., Lewis, R. D., and Evans, C. W. (2000). Analysis of the circadian clock gene *period* in the sheep blow fly *Lucilia cuprina*. *Genet. Res.* **75**, 257–267.

Weaver, D. R. (1998). The suprachiasmatic nucleus: a 25-year retrospective. *J. Biol. Rhythms* **13**, 100–112.

Weiner, J. (1999). Konopka's law. *In* "Time, Love, Memory," pp. 71–141. Knopf, New York.

Weitzel, G., and Rensing, L. (1981). Evidence for cellular circadian rhythms in isolated fluorescent dye-labelled salivary glands of wild type and an arrhythmic mutant of *Drosophila melanogaster*. *J. Comp. Physiol. A* **143**, 229–235.

Wheeler, D. A., Kyriacou, C. P., Greenacre, M. L., Yu, Q., Rutila, J. E., Rosbash, M., and Hall, J. C. (1991). Molecular transfer of a species-specific behavior from *Drosophila simulans* to *Drosophila melanogaster*. *Science* **251**, 1082–1085.

Wheeler, D. A., Hamblen-Coyle, M. J., Dushay, M. S, and and Hall, J. C. (1993). Behavior in light–dark cycles of *Drosophila* mutants that are blind, arrhythmic, or both. *J. Biol. Rhythms* **8**, 67–94.

White, L., Ringo, J., and Dowse, H. (1992a). A circadian clock of *Drosophila*: effects of deuterium oxide and mutations at the *period* locus. *Chronobiol. Int.* **9**, 250–259.

White, L., Ringo, J., and Dowse, H. (1992b). The effects of deuterium oxide and temperature on heart rate in *Drosophila*. *J. Comp. Physiol. B* **162**, 278–283.

Whitmore, D., and Block, G. D. (1996). Cellular aspects of molluskan biochronometry. *Semin. Cell Dev. Biol.* **7**, 781–789.

Williams, J. A, Su, H. S., Bernards, A., Field, J., and Sehgal, A. (2001). A circadian output in *Drosophila* mediated by *Neurofibromatosis-1* and Ras/MAPK. *Science* **293,** 2251–2256.

Williams, K. D., and Sokolowski, M. B. (1993). Diapause in *Drosophila melanogaster*: a genetic analysis. *Heredity* **71,** 312–317.

Winfree, A. T., and Gordon, H. (1977). The photosensitivity of a mutant circadian clock. *J. Comp. Physiol. A* **122,** 87–109.

Wise, S., Davis, N. T., Tyndale, E., Noveral, J., Folwell, M. G., Bedian, V., Emery, I. F., and Siwicki, K. K. (2002). Neuroanatomical studies of *period* gene expression in the hawkmoth, *Manduca sexta*. *J. Comp. Neurol.* **447,** 366–380.

Wood, K. V. (1995). Marker proteins for gene expression. *Curr. Opin. Biotech.* **6,** 50–58.

Yang, H. Q., Wu, Y. J., Tang, R. H., Liu, D., Liu, Y., and Cashmore, A. R. (2000). The C termini of *Arabidopsis* cryptochromes mediate a constitutive light response. *Cell* **103,** 815–827.

Yang, Z., and Sehgal, A. (2001). Role of molecular oscillations in generating behavioral rhythms in *Drosophila*. *Neuron* **29,** 453–467.

Yang, Z., Emerson, M., Su, H. S., and Sehgal, A. (1998). Response of the *timeless* protein to light correlates with behavioral entrainment and suggests separate pathways for visual and circadian photoreception. *Neuron* **21,** 215–223.

Yao, K.-M., and White, K. (1994). Neural specificity of *elav* expression: defining a *Drosophila* promoter for directing expression to the nervous system. *J. Neurochem.* **63,** 41–51.

Yasuyama, K., and Meinertzhagen, I. A. (1999). Extraretinal photoreceptors at the compound eye's posterior margin in *Drosophila melanogaster*. *J. Comp. Neurol.* **412,** 193–202.

Yoshida, T., and Kimura, M. T. (1995). The photoperiodic clock in *Chymomyza costata*. *J. Insect Physiol.* **41,** 217–222.

Young, M. W., and Judd, B. H. (1978). Nonessential sequences, genes and the polytene chromosome bands of *Drosophila melanogaster*. *Genetics* **88,** 723–742.

Young, M. W., Jackson, F. R., Shin, H. S., and Bargiello, T. A. (1985). A biological clock in *Drosophila*. *Cold Spring Harbor Symp. Quant. Biol.* **50,** 865–875.

Yu, Q., Colot, H. V., Kyriacou, C. P, Hall, J. C., and Rosbash, M. (1987a). Behaviour modification by *in vitro* mutagenesis of a variable region within the *period* gene of *Drosophila*. *Nature* **326,** 765–769.

Yu, Q., Jacquier, A. C., Citri, Y., Hamblen, M., Hall, J. C., and Rosbash, M. (1987b). Molecular mapping of point mutations in the *period* gene that stop or speed up biological clocks in *Drosophila melanogaster*. *Proc. Natl. Acad. Sci. USA* **84,** 784–788.

Zars, T. (2000). Behavioral functions of the insect mushroom bodies. *Curr. Opin. Neurobiol.* **10,** 790–795.

Zatz, M. (1996). Melatonin synthesis: Trekking toward the heart of darkness in the chick pineal. *Semin. Cell. Dev. Biol.* **7,** 811–820.

Zehring, W. A., Wheeler, D. A., Reddy, P., Konopka, R. J., Kyriacou, C. P., Rosbash, M., and Hall, J. C. (1984). P-element transformation with *period* locus DNA restores rhythmicity to mutant arrhythmic Drosophila melanogaster. *Cell* **39,** 369–376.

Zeng, H., Hardin, P. E., and Rosbash, M. (1994). Constitutive overexpression of the *Drosophila period* protein inhibits period mRNA cycling. *EMBO J.* **13,** 3590–3598.

Zeng, H., Qian, Z., Myers, M. P., and Rosbash, M. (1996). A light-entrainment mechanism for the *Drosophila* circadian clock. *Nature* **380,** 129–135.

Zerr, D. M., Hall, J. C., Rosbash, M., and Siwicki, K. K. (1990). Circadian fluctuations of *period* protein immunoreactivity in the CNS and the visual system of *Drosophila*. *J. Neurosci.* **10,** 2749–2762.

Zhang, X., McNeil, G. P., Hilderbrand-Chae, M. J., Franklin, T. M., Schroeder, A. J., and Jackson, F. R. (2000). Circadian regulation of the Lark RNA-binding protein within identifiable neurosecretory cells. *J. Neurobiol.* **45,** 14–29.

Zhang, Y. Q., Bailey, A. M., Matthies, H. J., Renden, R. B., Smith, M. A., Speese, S. D., Rubin, G. M., and Broadie, K. (2001). *Drosophila* fragile X-related gene regulates the MAP1B homolog Futsch to control synaptic structure and function. *Cell* **107**, 591–603.

Zheng, X. Z., Zhang, Y. P., Zhu, D. L., and Geng, Z. C. (1999). The *period* gene: High conservation of the region coding for Thr–Gly dipeptides in the *Drosophila nasuta* species subgroup. *J. Mol. Evol.* **49**, 406–410.

Zhu, L., McKay, R. R., and Shortridge, R. D. (1993). Tissue-specific expression of phospholipase C encoded by the *norpA* gene of *Drosophila melanogaster*. *J. Biol. Chem.* **268**, 15994–16001.

Zilian, O., Frei, E., Burke, R., Brentrup, D., Gutjahr, T., Bryant, P. J., and Noll, M. (1999). *double-time* is identical to *discs overgrown*, which is required for cell survival, proliferation and growth arrest in *Drosophila* imaginal discs. *Development* **126**, 5409–5420.

Zimmerman, W. F., and Goldsmith, T. H. (1971). Photosensitivity of the circadian rhythm and of visual receptors in carotenoid-depleted *Drosophila*. *Science* **171**, 1167–1169.

Zitnan, D., Ross, L. S., Zitnanova, I., Hermesman, J. L., Gill, S. S., and Adams, M. E. (1999). Steroid orchestration of a peptide hormone gene leads to orchestration of a defined behavioral sequence. *Neuron* **23**, 523–535.

Zordan, M., Osterwalder, N., Rosato, E., and Costa, R. (2001). Extra ocular photic entrainment in *Drosophila melanogaster*. *J. Neurogenet.* **15**, 97–116.

Zwiebel, L. J., Hardin, P. E., Liu, X., Hall, J. C., and Rosbash, M. (1991). A post-transcriptional mechanism contributes to circadian cycling of a per-β-galactosidase fusion protein. *Proc. Natl. Acad. Sci. USA* **88**, 3882–3886.

GLOSSARY

5′-flanking region—DNA sequences that flank the part of a gene that is transcribed into RNA and (usually) encodes a protein (except for ribosomal-RNA genes, etc.); the so-called 5′ end of a gene is stated to correspond to the 5′ end of the mRNA, not too far away from which (see **UTR**) is the codon where translation is initiated; the 5′-flanking region of a gene usually is taken to correspond to nontranscribed sequences "upstream" of (in the 5′ direction from) the transcription-start site; therefore this flanking DNA contains "*cis*-acting" (covalently bonded) gene-regulatory sequences (e.g., promoters; see **promoter**) that are bound by soluble ("*trans*-acting") proteins involved in activating transcription or keeping it relatively silent.

α—The "alpha" portion of an animal's daily cycle of rest versus activity, when it is locomoting.

abundance—A piece of molecular jargon that basically decodes as concentration, referring most commonly to varying relative levels of macromolecules (mRNAs, proteins); if a given such entity is at low concentration (at least episodically or throughout some stage of the life cycle), this jargon degenerates to usage of the paradoxical phrase "low abundance."

actogram—A plot of an animal's locomotor activity; the time base is usually a 24-hr period or multiple thereof; in the latter case, typically successive days of activity are plotted horizontally as well as vertically; thus days 1 and 2 on the top line of the plot, days 2 and 3 on the next line down, and so forth; if the animal behaved in a periodic manner, a systematic fluctuation between activity and rest can be appreciated by scanning a given line within the plot from left to right; scanning vertically (at roughly the left edge of the plot or the middle of it), one discerns the time (or phase) at which a given cycle of activity begins, peaks, or ends; therefore, if such phases are defined by a straight vertical line, the behavioral cycle duration is 24 hr; however, if (for example) the active portions of a cycle slide to the left (as one looks vertically), accentuated locomotion commenced earlier on successive days, and the period is less than 24 hr; conversely for a >24-hr behavioral rhythm in which such an actogram displays activity starting systematically later, day after day; aperiodic or arrhythmic behavior appears in an actogram as arbitrary time-dispersed segments of activity versus rest, whereby, if one puts a pencil point on a given portion of the plot, it is impossible to predict whether the placement would encounter a bout of locomotion or the absence thereof.

allele—A given form of a gene, such as one that has suffered a mutation [most commonly a base-pair substitution (see **bp**)]; in certain cases an intragenic deletion of several adjacent base pairs); a given mutant allele is symbolized by an italicized superscript, which can be any combination of letters and numbers (although, if the geneticist is being helpful, that person might chose a simple allelic designation which refers to the phenotypic effects of the mutation, e.g., per^{01}, the first "zero," or putatitvely null, mutant allele induced at the period locus of *Drosophila melanogaster*; or dbt^{ar}, a behavioral arrhythmia-inducing mutation at the double-time locus of that insect species); "wild-type allele" is the so-called normal form of the gene—symbolized with a "+" superscript—but more than one such allele is possible (e.g., among laboratory strains or different individual organisms existing in the wild); nevertheless, the different wild-type alleles, which may vary with respect to the gene's coding region (see **ORF**), are such that the alternative forms of the protein produced (see **isoform**) are all solidly functional; in contrast, the effects of mutant alleles typically are to produce subfunctional proteins, or aberrantly functioning ones, or none at all (depending on the nature of the changed base pair, if such a substitution is the cause of the mutation).

bHLH—Basic helix–loop–helix; a stretch of amino acids (aa; a.k.a. "residues") typically located relatively near the amino-(N-)terminal end of a protein—one that in this case functions as a transcription factor; among the many categories of such factors are those containing a α-helical stretch of mostly basic amino acids (ca. 15 in number), immediately upstream of a set of ca. 35 aa whose N-terminal portion consists of residues that are relatively devoid of secondary structure and "loop out," and whose C-terminal portion contains hydrophobic amino acids spaced at intervals characteristic of an amphipathic α-helix; the basic residues are primarily involved in binding to DNA, a relatively short stretch of nucleotides, or base pairs (see **bp**), in the 5′-flanking region (see that entry) of the gene in question; the quasi-hydrophobic C-terminal portion of the bHLH motif is concerned in part with interactions between the polypeptide that contains it and other protein molecules involved in transcriptional regulation (forming a dimer consisting of two heterologous polypeptides that partner in order that they might mutually regulate the relevant target gene); however, in some bHLH-containing proteins, additional amino-acid motifs located downstream of [in the carboxy (C)-terminal direction from] the basic helix–loop–helix region also participate in the requisite protein–protein interactions (see **PAS**).

Bolwig's organ—A photoreceptive structure subserving visual responses of insect larvae; this organ begins to form in early development from a neurectodermal placode in the embryonic head and extends afferent axons, which will become Bolwig's nerve (BN), into the developing central brain; among the BN targets in *Drosophila* are rhythm-regulating lateral neurons (see **LNs**); these contacts are first formed in late embryos and very likely maintained through the larval periods, at least in the sense that BN–LN contacts are observed in third-instar larvae, when a mutational effect on larval photoreception causes nonsynchrony of clock-gene expression in LN cells; among the roles played by BN is that of a "pioneer nerve" that facilitates ingrowth toward the brain of axons from photoreceptors in the developing imaginal eyes. BN disappears during an early pupal stage; during metamorphosis the LNs increase in number, or more such cells begin to express rhythm-related genes; concomitantly (as of the mid-pupal stage), there is a takeover of connections between photoreceptor afferents and the LNs by formation of contacts between imaginal eyes and these lateral-brain "clock neurons;" so far, the only such contacts experimentally specified are those involving axons ingrowing from the **H–B eyelet** (see entry) and LN cells located in relatively ventral regions of the metamorphosing brain.

bp—Base pair, that is, two DNA nucleotides (nt's) that participate in hydrogen-bonding-mediated association with one another across the two strands of a DNA double helix; a stretch of base pairs is quoted with respect to its integer number in a case where that is relatively small (e.g., 6 bp or 100 bp) compared with a situation in which several hundreds or thousands of base pairs/nucleotides are at issue (see **kb**).

bZIP—A category of DNA-binding protein that acts as a transcriptional regulator; the bZIP motif that can associate with intragenic regulatory nucleotide sequences is an extended α-helix whose N-terminal portion is enriched in basic amino acids (see **bHLH**); the C-terminal part of the helix contains hydrophobic residues (spaced every 7 aa; again, see **bHLH**); these are required for dimerization between two bZIP polypeptides; a pair of such molecules can be thought to come together by zipping up in the "7-aa hydrophobic" region; the residues of this type, spaced in this manner, are all leucines in some bZIP-containing proteins (hence the original name for this motif: leucine zipper), but other DNA-binding proteins in this category, containing different kinds of hydrophobic residues at the requisite positions within the ZIP region, were subsequently identified; a classification of transcription factors within the bZIP family is given the prefix PAR; this refers to intrapolypeptide segments rich in prolines and acidic amino acids, whose sequences are similar among members of the PAR bZIP subfamily.

CDL—Critical day length, the amount of daytime during a 24-hr period, below which an animal switches to a different biological mode; for example, in short photoperiods, females of various insect species exhibit ovarian diapause (shutdown of their reproductive system); as is probably obvious, this phenomenon correlates with ceasing performance of the biology in question as winter approaches (why waste energy on reproduction when one's offspring would encounter too-harsh conditions?).

cDNA—Complementary DNA, referring to double-stranded polynucleotide material that originated from RNA which was (first) copied into an RNA–DNA duplex by action of (animal-virus-derived) "reverse transcriptase"; this duplex is then used as a template to copy its DNA half into a complementary DNA-nucleotide sequence, resulting in a double-stranded cDNA molecule; the starting RNA is typically that which had been not only transcribed from a gene, but also processed such that stretches of nucleotides corresponding to intragenic noncoding introns were removed (by RNA processing, in particular splicing); it follows that a given cDNA molecule corresponds to a portion of the gene that is "expressed" (see **EST**), that is, transcribed, on the way to production of the protein or oligopeptide product; however, a subset of the base pairs contained within a cDNA molecule typically include, as well, stretches of nucleotides that are (naturally) not translated into amino-acid sequences (see **UTR**).

CNS structures—A limited array of named subsets of the central nervous systems of certain insects (here, chronobiologically studied ones); those structures most frequently mentioned include the brain's mushroom bodies (MBs) and their calyces, which are the MB portions located relatively near the cortex of the dorsal brain; cortical regions of CNS ganglia form a rind of neuronal and glial cell bodies that surround the neurite-containing neuropil; also referred to (or depicted in diagrams) are the supraesophageal ganglion, referring essentially to the entirety of the brain dorsal to the esophagus (an anterior portion of the alimentary system that runs roughly through the middle of the brain, aimed at more posterior parts of this system in the anterior thorax); also the subesophageal ganglion, usually smaller than the supraesophageal one; a cervical connective connects the subesophageal ganglion to the anterior portion of the thoracic–abdominal ganglionic complex (a.k.a. ventral nerve cord or VNC); the optic lobes (flanking the central brain) are given a separate entry; whether or not one regards these visual-system ganglia as CNS structures, they figure prominently in neurobiological studies of rhythms in various insect forms other than *Drosophila*.

crepuscular—A type of daily rhythmicity (here, always in the context of locomotor cycles) characterized by behavioral maxima occurring near the times that the lights come on and when they go off in light:dark (LD) cycles.

DD—Constant-dark conditions, in which both portions of a given earth-day involve subjecting the organism to D (see **LD**).

Df—A deletion, a.k.a. deficiency, which removes part of a chromosome; this type of genetic aberration is the one most commonly referred to in the context of analyzing rhythm-related genes and their mutations (here called *mut*'s); for example, the phenotypic effects of a fruit fly heterozygous for a *Df* and a mutation (*Df/mut*, whereby the latter is an intragenic change within a chromosomal region defined by material missing in the *Df*) might be compared with those of the mutant homozygote (*mut/mut*); or a fly heterozygous for a *Df* and the normal allele of the gene (i.e., one of the genetic loci deleted by this *Df*) would be compared phenotypically to the wild-type fly (homozygous for the gene's normal allele); if a *Df* is definitively known to be deleted of some portion of a chromosome (typically, ca. 1/20 to 1/50 of it)—by cytological examination of this chromosomally aberrant type—then it is almost certainly the case that several genes are missing (otherwise, if only a pair or quartet or so of chromosome "bands" is deleted, the *Df* would be invisible in the light microscope); therefore, a given *Df/Df* homozygote almost always has lethal effects on the organism's development (because nearly all chromosomal regions encompassing about a half-dozen adjacent genes or more contain at least one vital gene).

diel—An adjective describing daily cycles (in this case of locomotion) that occur in environmental oscillating conditions (usually meaning LD cycles in this review).

DNs—Dorsal neurons, located in the brain of *Drosophila* larvae, pupae, or adults, in which clock genes and others encoding rhythm-related factors are known to generate their mRNAs and protein products; the separate clusters of DNs (nominally three such neuronal groups) compose a small subset of the thousands of neurons located in the dorsal brain, such that "the DNs" represent only a few dozen CNS neurons exhibiting this particular feature of chronobiological cellular differentiation.

dosage, gene—Number of copies of a gene of interest, typically as produced by applying chromosome aberrations or transgenes within which are contained the normal (wild-type) allele of the genetic locus; thus, for example, a deletion (*Df*)/normal(+)-allele heterozygous type is generated and tested for its phenotype, with reference to that influenced by the +/+ genotype (for any autosomal gene; or for an *X*-chromosomal locus whose normal genetic constitution in a female is assessed with respect to a nonnormal phenotype that might be caused by the "one-dose" genotype); another example would involve comparison of the effects of an extra dose of the normal gene to the standard dosage of the locus (hence, a two-dose male compared with one that is genetically normal with regard to its *X*-chromosomal loci; or a three-dose type that carries a so-called duplication of the + allele, mediated by application of a chromosome rearrangement or a transgene, and is compared with the phenotype associated with the standard two doses of an autosomal locus or of an *X*-chromosomal one in females).

double mutant and double transgenic—Genetic constitutions involving a combination of two mutations or of two genes separately transformed into the organism; whereas the verbal appellation for these genotypes may reflect obvious genetic types, the potentially occult manner by which they are symbolized needs to be borne in mind; thus, a double mutant involving two mutations on the same chromosome is indicated simply by writing the two mutational symbols with a space between them (e.g., *gl cry^b*), but if the two mutations are on separate chromosomes, their symbols are separated by a semicolon (e.g., *norpA;cry^b*); the same conventions are applied (or should be) with reference to doubly transgenic types for which the transformed-in pieces of DNA happened to land on the same chromosome or on separate chromosomes.

E-box—A stretch of six nucleotides that forms a "consensus sequence" at which certain categories of transcription-factor proteins bind—in particular bHLH-containing gene-regulatory factors; a *cis*-acting sequence of the type involved in *Drosophila* clock-gene regulation is composed of the bases CACGTG (on one strand of the double helix), but sometimes the E-box designator is used to refer to a slightly longer stretch of nucleotides and/or one whose sequence varies mildly from the canonical six bases's worth; E-boxes that are potentially targets of these transcription-factor types (see **bHLH**) are usually located within the 5′-flanking region of the gene, but candidate E-box regulatees are also found in other parts of the gene's untranslated regions (see **UTR**), such as within an intron.

enhancer—A sequence of DNA base pairs (such as an E-box) that is recognized by gene-regulatory proteins, resulting in enhancement of the rate at which the gene is transcribed *and*, as is frequently observed, production of such gene products in certain tissues or cell types; a given enhancer can be located essentially anywhere within a genetic locus or even outside its confines (transcription enhancement from a distance, whereby the bound, positively acting regulatory protein is brought closer to the transcription unit essentially by DNA looping); nearly all the enhancers referred to in this work represent sequences in 5′-flanking DNA or within introns, representing in turn nontranscribed or nontranslated regulatory sequences; the 5′-flanking enhancers are found within what is loosely referred to as the "promoter region" of the gene (see **promoter**), but the promoter per se is defined by a short stretch of base pairs at which RNA polymerase promotes transcription of the primary gene product (RNA) in the proximate sense; thus, an enhancer located within the promoter region (typically upstream of it in the

5′ direction) is involved in communicating the effects of a bound transcription factor (e.g., a bHLH–PAS protein), to the polymerase, enhancing its catalytic effects in terms of synthesizing a gene-encoded sequence of RNA nucleotides.

enhancer trap—A phenomenon and also a type of genetic variant involving a transposon (loosely synonymous with a transgene) inserted at a genetic locus, such that the site of chromosomal insertion causes expression of transposon-contained DNA sequences to be expressed; the sequences might be a marker gene (see *lacZ*) or a gene-regulatory factor (see **GAL4 and** *gal4*); that such sequences are expressed is inferred to be the result of the transposon-contained material having come under the gene-regulatory control of an **enhancer** (see entry) near or within a genetic locus; such a positive effect can cause the transposon-contained sequences to be expressed in a developmental and/or tissue pattern which reflects that of the nearby "trapped" gene, that is, as controlled by the enhancer sequence(s) which indeed helps control when and where the normal genetic locus is expressed; however, the trap-mediated pattern of tissue expression may not mimic perfectly the places where the products of the nearby gene are normally found; an analogous problem can arise when one desires transformed-in DNA (see **transgenic**) to be controlled only by its intrinsic quality; yet, unwanted enhancer trapping may influence expression of the transgene that happens to land near a random regulatory sequence on some chromosome and cause extra cells and tissues to produce the protein encoded by the incoming transgene, whereas this DNA construct was designed to include its own enhancers and a promoter and thus mediate the normal tissue expression pattern; this setup might, for example, be aimed at "rescuing" the effects of a mutation (see **transgenic**).

entrainment—A phenomenon involving daily rhythms that are caused to occur in synchrony with fluctuating environmental conditions; thus, for example, an entrained animal will exhibit locomotor cycles whose durations match that of the environmental cycle, exemplified in turn by cases of 24-hr behavioral cycle durations that are usually observed in 12-hr light: 12-hr dark cycles; if the organism's "natural, endogenous" periodicity, underpinned by circadian-clock function, is in fact only "circa" 24-hr (see **free-running** and τ), it follows that the organism is reset on a daily basis to exhibit an entrained, synchronized rhythm in environmentally cycling conditions; underlying the daily resets are adjustments of the clock's phase (otherwise, the peak time for locomotion would, for example, drift later and later on successive days for an animal whose free-running period is longer than 24 hr); an entrained rhythm frequently exhibits anticipations of environmental transitions (e.g., upswings of locomotions before the lights come on in LD cycles), which implies an underlying clock function, but not necessarily (see Section IV.C.1); defining a daily rhythm as a bona-fide entrained one often includes the demand that the phase of at least one rhythmic component observed in environmentally fluctuating conditions is extended into the rhythmicity seen in subsequent constant conditions; therefore, if a morning peak of activity anticipates the time of lights-on (let us say), the "LD rhythm" may or may not be entrained, but it is apprehended more firmly to have that attribute if the LD morning peak "takes off into free run"; subsequently, as is typical, this "subjective morning peak" (see **PRC**) will occur later or earlier on successive 24-hr days in constant conditions.

EST—Expressed sequence tag, stemming from cDNAs generated from mRNAs taken from an organism at (for example) some stage of its life cycle or from a particular tissue taken from it; such cDNAs, which usually do not represent the fully transcribed RNA deriving from a given gene (i.e., all of its UTRs and its ORF; see **UTR** and **ORF**), are partially sequenced; these data create "tags" for collections of molecules representing genes that are expressed according to one's biological interests (in terms of the source of the mRNAs); the sequence data from ESTs are relentlessly archived, permitting investigators to search such databases for information corresponding to expressed genes, to ask, for instance, whether a particular category of protein is produced by the embryo or within the adult head (for example); a successful search of this sort depends upon translated sequences (see **ORF**) being contained within the ESTs of interest; ESTs successfully identified by the requisite computer searches can be obtained as physical entities ("clone by phone"), followed by experimental completion of their sequencing and use of them as molecular probes (e.g., to obtain additional, overlapping cDNAs from libraries, allowing for complete determination of the ORF; or to obtain fragments of genomic DNA in other kinds of libraries, such that the gene itself can be isolated).

extraocular—An adjective referring to photoreceptive structures and processes that do not operate within the organism's ("standard") external eyes; extraocular photoreception may occur in "excitable tissues" (e.g., within the brain or in appendages that take in sensory stimuli other than light) and in others (e.g., excretory structures).

free-running—An adjective applied to daily rhythmicity that operates in constant environmental conditions.

GAL4 and *gal4*—Respectively, a transcription factor (protein) and the nucleotide sequences encoding it in *Saccharomyces cerevisiae*; this factor is involved in galactose metabolism in this species of yeast; GAL4 "works" in higher eukaryotes, meaning—in *D. melanogaster*, for example—that covalent fusion of *gal4* downstream of regulatory sequences (here called *reg*), previously identified (and cloned) from a given metazoan gene, will (one hopes) mediate GAL4 production in the cells and tissues where that gene is normally expressed; the *cis*-acting DNA targets of GAL4 (see **UAS**) have been fused upstream of a variety of protein-encoding sequences (here called *pes*), such that a *reg-gal4*;UAS-*pes* doubly transgenic combination will result in this protein's production under the spatial (tissue) and developmental (life-cycle-stage) control of the *reg* sequence; this protein may be a marker factor that possesses intrinsic activity (see **lacZ**) or is experimentally known to be antigenic; alternatively, *pes* could be a tissue-disruptive agent that might mediate genetic ablation or physiological inactivation of cells.

gene or *Gene*—Words used here generically to describe how such factors are formally designated; the full gene name is italicized and begins with a lower-case letter for a genetic **locus** (see entry) that was initially defined (or subsequently altered) by a recessive mutation (see the entry **semidominant,** however); a dominant mutation, which usually causes a phenotypic defect not as salient or severe as when the mutation is homozygous, begins with an upper-case letter; some dominant-viable mutations are recessive lethals; genes originally identified by virtue of the protein product they encode (and in several cases also mutated in such coding sequences) typically are symbolized by would-be words (or other alphanumeric designators) that begin with an upper-case letter (this is in part because a null mutation in an enzyme-encoding gene, for example, leads to about half-normal activity when it is heterozygous with the wild-type, a.k.a. +, allele).

GFP—Green fluorescent protein; material involved in the bioluminescence of cnidarian coelenterates; cDNA encoding GFP cloned from the jellyfish, *Aequorea victoria* (and here called *gfp*), is widely used in cellular-marking transgenics; a molecular construct involving a driver factor is fused to *gfp*; the protein so encoded creates, in apparently any kind of heterologous cells (in this work, those of *D. melanogaster*), convenient marking all by itself (i.e., no substrate necessary, nor are any of the known factors with which GFP naturally interacts when excited in *A. victoria*); thus, for example, a UAS-*gfp* transgenic fruit fly also carrying a *gal4*-containing transgene (see **UAS** and **GAL and gal4**) can itself (the whole organism) be exposed to long-wave UV light, or a tissue-specimen taken from the animal can be so excited; this leads to emission of green light; in addition to the inherent, standalone marking properties of GFP-mediated signals, they routinely fill the entire cell (e.g., neuronal cell bodies along with their projecting neurites), whereas other transgenically engineered markers may not (e.g., *E. coli*-derived β-GAL, whose general properties are noted in the **lacZ** entry); moreover, GFP works when present in a fusion protein (e.g., all or part of a higher eukaryotic one, engineered to be cotranslated with sequences encoded by *gfp* cDNA); derivatives of wild-type *gfp* itself have been created (by *in vitro* mutagenesis) such that excitation leads to enhanced emission; or can be (more) conveniently effected using commonly available filters (fluorescein isothiocyanate ones, re 450- to 490-nm excitation); or that such stimulation leads to alternative colors of emission (e.g., "blue fluorescent protein"); or that (intra-GFP) chromophore formation is more rapid than the usual 4 hr; or that the GFP molecule has a shorter-than-normal half-life (see *luc*); some of the *Drosophila* transgenic cellular-marking studies mentioned in this work took advantage of an "enhanced" GFP form.

H–B eyelet—A structure contained within the visual system of *D. melanogaster*, located roughly between the compound eyes (in both sides of the head) and the distalmost **optic lobe** (see entry), called the lamina; the H–B eyelet was named after its discoverers, A. Hofbauer and E. Buchner, based on their demonstration that an antibody which leads to *in situ* staining of all known photoreceptors in the fly head also stains the extra structure in question, which was thus surmised to be an "extraocular" photoreceptor (viz., eyelet).

hybridization, *in situ*—A technique used to label structures contained within cells or tissues, whereby single-stranded nucleic-acid probes are synthesized such that they contain intrinsic label or entities that can lead to labeling (e.g., by their antigenicity) of the structures of interest; such structures include cells composing a certain tissue, or sites on intracellular chromosomes; the probes can consist of single-stranded RNA or DNA, designed to be complementary to mRNA within tissues where the gene in question is expressed or to be complementary to one strand of the double helix that makes up part of a chromosomal locus of interest; thus, one can determine, on the one hand, the tissues in which a gene is naturally transcribed and, on the other, to where within the genome a molecular clone of interest corresponds (i.e., the site on a given chromosome to which the probe anneals, thus marking the genetic locus in question, one that might have been defined only in molecular terms, not by a genetically mappable mutation); that this strategy and method involves "hybridization" may or may not be what occurs in actuality: a DNA probe applied for genetic-locus determination leads to DNA:DNA annealing, as opposed to formation of a molecular hybrid; the same misnomer obtains in a case where a labeled RNA probe is applied to determine where the complementary mRNA is present in one or more tissues.

hypomorph—A form of a gene, usually a mutated one, that leads to lowered functionality of the genetic locus; a hypomorph (with reference to the organism as opposed to the gene in question) is therefore a mutant type in which, for example, substantially lower than normal concentration of the normal gene products (mRNA and/or protein) are produced because of the mutation (as discussed shortly); alternatively, normal such amounts may be generated in a hypomorphic mutant, but the final (protein) product would carry out subnormal function (e.g., abnormally slow catalysis) on a molecule-by-molecule basis; some hypomorphs are so named solely by virtue of inferences stemming from phenotypic scrutiny of a mutant at the whole-organismal level, whereby it seems as if the mutation allows for "leaky" expression of the gene, but in the "normal direction," as opposed to the mutation causing some sort of deranged feature of protein function; the latter kind of mutation could cause a phenotypic defect that is worse than the effects of gene removal; for example, *mut*/+ would lead to a more severe alteration compared with the phenotype associated with *Df*/+; whereas, in a hypomorphic case, *Df*/*mut* leads to a phenotype "farther from wild type," compared with the effects of *mut*/*mut*—implying something like this: nothing over 5% normal causes a worse defect than does 5% normal/5% normal (total, 10%); in such a hypomorphic mutant example, actual biochemical analysis of the putatively subnormal product level or inherently subfunctional protein molecules might not yet have been possible; but, should study of the gene and its mutations at the molecular level ensue, the concrete reason for hypomorphy is not infrequently realized; in such cases, "subnormal" usually turns out to mean much less than the wild-type level of product concentration or protein function (as implied in the hypothetical case given above); therefore, a situation in which (biochemical) hypomorphy corresponds to something like three-fourths normal tends not to cause a phenotypic abnormality.

isoform—The form of a molecular entity, usually a gene product, that is the same as other forms—but not really; thus, for example, certain genes generate a set of strongly related mRNA species that are therefore called isoforms because each member of the set encodes a protein type similar to those specified by other members; but something like alternative splicings of the primary transcript, or usage of alternative translation-initiating codons by a given mRNA molecule, will lead to the synthesis of proteins that differ from one another within certain subsets of such polypeptides; one example of a small "family" of proteins, all of whose members are specified by a given gene, would be enzyme isoforms, all of which might contain the same core catalytic region(s) (defined by a stretch of amino acids or some number of amino-acid clusters), but these proteins would vary with respect to other regions of the polypeptides that are involved in regulating the nature of the catalysis, for instance.

kb—Kilobases, a convenient abbreviation for indicating 1,000 bp (see **bp**) or nucleotides, for example, a 13.2-kb genomic-DNA sequence, or a 4.5-kb mRNA species.

lacZ—A gene (whose full named is *lactose-Z*) contained within the "lactose operon" of the bacterium *Escherichia coli*; this one of the three enzyme-encoding genes within the operon that encodes β-galactosidase (β-GAL); catalysis mediated by it, solely by application of a convenient artificial substrate (known as X-gal), leads immediately to deposition of material that reflects blue light (i.e., no other reagents need be applied to elicit the color reaction); thus, *lacZ* is widely used in eukaryotic molecular genetics (implying that β-GAL in nonbacterial cells is nicely functional) as a marker; for example, gene-regulatory sequences cloned from *D. melanogaster* could be fused upstream of *lacZ* and such a molecular construct transformed into the germ line of the animal; the resulting *Drosophila* strain would exhibit blue coloration in cells and tissues where the normal gene is expressed, if all the normal regulatory sequences are contained in or trapped by the transgene; in the latter situation, β-GAL activity or antigenicity comes into play in transposon-mobilization procedures designed to detect enhancers of genes that might prove biologically interesting (here, by virtue of *lacZ* "reporting" of where and when they are expressed).

LD—Light:dark cycling conditions, usually (but not always) those made to fluctuate experimentally over the course of 24-hr cycles; crude approximations of natural such cycles are (in turn) typically employed by applying 12 hr of L followed by 12 hr of D; but in some experiments (or rhythm-monitoring "runs"), conditions mimicking "Northern-hemisphere winter," for example, are used (such as 8-hr L:16-hr D cycles).

limits of entrainment—**T cycles** (see entry) beyond which the organism no longer synchronizes to the period of an environmental oscillation; for example, a fruit fly entrains to T cycles that are ≥ 8 hr different from 24 hr, but at some point (e.g., $T \lesssim 14$ hr) the animal no longer behaves in synchrony and instead exhibits its free-running periodicity.

LL—Constant-light conditions, in which both portions of a given earth-day involve subjecting the organism to L.

LNs—Lateral-brain neurons, which are the companions of more dorsally located *Drosophila* cells (see **DNs**) in which clock genes (and others) generate their products; the separate clusters of LNs are indeed located in relatively lateral regions of the larval, the pupal, or the adult brain; but at the latter stage, one LN cluster is in a fairly dorsal location, and two additional such clusters are found within a more ventral region, which, however, is still dorsal to the esophagus (see **CNS structures**).

locus—A genetic term roughly synonymous with the word gene, but the latter term is sometimes used to mean the core of the genetic locus, its subset that contains protein-coding information (see **ORF**) and perhaps as well the principal gene-regulatory components (see **enhancer** and **promoter**); locus, if appreciated more broadly, could refer to the sum total of transcribed and nontranscribed DNA sequences present at this chromosomal site, required to express the gene products correctly; alternatively, the site of a genetic locus might be referred to in a rather formalistic manner (e.g., the intrachromosomal location to which a mutation got mapped by meiotic recombination, by application of *Df*'s, or by both), irrespective or in advance of acquiring any molecular-genetic information about the locus and its various categories of DNA contents.

luc—A cDNA originating from the beetle (firefly) *Photinus pyralis,* which encodes luciferase (LUC); upon delivery of luciferin to an insect transformed with a *luc*-expressible molecular construct (or, given internal manufacture by *P. pyralis* of this enzyme substrate), the relevant cells and tissues emit yellow light as the result of lucferin oxidation; if enough such light escapes from the tissues (which could be cultured specimens that had been dissected away from the transgenic animal) or from the body of the insect as a whole, the signals can be captured by a charge-coupled-device camera or, more conveniently, by a machine that in effect counts emitted photons and records such numbers, either for a one-time specified number of seconds or minutes (usually) or against a longer ongoing time base; therefore, molecular constructs (notably) in which regulatory sequences of rhythm-related genes (such as *per*, as cloned from *D. melanogaster*) are fused to *luc* allow for nicely quantitative reporting of the action of such sequences (see **5′-flanking**) in transfected cultured cells; also, the construct can be transformed into the germ line of fruitfly strains (see **transgenic**), such that real-time reporting of fluctuating levels of gene-product production are monitorable in live animals; that the LUC enzymatic function can be engineered to fluctuate, reflecting oscillations of clock-gene products like PER (see **PROTEIN**), is based on the fact that this reporter activity has a short half-life in heterologous cells (if not in all organisms other than firefly, then at least in higher plants, *D. melanogaster,* and mammals).

MAPK—Mitogen-activated protein kinase (a.k.a. microtubule-associated protein kinase, although this factor functions far more widely); this signal-transducing enzyme is a threonine/serine kinase whose phosphorylation (creating phospho-MAPK, here called P-MAPK) is the culminating event in a kinase cascade (although P-MAPK goes on to phosphorylate many different proteins and modify their functions); this cellular/biochemical pathway starts when a membrane-bound receptor is activated by a ligand; the resulting conformational change (e.g., dimerization and autophosphorylation of monomeric receptor polypeptides) allows for receptor biding to an SH2-containing adapter protein; this factor associates in turn with a guanine (G) exchange factor (called SOS), which binds to the so-called oncogene product Ras; this promotes dissociation of GDP from the latter and replacement by GTP binding to Ras, which is thereby activated; this permits Raf (another thereonine/serine kinase) to be bound by GTP–Ras, which is activated and proceeds to phosphorylate MEK, a kinase that can phosphorylate both tyrosine and serine residues in its protein substrates; one of them is MAPK whose modification to P-MAPK activates this enzyme, allowing it to "fan out" and phosphorylate a variety of farther-downstream protein substrates (see earlier discussion), including nuclear transcription factors and a certain kinase called sk6II (see Section X.B.2).

masking—A difficult-to-apprehend piece of rhythm jargon, referring to features of biological cycles that are only in part (or even not at all) underpinned by clock function; typically, some feature of the environmental conditions masks the wherewithal for the clock to be the only regulator that contributes to the rhythmicity; masking, however, does not mean only that an inhibition of the biological attribute is operating (e.g., an environmental stimulus impinging on locomotion that usually occurs during some portion of the daily cycle); in addition, a stimulus such as light (which is often cyclically present in rhythm-monitoring experiments) could accentuate "levels of the parameter"; what if, for example, short LD cycles (see T **cycles**) were causing the animal's locomotor rhythm to be mediated by its circadian clock? (see **limits of entrainment**)?:

the behavioral cycle durations should "look like" they contain only ca. 24-hr components, but light-induced increases in locomotion might also be operating (e.g., when the lights come a few times *per* day in something like $T = 8$ hr conditions); thus, a relatively high frequency rhythm would overlay the circadian fluctuation and might altogether obscure its presence within a plot of the animal's locomotion.

mosaic—A genetically mixed organism whose various tissues are composed of cells of different genotypes; for example, part of a mosaic individual might carry only a mutant allele of a rhythm-related gene (here called *rhy*), with the remaining tissues in this animal carrying the (recessive) mutation (superscript *mut*) in heterozygous condition with the normal allele; a series of such mosaics can be tested for biological rhythmicity (e.g., rest/activity locomotor cycles), followed by determination of which tissues in a given mosaic expressed the normal form of the gene versus only its *mut* allele; this might require that expression of the rhy^+ allele create its own tissue marking (such as by antigenicity of the encoded RHY protein), or that that allele would be genetically linked to a marker factor; in this regard, the type of mosaic applied (so far) in chronobiological experiments results from the progeny of certain genetic crosses designed to cause loss of an *X* chromosome that carries rhy^+ along with the dominant marker; the other homologous chromosome in the zygote (set up to develop into an adult mosaic) carries the *rhy* mutant allele, and elsewhere in the genetic background of the mosaic is a mutated marker gene—although such a *Drosophila* mutation is unnecessary if a transgene including bacterially derived *lacZ* marker is genetically linked to rhy^+; loss of the rhy^+ plus dominant-marker chromosome, occurring in a nucleus of the zygote early in development, simultaneously uncovers the effects of rhy^{mut} and leads to no marker expression in the tissues developing from that embryonic nucleus; one way of indicating the genotype of a mosaic is $rhy^+/rhy^{mut}//rhy^{mut}$; the double slash indicates the alternative tissue genotypes (to either side of this indicator) in the genetically mixed animal; so the question is: among all the separate mosaic individuals analyzed phenotypically and histologically, what tissues carrying rhy^+ are correlated with normal biological rhythmicity, and what ones that are solely rhy^{mut} correlate with the mutant version of this phenotype?

mushroom bodies—Paired neural structures located in a dorsal, posterior region of insect brains; various tissue preparations thereof lead to the appearance of mushroom-shaped bodies (MBs) in which the cap is formed by MB cell bodies and dendrites, with the stalk-appearing substructures being formed by MB axons (intrinsic and extrinsic such fibers); the MBs receive sensory information (eventually) from olfactory inputs (in terms of current knowledge for *Drosophila*), transmitted to the central brain by interneurons connecting sensory afferents, notably those elaborated by the antenna; various experimental manipulations of the MBs indicate that they are involved in "processing" sensory cues subserving insect learning and memory; also, axons of certain circadian-pacemaker neurons located in a ventrolateral brain region (see **LNs**), terminate in the vicinity of the MBs in *Drosophila*, suggesting that these structures could participate in the regulation of rhythmic behavior; however, the results of damaging the MBs mutationally and by chemical treatment were mostly negative in terms of establishing a chronobiological role for these dorsoposterior brain structures.

neurofribromin—A protein originally identified by molecular identification of neurofibromatosis-1 mutations in humans (which cause multiple tumors in the peripheral nervous system); sequencing the gene that was genetically so defined implied that it encodes a previously unknown polypeptide (NF1); a potion of NF1 is similar to part of a protein called GAP, which binds to certain phosphotyrosines in activated receptors; this in turn promotes GAP binding to the Ras polypeptide (see **MAPK**); indeed, HF1 interacts with Ras, functioning as a GTP-activator of the latter and contributing to "Ras-cycling" between its inactive and active forms (respectively, as bound by GDP or GTP).

Northern blotting—A way of visualizing RNAs on a gel; the geographically based adjective invoked to signify this procedure stems from a smarmy jape foisted on the scientific community by virtue of a biochemical "blotting" tactic; in executing the original version of this technique, fragments of DNA (often created by digestion with restriction endonucleases) are separated by size electrophoretically, denatured, then transferred by capillary action onto a sheet of nitrocellulose paper to which single-stranded such molecules adhere; the paper is then incubated in a solution containing a molecular probe (usually radiolabeled, single-stranded DNA from some known source, typically derived from a single gene), which will anneal to the portion of paper containing complementary nucleotide sequences; this method for detecting a tiny subset of all the size-separated DNA fragments (i.e., one or a few "bands," corresponding to locations on the electropherogram annealed by the probe and detected by autoradiography) was invented by one Edward Southern, thus "Southern blotting"; by analogy (its inherent gibberish notwithstanding), Northern blotting involves separation of RNA species via agarose-gel electrophoresis; this is followed by blotting the "ladder" of such molecules (whereby small mRNA species had moved more rapidly through the gel than large ones) into nitrocellulose paper or a nylon membrane and detection of the sizes and relative amounts of complementary RNA molecules, again by application of a labeled DNA probe.

null mutant or allele—An organism exhibiting the biological effects of a mutation that has lost all function vis à vis that of the normal allele of some gene; the "null phenotype" may therefore be no rhythmicity (in the context of the subject of this review), developmental lethality, or whatever is the consequence of a mutation that behaves as if the gene is completely deleted; such "null alleles" can be caused by mutations that (indeed) delete all or part of the gene, or others in which an amino-acid-specifying codon is changed by a base-pair substitution to a premature stop (a.k.a. "nonsense") codon; another type of null mutation might result from a base-pair substitution that leads to an amino-acid substitution, one that causes thoroughgoing loss of the protein's function (even though the mutated polypeptide might be present in normal amounts).

optic lobes—The ganglia within the proximal (brain-adjacent) portions of an insect's visual system, or, from another perspective, the relatively lateral portions of its anterior CNS; there are typically four bilaterally paired such lobes, composed of the distally located lamina (right underneath the compound eye), the medulla (underneath the lamina), and the lobula plus lobula plate; the latter two lobes are closest to the central brain proper, the lobula being located anterior to the lobula plate; several of the LN clock-gene expressing neurons in *D. melanogaster* (see **LNs**) are located at the anterior rim of the medulla, which can be thought of as part of that optic lobe; indeed this region of the CNS is called the "accessory medulla" (AMe) in some of the relevant reports; other insects, such as cockroach, possess circadian-pacemaker activity within this structure, with these clock functions being output anatomically from well-described neurites emanating from the AMe.

ORF—Open-reading frame, a sequence of nucleotides (usually hundreds in a row) that corresponds to an amino-acid sequence; that is, a "frame" to start the on-paper "reading" of such a base sequence can be found such that a long sequence of codons each corresponding to an amino acid is located; typically, ORFs are searched for in base-pair sequences contained within cDNAs (see **cDNA**) because such clones are derived from mRNAs, most of which were transcribed from protein-encoding genes; however, a cDNA sequence may contain UTRs (see **UTR**) as well as ORFs, whereby, for example, a reading frame "open" for translation into a stretch of many amino acids would start some number of nucleotides in from the 5′ end of one strand of the double-stranded cDNA; therefore, 5′ of the ORF's beginning are untranslatable sequences, signified by the fact that any of the three possible reading frames of the UTR are such that one or more stop codons are encountered, but the relatively 3′-located ORF by definition leads to no stop until the natural translation-terminating codon is encountered at the extreme 3′ end of the ORF (after which—in the gene sequence and that of a given corresponding cDNA—"3′ UTR" sequences may be present, representing the tailend of the mRNA); long ORFs (as mentioned) are in a way the rule, in that most of the gene products considered in this work are macromolecular proteins; however, short ORFs are properly a feature of the informational content of certain cDNAs and genes giving rise to them—those that naturally encode one or more oligopeptides as their final products (e.g., neuropeptides comprising only 10 aa or even fewer, up to perhaps 100 such residues).

PAS—A domain or motif contained within a certain category of proteins involved in, among many other biological processes, transcriptional control and the intracellular molecular processing of incoming signals; the PAS region of

such a polypeptide is composed of ca. 250–300 contiguous amino acids, usually located relatively near the N-terminus; the PAS acronym takes its letters
from the three historically founding members of this protein family, PERIOD
(then a putative circadian clock factor known in *D. melanogaster* only), the
ARNT protein (a mammalian nuclear translocator involved in the intracellular trafficking of a transcription factor that binds to aryl hydrocarbons), and
SINGLE-MINDED (a transcription factor discovered in *Drosophila* by virtue
of certain neural-lethal developmental mutations in the *sim* gene); protein
families in general are not infrequently named (with three-letter acronyms)
from this kind of investigatory and historical perspective; the PAS region can
be involved in interpolypeptide molecular interactions (e.g., PASs contained
within heterologous proteins might mediate dimer formation) and may or may
not bind a ligand (depending on the nature of the gene product's function);
one large subcategory of PAS proteins are those in which **bHLH** motifs (see
entry) are located not far upstream, nearer to the N terminus.

PCR—Polymerase chain reaction, in which tiny amounts of DNA in solution can
be amplified to manageable concentrations (e.g., for cloning into microbial-
genetic vectors, aimed at nucleotide sequencing or preparation of molecular
probes); primers complementary to portions of the complementary strands of
DNA to be amplified are prepared (these may be "degenerate" primers containing a host of nucleotide sequences among them, if, for example, sequences
of desired amplifiees are unknown or nondefinitively deduced); heat-resistant,
bacterially derived DNA polymerase will synthesize nucleotide sequences between the primers; that is, at the beginning of a series of PRCs, a given piece
of DNA molecule will denature to single-stranded material by heating and
copied into molecules identical to the starting one; a series of further denatu-
ration/polymerization cycles leads to an exponential increase in concentration
of the starting material; PCR is applied essentially all day/every day in molecular genetic investigations; exemplified in this work are PCRs used to amplify
cDNA copies of RNAs extracted from an insect (or certain of its tissues) of
chronobiological interest, such that material corresponding to a gene known
to function in this species (by virtue of identifying its protein) can be cloned,
characterized, and manipulated.

penetrance—A genetic term referring to the proportion of genotypically uniform individuals that exhibit a given phenotype; for example, all individuals
within a group of fruit flies might be homozygous for a recessive mutation or
heterozygous for a **semidominant** one (see entry), but perhaps only one-third,
one-half, or three-fourths or so of such individuals will display the mutant character in question (e.g., such a fraction would be arrhythmic behaviorally, with
the remainder of these insects behaving normally); "impenetrant" phenotypes
are a mystery as to why the same interorganismal genotype leads to different
phenotypic outcomes among the individuals that get assayed separately.

phenocopy—An organismic state induced by treating a genetically normal organism in some manner (environmental condition, chemical), leading to a phenotypic defect that mimics the effects of a mutation; cf. "genocopy" (a cute term coined in the text), which involves two unrelated kinds of genetic variants (e.g., a mutation at some locus, vis à vis a transgene manipulation of a separate gene) that end up causing similar phenotypes.

promoter—A regulatory region of a gene, located near the transcription-initiating nucleotide (a.k.a. the "+1 nt") and necessary as a *cis*-acting sequence for such mRNA production to occur (many but not all promoters are located upstream of +1); the promotor per se is typically a short stretch of nucleotides, largely concerned with the action of RNA polymerase; but "promoter region" of a gene is more loosely ascribed to its 5′-regulatory sequences that are involved in broader features of controlling the gene's expression, for example, life-cycle stages during and tissue types within which the encoded transcripts and protein molecules are produced (also see **5′-flanking region** and **enhancer**).

PRC—Phase–response curve, which plots shifts of a circadian rhythm induced by pulsatile environmental stimuli; thus, for example, the usual peak times of a locomotor rhythm are shifted to earlier or later cycle times, depending on the time during a free-running daily cycle when a pulse of light or altered temperature is delivered; such stimuli are the main ones applied, although a PRC can also be generated by things like pulses of chemical treatments; moreover, the stimuli in question are typically delivered for a few minutes, but sometimes pulses of 1 hr or more are administered; consider that a PRC determination typically occurs for organisms that were in LD cycling conditions before proceeding into constant ones; after proceeding into DD, the "first half" of a given cycle (defined now by time per se instead of the environmental cycle duration) is called the "subjective day" and the second half the "subjective night"; classically (including that the following pertains to PRCs determined for all kinds of organisms), light pulses delivered during the subjective day lead to minimal phase shifts, whereas those given during the early subjective night induce phase delays, and light-subjective-night pulses cause phase advances; respectively, these phase shifts are observed as later occurrences of (for example) a locomotor peak compared with its time of occurrence in an undisturbed, free-running individual; whereas the opposite kind of (advance) shift is observed as a "unexpectedly" earlier peak (post-pulse); with reference to the **entrainment** entry, note that these features of the PRC can explain daily resets of the circadian clock and a revealed rhythm that are observed in naturally fluctuating conditions: if an organism's free-running period is shorter than 24 hr, it will come to the end of its "internal day" when the lights are still on, such that an "early-internal-night" stimulus will cause a daily delay of the clock that can put this specimen into synchrony with the natural cycle (otherwise, the phase of its biological rhythm would go out of synch by drifting to an earlier time each day); conversely, an organism whose τ is longer than 24 hr will encounter light as it approaches the end of its internal night, and that stimulus can effect an entraining daily advance of its clock.

PROTEIN—A term whose meaning is obvious, but is entered to denote that this "level" of gene product, compared with italicized designators for the encoding gene or its mRNA, is signified by an all-upper-case version of the word or the abbreviation that names the gene; thus, for example, the *period* (*per*) gene encodes PERIOD (PER) protein.

ρ—The "rho" portion of an animal's daily cycle of rest versus activity when it is locomotor-quiescent.

semidominant—An allelic form of a gene, usually referring to a mutation, which causes a phenotype somewhere in between those influenced by related genotypes; for example, a semidominant mutation (the *Jrk* allele) at the *Clock* (*Clk*) locus of *D. melanogaster* causes ca. one-half to two-thirds of individually monitored $Clk^{Jrk}/+$ flies to be arrhythmic in terms of daily fluctuations of locomotion, whereas nearly all Clk^+/Clk^+ individuals behave rhythmically and all Clk^{Jrk}/Clk^{Jrk} flies are arrhythmic; in other examples of semidominant mutations, gene names and abbreviations were invoked (perhaps unfortunately) that begin with lower-case letters—such as for the *period*, or *per*, gene of *D. melanogaster*, essentially all of whose mutations are semidominant; for example, in the case of the *Short*, or *S*, mutant allele, $per^S/+$ flies exhibit 21.5-hr free-running periods compared with those for per^+/per^+ of ca. 24 hr and those for per^S/per^S of 19 hr.

τ— The "tau" value in hours that defines the free-running period of a circadian rhythm, operating in environmentally constant conditions.

T cycles—Time periods during which the environmental conditions are oscillating; for example, T-cycle variations are imposed experimentally by placing the organism in light:dark conditions that might vary both downward and upward from "standard" 12-hr L:12:hr D cycles; also, the upper-case T used here is indirectly pitted against lower-case τ, the latter referring to periods observed in constant conditions.

temperature compensation—A clock phenomenon in which biological cycle durations remain similar as the temperature is varied; that such clocks do not, for example, speed up as the temperature goes up is taken to mean that some mechanism is actively involved (thus, invocation of a more loaded term than "temperature independence," in the context of biochemistry not having that attribute).

T–G repeat—A series of repeated threonine–glycine dimers present within many types of PER protein; the presence and length of the repeat vary widely among species and even within them; for example, in *D. melanogaster* the *per* gene exists in several different T–G (intraprotein)-encoding forms among laboratory strains and among individuals taken from natural populations; the maximum T–G repeat length is only about a score of such tandemly arranged amino-acid pairs, thus forming at most ca. 3% of the *D. melanogaster* polypeptide; other *Drosophila* or nondipteran species of insect produce PERs with much shorter T–G repeats or none at all.

transgenic—A strain of organism carrying a piece of DNA that was stably transformed into the chromosomes of its germline; such a strain is established so that all animals, down the generations, carry the transgene in question; the incoming material (originally transformed in) may be relatively simple (e.g., a fragment of genomic DNA to be tested as to whether it contains all the information necessary for biologically meaningful expression of some gene, including generation of a largely normal product), or it might have been heavily engineered (e.g., such that stretches of gene-regulatory sequences were fused to more than one kind of protein-coding sequences, whether or not the latter originated within the to-be-transformed species); "a transgenic" (sometimes designated "transformant") also refers to an individual organism carrying a given transgene; note that, in *Drosophila* experiments, even a simple transgene (as just described) needs to be manipulated such that it is fused to special DNA sequences derived from transposons (see **transposon**), which flank the incoming DNA of interest and are contained within an appropriate bacterially derived cloning vector, in order that insertion of such incoming (embryo-injected) DNA can be catalyzed into a site on one of the fly's chromosomes (those present in precursors of the eventual adults' gametes); such genomic "insert sites" (see **transposon**) are arbitrary—for example (unless special precautions are taken), almost never near the locus corresponding to an intraspecific gene whose DNA was used to produce the molecular construct for transformation.

transposon—A potentially mobile genetic element; a given transposon is composed of a few kilobases of DNA and might "sit" inserted at some chromosomal location forever; but if the germ cells of the transposon-bearing organism produce a DNA-metabolizing "transposase" enzyme (which is naturally encoded by full-blown versions of the transposon and is here called XPOS), then removal of the inserted material (which is often an incomplete version of the transposon, that is, lacking XPOS-encoding sequences) is catalyzed; this can be followed by reinsertion of the genetic element at a new chromosomal location (likely near the original insertion site, but not uncommonly far away, even on another chromosome); one of the best-known transposon types is the P-element in *D. melanogaster*; such P's, or there derivatives, are the only transposons dealt with in this work; any transgene (see **transgenic**) is in a way a potential transposon, especially in light of engineered such factors being heavily modified P-elements; a basic such element includes only transposon sequences per se, at least those necessary for insertion into chromosomal sites; a simple P-element modification would be to ligate in marker-making DNA; often two stretches of DNA are inserted between the sequences required for chromosomal insertion: a visible marker (such as an eye-color one) that permits tracking of the transposon's inheritance in genetic crosses, along with **lacZ** (see entry) for internal-tissue-marking purposes; the insertion-requiring sequences and the marker ones can be exploited in the following kinds of genetic strategies and crosses: fruit flies, each carrying the starting element on some chromosome, are mated to others possessing an XPOS-encoding transgene (which happens to sit permanently in its genomic location); in the germ cells of the resulting doubly transgenic offspring, the marker-bearing element goes mobile, such that a substantial proportion of the progeny in the next generation carries the mobilized factor in a novel genomic location; genetic inbreeding can lead to homozygosity of a given new insertion. Might such a molecular change at this chromosomal site cause subnormal or otherwise abnormal expression of a nearby gene, leading to an anomalous biological phenotype? If so, it is likely that the gene responsible for the abnormality is transposon-tagged and readily clonable (transposon-sequence-containing clones are recovered from the new mutant in a manner designed to "clone out" adjacent chromosomal sequences). Moreover, might expression of the transposon-contained cellular marker (again: see **lacZ**) suggest the locations of tissues related to the phenotypic anomaly?

UAS–Upstream-activating sequence, a stretch of DNA that comprises the *cis*-acting regulatory material targeted by the GAL4 protein in *S. cerevisiae* (see **GAL4/gal4;** bear in mind that UAS is also used to designate gene-regulatory sequences bound by other kinds of transcription factors); experimentally, *Drosophila* transgenes are constructed in which a tandem array of the (yeast-derived) UASs is fused to sequences encoding a protein that has some practical utility (DNA that is here called *ppu*), for example, a cellular marker or a factor that eliminates or physiologically disrupts the cells in which the UAS-*ppu* transgene has its expression driven by GAL4, whose production is in turn set up to be under the control of regulatory sequences cloned or enhancer-trapped (see **enhancer trap**) in the doubly transgenic organism (again, and for the purposes of this monograph, *D. melanogaster*).

ultradian—Rhythmicity whose cycle durations are quite a bit less than one day, for example, 5–10 hr (thus not near the 24-hr "circa-dian" range), or even shorter (in the range of minutes).

UTR—Untranslated regions of an mRNA molecule; one such region flanks the mRNAs **ORF** (see entry) on its 5' (upstream) side, which means that nearly all transcripts contain "leader" sequences (corresponding to a portion of the gene downstream of its transcription-start site, but upstream of its amino-acid-coding information); the leader is involved in things like binding of the mRNA to the ribosome; there is also (almost always) a 3' UTR downstream of the transcript's ORF; this "trailer" sequence contains sites for poly-adenylic acid (poly-A) attachment, which is nearly ubiquitous feature of mature mRNAs in eukaryotes; 3' UTRs can also be involved in transcript stability (the wherewithal for the mRNA to accumulate to relatively robust levels as opposed to being turned over in a rapid manner).

Western blotting—A way to detect polypeptides on gels; the name for this method is an extension of the molecular-biological japery introduced in the Northern-blotting entry; thus, an intrinsically bizarre metaphor was invoked to describe a series of procedures in which tissue-extracted proteins are first denatured (by solubilization in sodium dodecyl sulfate), then separated by molecular weight electrophoretically (in a polyacrylamide gel); this material is subsequently transferred (blotted) into a sheet of nitrocellulose paper (by analogy to how size-separated RNA species are pulled out of gels used in Northern blottings); application of reagents to the Western blot leads to detection of a linear array of polypeptide species (i.e., those distributed along a given lane of the electrophoresed gel); the substance applied could be a general protein-staining reagent (although that can be applied directly to the gel, pre-blotting), or, more usefully, the blotted-onto paper is incubated with antibody (e.g., anti-PER) raised for the purpose of detecting a particular type of protein.

yeast two-hybrid—A "system" (strategy) in which certain pieces of DNA from *S. cerevisiae*—along with many such pieces derived from any organism that is the focus of molecular inquiry, and (in one version of the tactics devoted to this system) certain bacterial sequences—all are combined in multiply transformed yeast cells; thus, a molecular-genetic construct is introduced into such cells, whereby the transgene constitutively generates a protein composed of two fused parts: the DNA-binding subset of a (yeast or bacterial) transcription factor, covalently bound to a "bait" protein fragment; the latter is derived from material suspected to interact with other protein molecules naturally produced in the organism of interest (for the purposes of this work, *D. melanogaster*); a vast array of such fungal cells is then transformed individually with constructs that include cDNAs cloned (starting with mRNAs) from *Drosophila*; this array of clones is designed such that each has the capacity to generate a fusion protein consisting of a fragment of fly protein joined to the transcriptional-activation domain of the microbial transcription factor; a small subset of the cloned-in-yeast library of fusion DNA sequences—a library known as the "fish" or "prey"—has the potential to generate chimeric proteins that would bind to the bait one; if this occurs (in a given multiply transformed yeast clone), the bait-with-prey interaction brings the separate subsets of the microbial transcription factor into physical proximity; it was predetermined that such non-covalently bound (composite) polypeptides are able to function as does the natively active transcription factor; in terms of the two-hybrid screening, the array of yeast clones is set up to require activation of a nutritional factor by the composite transcription one (i.e., with the proviso that the bait–prey interaction led to formation of the composite); if and only if the nutritionally related gene is so activated does a surviving yeast colony form (the genetic background of which disallows such growth in the absence of the composite transcription factor's activity); therefore, yeast subclones are selected that contain the (*Drosophila*) library-derived sequences encoding polypeptide fragments that can bind to the bait-contained protein fragment; potentially, such prey proteins function in part by naturally occurring (intrafly) interactions with the one that was used to create the bait; further applications of the yeast two-hybrid system can mediate (intrafungal) marker expression that reflects the strength of bait–prey interactions; for this, the composite transcription factor would activate a transgene designed to contain its *cis*-acting regulatees fused to a marker-encoding sequence, typically **lacZ** (see entry); thus, addition of β-GAL substrate (X-gal) to the multiply transformed yeast cells leads to a given intensity of (blue) reaction product, depending on the tightness of transcription-factor reconstitution that is mediated by bait–prey binding.

Zeitgeber—A time giver that is often used as an environmental condition to influence clock functioning and the rhythms so regulated; usually, LD cycles are employed to give time information to the organism; if such environmental cycles are 24 hr in duration, Zeitgeber Time 0 is defined as the beginning of the 12-hr photic period; in the most simplistic such LD cycles they are divided in half, such that ZT12 specifies when the lights go off (this simplicity is routine in chronobiological experiments, especially those devoted to the daily rhythms of locomotion and abundance oscillations of molecules in insects).

Index